MEMBRANE TECHNOLOGY FOR WATER AND WASTEWATER
TREATMENT, ENERGY AND ENVIRONMENT

T0262830

Sustainable Water Developments
Resources, Management, Treatment, Efficiency and Reuse

Series Editor

Jochen Bundschuh
University of Southern Queensland (USQ), Toowoomba, Australia
Royal Institute of Technology (KTH), Stockholm, Sweden

ISSN: 2373-7506

Volume 3

Membrane Technology for Water and Wastewater Treatment, Energy and Environment

Editors

Ahmad Fauzi Ismail

Advanced Membrane Technology Research Centre (AMTEC),
Universiti Teknologi Malaysia (UTM), Skudai, Johor, Malaysia

Takeshi Matsuura

Department of Chemical and Biological Engineering,
University of Ottawa, Ottawa, Ontario, Canada

CRC Press
Taylor & Francis Group
Boca Raton London New York

CRC Press is an imprint of the
Taylor & Francis Group, an **informa** business

A BALKEMA BOOK

Co-published by IWA Publishing
Alliance House, 12 Caxton Street, London SW1H 0QS, UK
Tel: +44 (0)20 7654 5500, Fax: +44 (0)20 7654 5555
publications@iwap.co.uk
www.iwapublishing.com
ISBN: 9781780407951

Cover photo
The cover shows the asymmetric structure of a hollow fiber membrane configuration which is widely used for fluid separation processes. Fine structural details of the membrane are important to ensure effective and efficient separation processes. Thus the cover shows an integral part of the membrane process if optimization of the membrane system for water and waste water treatment, energy and environment is to be envisaged.

First issued in paperback 2021

CRC Press/Balkema is an imprint of the Taylor & Francis Group, an informa business

© 2016 Taylor & Francis Group, London, UK

Typeset by MPS Limited, Chennai, India

Library of Congress Cataloging-in-Publication Data

Names: Ismail, Ahmad Fauzi editor. | Matsuura, Takeshie editor.
Title: Membrane technology for water and wastewater treatment, energy and
    environment / editors, Ahmad Fauzi Ismail, Takeshi Matsuura.
Description: Boca Raton : CRC Press, 2016. | Includes bibliographical
    references and index. | Description based on print version record and CIP
    data provided by publisher; resource not viewed.
Identifiers: LCCN 2015048706 (print) | LCCN 2015046520 (ebook) | ISBN
    9781315645773 (ebook) | ISBN 9781138029019 (hardcover : alk. paper)
Subjects: LCSH: Water—Purification—Membrane filtration.
Classification: LCC TD442.5 (print) | LCC TD442.5 .M4735 2016 (ebook) | DDC
    628.1/64–dc23

Published by: CRC Press/Balkema
                P.O. Box 11320, 2301 EH Leiden, The Netherlands
                e-mail: Pub.NL@taylorandfrancis.com
                www.crcpress.com – www.taylorandfrancis.com

ISBN 13: 978-1-138-61199-3 (pbk)
ISBN 13: 978-1-138-02901-9 (hbk)

Publisher's Note
The publisher has gone to great lengths to ensure the quality of this reprint but points out that some imperfections in the original copies may be apparent.

About the book series

Augmentation of freshwater supply and better sanitation are two of the world's most pressing challenges. However, such improvements must be done economically in an environmentally and societally sustainable way.

Increasingly, groundwater – the source that is much larger than surface water and which provides a stable supply through all the seasons – is used for freshwater supply, which is exploited from ever-deeper groundwater resources. However, the availability of groundwater in sufficient quantity and good quality is severely impacted by the increased water demand for industrial production, cooling in energy production, public water supply and in particular agricultural use, which at present consumes on a global scale about 70% of the exploited freshwater resources. In addition, climate change may have a positive or negative impact on freshwater availability, but which one is presently unknown. These developments result in a continuously increasing water stress, as has already been observed in several world regions and which has adverse implications for the security of food, water and energy supplies, the severity of which will further increase in future. This demands case-specific mitigation and adaptation pathways, which require a better assessment and understanding of surface water and groundwater systems and how they interact with a view to improve their protection and their effective and sustainable management.

With the current and anticipated increased future freshwater demand, it is increasingly difficult to sustain freshwater supply security without producing freshwater from contaminated, brackish or saline water and reusing agricultural, industrial, and municipal wastewater after adequate treatment, which extends the life cycle of water and is beneficial not only to the environment but also leads to cost reduction. Water treatment, particularly desalination, requires large amounts of energy, making energy-efficient options and use of renewable energies important. The technologies, which can either be sophisticated or simple, use physical, chemical and biological processes for water and wastewater treatment, to produce freshwater of a desired quality. Both industrial-scale approaches and smaller-scale applications are important but need a different technological approach. In particular, low-tech, cost-effective, but at the same time sustainable water and wastewater treatment systems, such as artificial wetlands or wastewater gardens, are options suitable for many small-scale applications. Technological improvements and finding new approaches to conventional technologies (e.g. those of seawater desalination), and development of innovative processes, approaches, and methods to improve water and wastewater treatment and sanitation are needed. Improving economic, environmental and societal sustainability needs research and development to improve process design, operation, performance, automation and management of water and wastewater systems considering aims, and local conditions.

In all freshwater consuming sectors, the increasing water scarcity and correspondingly increasing costs of freshwater, calls for a shift towards more water efficiency and water savings. In the industrial and agricultural sector, it also includes the development of technologies that reduce contamination of freshwater resources, e.g. through development of a chemical-free agriculture. In the domestic sector, there are plenty of options for freshwater saving and improving efficiency such as water-efficient toilets, water-free toilets, or on-site recycling for uses such as toilet flushing, which alone could provide an estimated 30% reduction in water use for the average household. As already mentioned, in all water-consuming sectors, the recycling and reuse of the respective wastewater can provide an important freshwater source. However, the rate at which these water efficient technologies and water-saving applications are developed and adopted depends on the behavior of individual consumers and requires favorable political, policy and financial conditions.

Due to the interdependency of water and energy (water-energy nexus); i.e. water production needs energy (e.g. for groundwater pumping) and energy generation needs water (e.g. for cooling), the management of both commodities should be more coordinated. This requires integrated energy and water planning, i.e. management of both commodities in a well-coordinated form rather than managing water and energy separately as is routine at present. Only such integrated management allows reducing trade-offs between water and energy use.

However, water is not just linked to energy, but must be considered within the whole of the water-energy-food-ecosystem-climate nexus. This requires consideration of what a planned water development requires from the other sectors or how it affects – positively or negatively – the other sectors. Such integrated management of water and the other interlinked resources can implement synergies, reduce trade-offs, optimize resources use and management efficiency, all in all improving security of water, energy, and food security and contributing to protection of ecosystems and climate. Corresponding actions, policies and regulations that support such integral approaches, as well as corresponding research, training and teaching are necessary for their implementation.

The fact that in many developing and transition countries women are disproportionately disadvantaged by water and sanitation limitation requires special attention to this aspect in these countries. Women (including schoolgirls) often spend several hours a day fetching water. This time could be much better used for attending school or working to improve knowledge and skills as well as to generate income and so to reduce gender inequality and poverty. Absence of in-door sanitary facilities exposes women to potential harassment. Moreover, missing single-sex sanitation facilities in schools and absence of clean water contributes to diseases. This is why women and girls are a critical factor in solving water and sanitation problems in these countries and necessitates that men and women work side by side to address the water and wastewater related operations for improvement of economic, social and sustainable freshwater provision and sanitation.

Individual volumes published in the series span the wide spectrum between research, development and practice in the topic of freshwater and related areas such as gender and social aspects as well as policy, regulatory, legal and economic aspects of water. The series covers all fields and facets in optimal approaches to the:

- Assessment, protection, development and sustainable management of groundwater and surface water resources thereby optimizing their use.
- Improvement of human access to water resources in adequate quantity and good quality.
- Meeting of the increasing demand for drinking water, and irrigation water needed for food and energy security, protection of ecosystems and climate and contribution to a socially and economically sound human development.
- Treatment of water and wastewater also including its reuse.
- Implementation of water efficient technologies and water saving measures.

A key goal of the series is to include all countries of the globe in jointly addressing the challenges of water security and sanitation. Therefore, we aim for a balanced selection of authors and editors originating from developing and developed countries as well as for gender equality. This will help society to provide access to freshwater resources in adequate quantity and of good quality, meeting the increasing demand for drinking water, domestic water and irrigation water needed for food security while contributing to socially and economically sound development.

This book series aims to become a state-of-the-art resource for a broad group of readers including professionals, academics and students dealing with ground and surface water resources, their assessment, exploitation and management as well as the water and wastewater industry. This comprises especially hydrogeologists, hydrologists, water resources engineers, wastewater engineers, chemical engineers and environmental engineers and scientists.

The book series provides a source of valuable information on surface water but especially on aquifers and groundwater resources in all their facets. As such, it covers not only the scientific and technical aspects but also environmental, legal, policy, economic, social, and gender aspects of groundwater resources management. Without departing from the larger framework of integrated groundwater resources management, the topics are centered on water, solute and heat transport in aquifers, hydrogeochemical processes in aquifers, contamination, protection, resources assessment and use.

The book series constitutes an information source and facilitator for the transfer of knowledge, both for small communities with decentralized water supply and sanitation as well as large industries that employ hundreds or thousands of professionals in countries worldwide, working in the different fields of freshwater production, wastewater treatment and water reuse as well as those concerned with water efficient technologies and water saving measures. In contrast to many other industries, suffering from the global economic downturn, water and wastewater industries are rapidly growing sectors providing significant opportunities for investments. This applies especially to those using sustainable water and wastewater technologies, which are increasingly favored. The series is also aimed at communities, manufacturers and consultants as well as a range of stakeholders and professionals from governmental and non-governmental organizations, international funding agencies, public health, policy, regulating and other relevant institutions, and the broader public. It is designed to increase awareness of water resources protection and understanding of sustainable water and wastewater solutions including the promotion of water and wastewater reuse and water savings.

By consolidating international research and technical results, the objective of this book series is to focus on practical solutions in better understanding groundwater and surface water systems, the implementation of sustainable water and wastewater treatment and water reuse and the implementation of water efficient technologies and water saving measures. Failing to improve and move forward would have serious social, environmental and economic impacts on a global scale.

The book series includes books authored and edited by world-renowned scientists and engineers and by leading authorities in economics and politics. Women are particularly encouraged to contribute, either as author or editor.

Jochen Bundschuh
(Series Editor)

Editorial board

Table of contents

Part VI Membranes for other applications

List of contributors

Ahmad Arabi Shamsabadi   Marun Petrochemical Company, Petzone, Mahshahr, Iran and Petroleum University of Technology, Ahwaz, Iran

Mohd Sohaimi Abdullah   Advanced Membrane Technology Research Centre (AMTEC), Universiti Teknologi Malaysia, Skudai, Johor, Malaysia

Zainan Abdullah   Sime Darby Latex Sdn. Bhd., Batu Anam, Segamat, Johor, Malaysia

Elmahdi M. Abousetta   Chemical Engineering Department, Computer Engineering Department, University of Tripoli, Tripoli, Libya

Abdul Latif Ahmad   School of Chemical Engineering Campus, Universiti Sains Malaysia, Seri Ampangan, Nibong Tebal Penang, Malaysia

Nor Aini Ahmad   School of Chemical Engineering, Engineering Campus, Universiti Sains Malaysia, Nibong Tebal, Penang, Malaysia

Ammar Mohd Akhir   Faculty of Chemical Engineering, Universiti Teknologi MARA, Shah Alam, Selangor, Malaysia

Abdulhakim M. Alamaria   Department of Chemical Engineering, Faculty of Chemical Engineering, Universiti Teknologi Malaysia, Johor Bahru, Malaysia

Yousof H.D. Alanezi   Department of Chemical Engineering Technology, College of Technological Studies (CTS), Public Authority of Applied Education and Training (PAAET), Adailiyah, Kuwait

Mohamed Kheireddine Aroua   Department of Chemical Engineering, University of Malaya, Kuala Lumpur, Malaysia

Gangasalam Arthanareeswaran   Membrane Research Laboratory, Department of Chemical Engineering, National Institute of Technology, Tiruchirappalli, India

M. Rezaei-DashtArzhandi   Advanced Membrane Technology Research Centre (AMTEC), Universiti Teknologi Malaysia, Skudai, Johor, Malaysia

Ahmad Jaril Asis   Processing & Engineering, R&D Centre, Sime Darby Research, Jalan Pulau Carey, Pulau Carey, Kuala Langat, Selangor, Malaysia

Masoud Bahrami Babaheidari   Petroleum University of Technology, Ahwaz, Iran

Gholamreza Bakeri   Advanced Membrane and Biotechnology Research Center, Faculty of Chemical Engineering, Babol Noshirvani University of Technology, Babol, Iran

Wan Ramli Wan Daud   Fuel Cell Institute, Universiti Kebangsaan Malaysia, UKM Bangi, Selangor, Malaysia

Norfadilah Dolmat   Faculty of Petroleum and Renewable Energy Engineering, Universiti Teknologi Malaysia, UTM Skudai Johor, Malaysia

Chutima Eamchotchawalit   Thailand Institute of Scientific and Technological Research, Khlong Luang, Pathumthani, Thailand

David Gethin   Civil and Computational Engineering Center, University of Wales, Swansea, UK

Azadeh Ghaee	Department of Life Science Engineering, Faculty of Interdisciplinary New Science and Technologies, University of Tehran, Tehran, Iran
Mostafa Ghasemi	Fuel Cell Institute, Universiti Kebangsaan Malaysia, UKM Bangi, Selangor, Malaysia
M.M. Nawawi Ghazali	Department of Chemical Engineering, Faculty of Chemical Engineering, Universiti Teknologi Malaysia, Skudai, Johor, Malaysia
Sina Gilassi	Department of Chemical Engineering, Universiti Teknologi PETRONAS, Bandar Seri Iskandar, Perak, Malaysia
Pei Sean Goh	Advanced Membrane Technology Research Centre (AMTEC), Universiti Teknologi Malaysia, Skudai, Johor, Malaysia and Faculty of Petroleum and Renewable Energy Engineering, Universiti Teknologi Malaysia, Skudai, Johor, Malaysia
Norlisa Harruddin	Centre of Lipids Engineering and Applied Research (CLEAR), Department of Chemical Engineering, Faculty of Chemical Engineering, Universiti Teknologi Malaysia, Skudai Johor, Malaysia
Zawati Harun	University Tun Hussien Onn, Parit Raja, Batu Pahat, Johor, Malaysia
S. Jalaledin Hashemi	School of Chemical Engineering, College of Engineering, University of Tehran, Tehran, Iran
Seyed A. Hashemifard	Chemical Engineering Department, Gas and Petrochemical Faculty, Persian Gulf University, Bushehr, Iran
Nidal Hilal	Centre for Water Advanced Technologies and Environmental Research (CWATER), College of Engineering, Swansea University, Swansea, UK and Qatar Energy and Environment Research Institute (QEERI), Doha, Qatar
Ani Idris	Institute of Bioproduct Development (IBD), Department of Bioprocess Engineering Faculty of Chemical Engineering, Universiti Teknologi Malaysia, Skudai Johor, Malaysia
Abudussalam O. Imdakm	Chemical Engineering Department, Computer Engineering Department, University of Tripoli, Tripoli, Libya
Ahmad Fauzi Ismail	Advanced Membrane Technology Research Centre (AMTEC), Universiti Teknologi Malaysia, Skudai, Johor, Malaysia and Faculty of Petroleum and Renewable Energy Engineering, Universiti Teknologi Malaysia, Skudai, Johor, Malaysia
Norafiqah Ismail	Advanced Membrane Technology Research Centre (AMTEC), Universiti Teknologi Malaysia, Skudai, Johor, Malaysia
Arisa Jaiyu	Thailand Institute of Scientific and Technological Research, Khlong Luang, Pathumthani, Thailand
Ali Kazemi Joujili	Advanced Membrane and Biotechnology Research Center, Faculty of Chemical Engineering, Babol Noshirvani University of Technology, Babol, Iran
Sharifah Aishah Syed Abdul Kadir	Faculty of Chemical Engineering, Universiti Teknologi MARA, Shah Alam, Selangor, Malaysia
Feras M. Kafiah	Mechanical Engineering Department, King Fahd University of Petroleum & Minerals, Dhahran, Kingdom of Saudi Arabia (KSA)
Khairul S.N. Kamarudin	Faculty of Petroleum and Renewable Energy Engineering, Universiti Teknologi Malaysia, Skudai Johor, Malaysia

Ali Kargari	Membrane Processes Research Laboratory (MPRL), Department of Petrochemical Engineering, Amirkabir University of Technology (Tehran Polytechnic), Mahshahr Campus, Mahshahr, Iran
Kanungnuch Keawsupsak	Thailand Institute of Scientific and Technological Research, Khlong Luang, Pathumthani, Thailand
Zafarullah Khan	King Fahd University of Petroleum & Minerals, Mechanical Engineering Department, Dhahran, Kingdom of Saudi Arabia (KSA)
Kok Keong Lau	Chemical Engineering Department, Universiti Teknologi PETRONAS, Bandar Sri Iskandar, Perak, Malaysia
Woei Jye Lau	Advanced Membrane Technology Research Centre (AMTEC), Universiti Teknologi Malaysia, Skudai, Johor, Malaysia and Faculty of Petroleum and Renewable Energy Engineering, Universiti Teknologi Malaysia, Skudai, Johor, Malaysia
Choe Peng Leo	School of Chemical Engineering, Engineering Campus, Universiti Sains Malaysia, Nibong Tebal, Penang, Malaysia
Jit-Kang Lim	School of Chemical Engineering Campus, Universiti Sains Malaysia, Seri Ampangan, Nibong Tebal, Penang, Malaysia and Department of Physics, Carnegie Mellon University, Pittsburgh, USA
Serene Sow Mun Lock	Chemical Engineering Department, Universiti Teknologi PETRONAS, Bandar Sri Iskandar, Perak, Malaysia
Siew-Chun Low	School of Chemical Engineering Campus, Universiti Sains Malaysia, Seri Ampangan, Nibong Tebal, Penang, Malaysia
Khalid Mahroug	Chemical Engineering Department, Computer Engineering Department, University of Tripoli, Tripoli, Libya
Kaveh Majdian	Marun Petrochemical Company, Petzone, Mahshahr, Iran
Zakaria Man	CO_2 Management, MOR, Universiti Teknologi PETRONAS, Bandar Sri Iskandar, Perak, Malaysia
Takeshi Matsuura	Department of Chemical and Biological Engineering, University of Ottawa, Ottawa, Ontario, Canada
Abdul Wahab Mohammad	Research Centre for Sustainable Process Technology (CESPRO), Faculty Engineering and Built Environment, Universiti Kebangsaan Malaysia (UKM), Bangi, Selangor, Malaysia and Department of Chemical and Process Engineering, Faculty of Engineering and Built Environment, Universiti Kebangsaan Malaysia (UKM), Bangi, Selangor, Malaysia
Hilmi Mukhtar	CO_2 Management, MOR, Universiti Teknologi PETRONAS, Bandar Sri Iskandar, Perak, Malaysia
Haikal Mustafa	Faculty of Chemical Engineering, Universiti Teknologi MARA, Shah Alam, Selangor, Malaysia
Be Cheer Ng	Advanced Membrane Technology Research Centre (AMTEC), Universiti Teknologi Malaysia, Skudai, Johor, Malaysia.
Ching Yin Ng	Department of Chemical and Process Engineering, Faculty of Engineering and Built Environment, Universiti Kebangsaan Malaysia (UKM), Bangi, Selangor, Malaysia
Law Yong Ng	Department of Chemical and Process Engineering, Faculty of Engineering and Built Environment, Universiti Kebangsaan Malaysia (UKM), Selangor, Malaysia
Qi-Hwa Ng	School of Chemical Engineering Campus, Universiti Sains Malaysia, Seri Ampangan, Nibong Tebal, Penang, Malaysia

Norul Fatiha Mohamed Noah	Centre of Lipids Engineering and Applied Research (CLEAR), Department of Chemical Engineering, Faculty of Chemical Engineering, Universiti Teknologi Malaysia, Skudai Johor, Malaysia
Ahmadilfitri Md Noor	Sime Darby R&D Center Carey Island – Downstream, Jalan Pulau Carey, Pulau Carey, Selangor, Malaysia
Tze Ching Ong	University Tun Hussien Onn, Parit Raja, Batu Pahat, Johor, Malaysia
Zing-Yi Ooi	Centre of Lipids Engineering and Applied Research (CLEAR), Department of Chemical Engineering, Faculty of Chemical Engineering, Universiti Teknologi Malaysia, Skudai Johor, Malaysia
Noor Hidayu Othman	Sime Darby R&D Center Carey Island – Downstream, Jalan Pulau Carey, Pulau Carey, Selangor, Malaysia and Faculty of Petroleum and Renewable Energy Engineering, Universiti Teknologi Malaysia, Skudai, Johor, Malaysia
Norasikin Othman	Centre of Lipids Engineering and Applied Research (CLEAR), Department of Chemical Engineering, Faculty of Chemical Engineering, Universiti Teknologi Malaysia, Skudai Johor, Malaysia
Julaluk Pannoi	Thailand Institute of Scientific and Technological Research, Khlong Luang, Pathumthani, Thailand
Thinesh Rachandran	School of Chemical Engineering, Engineering Campus, Universiti Sains Malaysia, Nibong Tebal, Penang, Malaysia
Mostafa Rahimnejad	Advanced Membrane and Biotechnology Research Center, Faculty of Chemical Engineering, Babol Noshirvani University of Technology, Babol, Iran
Nejat Rahmanian	School of Engineering, University of Bradford, Bradford, UK
Samad Sabbaghi	Faculties of Advanced Technologies, Nano Chemical Engineering Department, Shiraz University, Shiraz, Iran
Nur Aimie Abdullah Sani	Advanced Membrane Technology Research Centre (AMTEC), Universiti Teknologi Malaysia, Skudai, Johor, Malaysia
Ramesh Kumar Saranya	Membrane Research Laboratory, Department of Chemical Engineering, National Institute of Technology, Tiruchirappalli, India
Mojtaba Shariaty-Niassar	Transport Phenomenon and Nanotechnology Laboratory, School of Chemical Engineering, College of Engineering, University of Tehran, Tehran, Iran
Azmi Mohd Shariff	Chemical Engineering Department, Universiti Teknologi PETRONAS, Bandar Sri Iskandar, Perak, Malaysia
Biruh Shimekit	CO_2 Management, MOR, Universiti Teknologi PETRONAS, Bandar Sri Iskandar, Perak, Malaysia and Addis Ababa Institute of Technology, School of Chemical and Bio-Engineering, Addis Ababa, Ethiopia
Mohammad Mahdi A. Shirazi	Membrane Processes Research Laboratory (MPRL), Department of Petrochemical Engineering, Amirkabir University of Technology, Mahshahr Campus, Mahshahr, Iran
Passakorn Sueprasit	Thailand Institute of Scientific and Technological Research, Khlong Luang, Pathumthani, Thailand
Nik Meriam Nik Sulaiman	Department of Chemical Engineering, University of Malaya, Kuala Lumpur, Malaysia

Yogarathinam Lukka Thuyavan	Membrane Research Laboratory, Department of Chemical Engineering, National Institute of Technology, Tiruchirappalli, India
Jaya Kumar Veellu	Processing & Engineering, R&D Centre, Sime Darby Research, Jalan Pulau Carey, Pulau Carey, Kuala Langat, Selangor, Malaysia
Abu Zahrim Yaser	Chemical Engineering Programme, Faculty of Engineering, Universiti Malaysia Sabah, Jalan UMS, Kota Kinabalu, Sabah, Malaysia
Khairul Muis Mohamed Yusof	Processing & Engineering, R&D Centre, Sime Darby Research, Jalan Pulau Carey, Pulau Carey, Kuala Langat, Selangor, Malaysia
Mohd Suria Affandi Yusoff	Sime Darby R&D Center Carey Island – Downstream, Jalan Pulau Carey, Pulau Carey, Selangor, Malaysia
Mohammad M. Zerafat	Transport Phenomenon and Nanotechnology Laboratory, School of Chemical Engineering, College of Engineering, University of Tehran, Tehran, Iran

Foreword by Nidal Hilal

This book was written as one of the "Sustainable Water Developments" book series. Realizing, however, that water, energy and food are the three pillars to sustain the growth of human population in the future, the book deals with all the above aspects with particular emphasis on water and energy.

Despite the availability of several books on similar subject, this book is unique by including the results of research efforts mainly in the ASEAN and the Middle East Regions.

With a large population of the ASEAN nations (around 300 million) and their rich natural resources, sustainable development of the region based on suitable utilization of the valuable resources will have a strong impact on the growth of the entire global economy. However, with limited land resources, inadequate energy supply and growing water stress, the region faces the challenge of providing enough water and energy to grow enough food for the burgeoning population. These issues and challenges in the water, energy and food sectors are interwoven in many complex ways and cannot be managed effectively without cross-sectorial integration. Luckily, the technologies are currently available to address all the above issues simultaneously, and the membrane separation technology is one of such key technologies. Therefore, it is no wonder that strong membrane research and development activities are now being undertaken in the ASEAN region. It is also needless to say that, as the primary supplier of the fossil fuel to the global market yet with scarce water resources, the production of drinking water and the rational management of energy by the nations in the Middle East is equally crucial to maintain the peace in the whole world.

In this era of the global collaboration, it seems nowadays of primary importance to assemble the results of the individual research efforts in one place, thus promoting the networking among different research institutions in order to achieve the common goal of finding the solutions to the challenges we are facing. It was very fortunate therefore that the editors were successful to collect twenty eight papers; all related to water and energy, from the region as well the other parts of the world.

By highlighting the advances in the membrane technology, the editors were also successful in achieving the following main aims:

To promote interdisciplinary collaboration
To foster joint ventures
To promote the growth of scientific and technical development in membrane technology in the region and beyond

I am confident that the book will open up a new avenue for international collaboration not only within but also beyond ASEAN and the Middle East regions on the development of new membranes and membrane processes, thus contributing to the welfare of the entire human community.

Nidal Hilal
Professor and Director, CWATER
Swansea University
United Kingdom
and
Editor-in-Chief, Desalination
October 2015

Editors' foreword

As the book title "Membrane Technology for Water and Wastewater Treatment, Environment and Energy" indicates, the book is primarily concerned with environmental applications of membrane technology, particularly the treatment of water and wastewater.

Even though several books may have been written on a similar subject, this book is very unique for the reason that the chapters have been contributed, on invitation, mainly by the active researchers from the ASEAN and the Middle East regions.

With the large population of ASEAN nations (about 300 million) and their rich natural resources, e.g. agricultural products in Thailand, and petroleum, natural gas and water in Indonesia and Malaysia, sustainable development of the region based on suitable utilization of the valuable resources will have a strong impact on the growth of the entire global economy.

This book provides information to assess the quantity and quality of current research and development activities in the ASEAN region, especially on membrane science and technology, which are recognized nowadays as one of the most powerful tools to solve environmental and energy problems. The book also reflects upon those in the Middle East region, where the research on water treatment is growing due to scarcity of water. Thus, it is believed that the book will open up a new avenue for the establishment of global collaborations to achieve our common goal of the welfare of human society.

Part I is for the development of novel separation membranes for water and wastewater treatment. The membranes were fabricated either by blending of different polymers (Chapters 1 and 2) or incorporation of nanoparticles (Chapter 3). The membrane surface was modified and the membrane surface-pollutant interaction was studied by the QCM-D method (Chapter 4). Hollow fibers were spun from polylactic acid (PLA) (Chapter 5). Flat sheet isotactic polypropylene membrane was fabricated by the TIPS method (Chapter 6). Superhydrophobic membrane was prepared by grafting of saline compounds on an alumina surface (Chapter 7). The first 5 chapters are for the pressure driven membrane processes, whereas the last two chapters are for water treatment by membrane distillation (MD).

Part II is for various applications of membrane separation processes in water and wastewater treatment. Synthetic oily wastewater was treated by ceramic membrane (Chapter 8). Removal of dye solutions typical for palm oil mill, leather, textile, pulp and paper industries was attempted by nanofiltration (NF) (Chapter 9). Wastewater from palm oil (Chapter 10) and rubber industries (Chapter 11), typical for Malaysia, was treated by ultrafiltration (UF)/NF and ceramic UF membrane, respectively. Application of membrane separation technology for biodiesel processing is reviewed (Chapter 12).

Part III is for modeling concerning water and wastewater treatment. Mathematical models were developed for NF based deionization process (Chapter 13). The Monte Carlo method was applied to simulate MD process (Chapter 14). Finally, drying in the multilayer formation of ceramic membrane was simulated by a mathematical model (Chapter 15).

Part IV is for the preparation of membranes for energy and environment applications, more specifically for gas separation membranes. The first two chapters are for the preparation of polymeric flat sheet membranes from polyphenylene oxide (PPO) (Chapter 16) and polyetherimide (PEI) (Chapter 17). The third one is for the preparation of inorganic (perovskite) hollow fiber membranes (Chapter 18), which is followed by two chapters where the fabrication of mixed matrix membranes (MMMs) by filling clay nanoparticles (Chapter 19) into the polymer matrix is discussed. Two chapters (Chapters 16 and 19) are for removing CO_2, a powerful greenhouse gas, whereas others are for CH_4/H_2 (Chapter 17) and O_2/N_2 (Chapter 18) separation.

Part V is for energy and environmental applications. Two of them are for CO_2/CH_4 separation by using either commercially available PTFE membrane (Chapter 20) or emulsion liquid membrane (Chapter 21). The third one is a review of olefin and paraffin separation by membrane contactor (MC) (Chapter 22). The forth chapter discusses on the reliability of the gas permeation method which is often used to measure the pore size of membranes for membrane distillation (MD) and membrane contactor (MC) where Knudsen and viscous flow dominate (Chapter 23). The last chapter compares co-current and counter-current flow for the separation of gas mixtures by hollow fiber membranes (Chapter 24).

Part VI is dedicated to various applications of membrane separation processes concerning energy, environment and food. The first one is for microbial fuel cell application (Chapter 25). The second and third chapters are for membranes used in the organic environment. Polyphenyl sulfone membrane is used for the separation of dye from methanol (Chapter 26) and a review is made on degumming and deacidification of vegetable oils by membrane processes (Chapter 27). The last chapter is for the fabrication of electro-spun nanofiber membrane, a novel membrane that can be used for various applications (Chapter 28).

The editors believe that many innovative ideas on the development of membrane science and technology are assembled in this book. The readers may also have a glimpse into the rapidly growing R and D activities in the ASEAN and the Middle East regions that are emerging as the next generation R and D centers of membrane technologies, especially owing to their need of technology for water and wastewater treatment. As the table of contents indicates the book addresses applications of membrane science and technology for water and wastewater treatment, energy and environment. Hence, this book will be useful not only for the engineers, scientists, professors and graduate students who are engaged in the R & D activities of the field, but also for those who are interested in the sustainable development of these particular regions.

Ahmad Fauzi Ismail
Takeshi Matsuura
(Editors)
November 2015

About the editors

Professor Ahmad Fauzi Ismail is the Founding Director of Advanced Membrane Technology Research Center (AMTEC), Universiti Teknologi Malaysia (UTM). Professor Fauzi obtained a Ph.D. in Chemical Engineering in 1997 from the University of Strathclyde and M.Sc. and B.Sc. from Universiti Teknologi Malaysia in 1992 and 1989 respectively. He is the author and co-author of over 380 refereed journal articles. He has authored 5 books, 30 book chapters and 3 edited books, 3 Patents granted and 17 Patents pending. He has won more than 120 awards and among the outstanding awards are the Malaysia Young Scientist Award in 2000; ASEAN Young Scientist Award in 2001; Two times winner of the National Intellectual Property Award (Patent Category), 2009 and (Product Category), 2013; Two times winner of the National Innovation Award (Waste to Wealth Category), 2009 and (Product Category), 2011. Recently, he won the National Academic Award (Innovation and Product Commercialization Category) in August 2013 and the Malaysian Toray Science and Technology Foundation Award, in November 2013. He also won the IChemE Innovator of the Year 2014 awarded by IChemE UK in October 2014.

In December 2014, he was awarded the Merdeka Award in the Outstanding Scholastic Achievement Category.

He is a Fellow of the Academy of Sciences Malaysia. At present, he is the Editor of *Desalination* with impact factor of 3.96 and is editor of Jurnal Teknologi (Scopus Index). He is also an Editorial Board Member of the *Journal of Membrane Water Treatment* and the Advisory Editorial Board Member of the *Journal of Chemical Technology and Biotechnology*, just to name a few.

Professor Fauzi's research focuses on the development of polymeric, inorganic and novel mixed matrix membranes for water desalination, waste water treatment, gas separation processes, membrane for palm oil refining, photocatalytic membrane for removal of emerging contaminants and polymer electrolyte membrane for fuel cell applications. Professor Fauzi has been involved extensively in R&D&C for multinational companies related to membrane-based processes for industrial application.

In 2013, Ministry of Education unanimously awarded him the Innovative Action Plan for Human Capital Development Tertiary Level award. Currently Prof. Fauzi is the Deputy Vice Chancellor of Research and Innovation, Universiti Teknologi Malaysia.

Professor Matsuura received his B.Sc. (1961) and M.Sc. (1963) degrees from the Department of Applied Chemistry at the Faculty of Engineering, University of Tokyo. He went to Germany to pursue his doctoral studies at the Institute of Chemical Technology of the Technical University of Berlin and received the degree of Doktor-Ingenieur in 1965.

After working at the Department of Synthetic Chemistry of the University of Tokyo as an assistant and at the Department of Chemical Engineering of the University of California as a postdoctoral research associate, he joined the National Research Council of Canada in 1969. He came to the University of Ottawa in 1992 as a professor and the chairholder of the British Gas/NSERC Industrial Research Chair. He served as a professor of the Department of Chemical Engineering and the director of the Industrial Membrane Research Institute (IMRI) until he retired in 2002. He was appointed to professor emeritus in 2003. He served also at the Department of Chemical and Environmental Engineering of the National University of Singapore as a visiting professor during the period of January to December 2003, and at University Technology Malaysia (UTM), Skudai, Malaysia (currently at the Advanced Membrane Technology Research Centre (AMTEC) of UTM), as a distinguished visiting professor, in the years 2007 and 2009–2015. He stayed at the Department of Environmental Engineering and Biotechnology of Myongji University, Yongjin, Korea, from January to April 2008 by the support of the Brain Pool Program.

Professor Matsuura received the Research Award of the International Desalination and Environmental Association in 1983. He is a fellow of the Chemical Institute of Canada and is a member of the North American Membrane Society. He received the George S. Glinski Award for Excellence in Research from the Faculty of Engineering of the University of Ottawa in 1998. A special issue honouring his 75th birthday was published by *Desalination* in February 2012. He was appointed a member of the Advisory Board of the *Journal of Membrane Science* in October 2013.

Part I
Water and wastewater treatment
(novel membrane development)

CHAPTER 1

Preparation and characterization of blended polysulfone/polyphenylsulfone ultrafiltration membranes for palm oil mill effluent treatment

Norafiqah Ismail, Woei Jye Lau & Ahmad Fauzi Ismail

1.1 INTRODUCTION

The palm oil industry forms the backbone of the economy of Malaysia, where the country today is the world's second largest producer and exporter of palm oil after Indonesia. In Malaysia, the palm oil mill industry is identified as the one discharging the largest pollution load into the rivers and this has been reported for the last three decades (Hwang *et al.*, 1978). Currently, there are more than 250 palm oil mills distributed in peninsular Malaysia, generating over 20 million metric tons of effluent every year. In addition, one of the most crucial problems that the oil palm industry faces is waste disposal, since about 2.5–3.0 m³ of palm oil mill effluent (POME) are generated for every metric ton of crude palm oil produced (Borja and Banks, 1994). An efficient treatment system is therefore highly desirable in palm oil mills in order to control the effluent discharge and to meet the standard parameters set by the Malaysian Department of Environment (DOE).

There are several common POME treatment methods being adopted in Malaysia. One of the types of POME treatment is biological treatment of anaerobic and aerobic systems. According to Perez *et al.* (2001), the anaerobic process is a suitable treatment method due to the organic characteristics of POME and thus is widely applied in Malaysia. However, the biological treatment process needs proper maintenance and closed monitoring as the efficiency of the process strongly depends on microorganisms to break down the pollutants which usually require long retention time and skilled personnel to handle (Quah *et al.*, 1982). An efficient POME treatment has always been considered a burden to industrial players since no profit is gained from this activity. Although biogas can be potentially produced from biological treatment, it still requires an additional treatment process (Ahmad *et al.*, 2005). Besides biological treatment, pond systems have also been applied in Malaysia for POME treatment since 1982 and they are classified as waste stabilization pond (Onyia *et al.*, 2001). The main reason of using ponding systems is due to their low operating and maintenance costs. In general, ponding systems are easy to operate, but it is always associated with several drawbacks which include the vast amount of land required, relatively long hydraulic retention time (between 45 and 60 days), bad odor created and difficulty in maintaining the liquor distribution and collecting biogas. In order to tackle this problem, membrane-based treatment processes have become the main focus among researchers due to their excellent performance in treating POME and producing permeate of high quality, in addition to its much smaller treatment footprint. This phenomenon can be reflected by the increasing number of technical papers published (Abdurahman *et al.*, 2011; Ahmad *et al.*, 2006; Idris *et al.*, 2007).

Membrane technology, particularly ultrafiltration (UF), appears to be a potential treatment process to treat POME efficiently owing to its micrometer (μm) range pore size. Industrially, UF membranes have been successfully implemented in municipal and industrial wastewater treatments. It is gaining importance for water and industrial wastewater treatment especially in the recovery of chemicals from the industrial wastewater, desalination, drinking water purification, and removal of oil from oil-water emulsions (Tansel *et al.*, 2000). It is reported in the work of Mameri *et al.* (2000) that UF membrane could reduce up to 90% of the COD value of olive mill washing water. Lo *et al.* (2005) and Wu *et al.* (2007) on the other hand found that UF membranes made of polysulfone (PSF) were able to treat and reclaim the protein from poultry processing wastewater and POME, respectively, with significant efficiencies in reducing the value of several important effluent parameters. In recent years, the potential of using polyphenylsulfone (PPSF) membranes has been reported in different fields of membrane applications such as fuel cell fabrication, pervaporation, gas separation and water treatment processes. PPSF, which is a third member of the PSF family, comprises sulfone moieties, ether linkages and biphenyl group in its repeat group. The replacement of the bisphenolA group by biphenyl group in PSF structure increases the impact strength of PPSF polymer, making it more suitable for UF membrane preparation.

The main objective of the present work was to investigate the potential of the PSF membranes blended with different content of PSSU for POME treatment process. Prior to the effluent treatment process, the membranes were characterized using different instruments, i.e. scanning electron microscope (SEM), atomic force microscope (AFM) and contact angle goniometer, in addition to the water flux and BSA rejection tests. The performances of membranes for POME treatment process were evaluated in terms of BOD_3, COD and turbidity.

1.2   EXPERIMENTAL

1.2.1   *Materials*

Polysulfone (Udel-P1700, MW: 35,400 g mol^{-1}) obtained from Amoco Chemical was used in this study as membrane forming material. Another type of polymer – polyphenylsulfone (MW: 53,000–59,000 g mol^{-1}) purchased from Sigma-Aldrich was used together with PSF in preparing PSF/PPSF blend membranes. 1-methyl-2-pyrrolidone (NMP) with analytical purity of 99.5% purchased from Merck was used as solvent. Bovine serum albumin (BSA, 67 kDa) purchased from Sigma Aldrich was used as test solute by dissolving it in DI water.

1.2.2   *Membrane preparation*

Table 1.1 shows the composition of five different dope solutions that were used to prepare UF hollow fiber membranes of various properties. These membranes were prepared from a dope solution consisted of 20 wt% total polymer (with different PSF/PPSF ratio) and 80 wt% NMP solvent. The details of dry-jet wet spinning process for hollow fiber membrane fabrication can be found elsewhere (Ismail, 2014). In order to remove the residual solvent from membrane matrix, it was needed to immerse the membrane in water bath for at least 1 day. Afterwards, the membrane was air-dried naturally before it was used for module making.

1.2.3   *Membrane characterizations*

The morphologies of the membranes were observed by a tabletop scanning electron microscope (SEM) (TM3000, Hitachi).The cross sections of samples were prepared by fracturing the membranes in liquid nitrogen followed by a thin layer of platinum coating using sputter coater. The surface roughness of the dried hollow fiber membranes on the other hand was determined by AFM (Multimode Nanoscope, DI company) in a dynamic force mode (DFM) with a 20 μm scanner and SI-DF40 (spring constant = 42 N m^{-1}) cantilever. For this analysis, the membrane samples were required to be cut into lengths of approximately 1 cm and located on the top of the scanner tube by carbon tape. The outer surface of the hollow fiber membranes was observed within the scan size of 1 μm × 1 μm. In this study, R_a and RMS were used as the evaluation parameter to compare the roughness of membrane produced. The hydrophilicity of membranes was studied by measuring the angle between the membrane surface and the meniscus formed by the water using

Table 1.1.   Composition of the dope solution for blended PSF/PPSF
membranes preparation.

| | Dope solution compositions [wt%] | | |
| | Total polymer weight (20%) | | Solvent (80%) |
Membrane	PSF	PPSF	NMP
PSF	20	–	80
75PSF/25PPSF	15	5	80
50PSF/50PPSF	10	10	80
25PSF/75PPSF	5	15	80
PPSF	–	20	80

contact angle goniometer (OCA 15plus, DataPhysics). To determine membrane porosity, ε [%], the gravimetric method as expressed in the following equation can be employed:

$$\varepsilon = \frac{\dfrac{w_1 - w_2}{\rho_w}}{\dfrac{w_1 - w_2}{\rho_w} + \dfrac{w_2}{\rho_p}} \tag{1.1}$$

where w_1 is the weight of the wet membrane [g], w_2 is the weight of the dry membrane [g], ρ_p is the density of the polymer [g cm^{-3}] and ρ_w is the density of water [g cm^{-3}].

1.2.4   UF experiment

The permeation flux and BSA rejection of all resultant membranes were measured by a customized cross-flow UF system. In this experiment, the pressure was applied on the shell side (feed side) of the hollow fiber membranes and the permeate flux was created in outside-in filtration mode. For each membrane sample, a module consisted of 10 fibers in 21 cm long was used. The pure water flux, J_{PWF} [L m^{-2} h^{-1}], was measured at 0.2 MPa after a steady-state flux was achieved:

$$J_{PWF} = \frac{V}{AT} \tag{1.2}$$

where V is the permeate volume [L], A is the membrane surface area [m^2] and T is the time [h].

To determine membrane separation efficiency, a feed solution containing 1000 mg L^{-1} of BSA test solute was prepared. The membrane rejection against BSA solute, R [%], was determined by measuring the concentration in the feed and permeate using UV-VIS spectrophotometer (DR5000, Hach) at the wavelength of 280 nm:

$$R = \left(1 - \frac{c_p}{c_f}\right) \times 100 \tag{1.3}$$

where c_p is the concentration of the permeate [mg L^{-1}] and c_f is the concentration of the feed [mg L^{-1}].

1.2.5   POME sample and analysis

The POME sample used in this work was collected from Sime Darby East Oil Mill, Carey Island, Selangor, without undergoing any pretreatment process. Analytical tests were carried out to evaluate the quality of the POME before and after membrane treatment process. Chemical oxygen demand (COD) tests were carried out by putting 2 mL sample into the oxidizing acid reagent solution and then it was held at 150°C for 2 h. After it was cooled down to ambient temperature, the COD value of sample was analyzed using a colorimeter (DR/890, Hach) with readings taken at 435 nm wavelength. Biological oxygen demand (BOD) tests on the other hand were performed based on 3-day period, i.e. BOD_3. For this analysis, initial dissolved oxygen (DO) of each solution was first determined using a portable dissolved oxygen meter (PRODO, YSI). After 3 days of incubation in the dark condition at 20°C, the DO of the final samples was re-analyzed in order to calculate the change of DO value, i.e. BOD value. With respect to sample turbidity, a portable turbidimeter (2100Q, Hach) was used for measuring the turbidity of feed and permeate sample at room condition.

1.3 RESULTS AND DISCUSSION

1.3.1 *Effect of PSF:PPSF weight ratio on membrane properties*

1.3.1.1 *Morphological studies*
The cross-sectional morphology of membranes prepared in this study was examined using SEM. The effects of PPSF concentration on the structural properties of PSF is presented in Figure 1.1. As can be seen, the addition of PPSF in the dope solution has indeed played a role in altering membrane morphology. With increasing PPSF concentration in the PSF dope solution, it is found that the finger-like structures were extended from both the inner and outer skin layer of membrane by suppressing the number of big macrovoids formed at the intermediate layer. However, when the dope solution was prepared from PPSF only, the viscosity of the dope solution was increased, owing to the higher molecular weight of PPSF compared to PSF. Because of this, a relatively denser and thick active layer was formed in PPSF membrane as shown in Figure 1.1.e. In this PPSF membrane, a very thick dense top layer was observed followed by sponge-like layer supported a uniformly formed finger-liked inner layer. These results indicate that kinetically the overall diffusion between components in phase inversion system can be suppressed because of the increase in the rheological hindrance or delayed exchange between solvent and non-solvent. In the following section, the effects of the morphology change of PSF membrane due to the introduction of PPSF polymer at different weight ratio on the water flux and BSA removal are further discussed.

1.3.1.2 *Membrane filtration performance and its relation to membrane characteristics*
Figure 1.2 shows the pure water flux and BSA rejection of all the resultant membranes measured at operating pressure of 0.2 MPa. The water flux of membrane was improved with increasing the PPSF concentration from PSF:PPSF weight ratio of 1:0 to 0.5:0.5, recording 82.81 L m^{-2} h^{-1} in the 50PSF/50PPSF membrane compared to 11.09 L m^{-2} h^{-1} shown in the pristine PSF membrane, i.e. PSF. The flux enhancement is mainly due to the development of more finger-like structures as evidenced in the SEM image (Fig. 1.1) coupled with the increase in overall membrane porosity (Table 1.2), which reduces the transport resistance of water and increases the water permeation rate. Further increase in PPSF concentration to weight ratio of 0.25:0.75, however resulted in lower pure water flux as evidenced in 25PSF/75PPSF membrane. This is probably due to the decrease in overall membrane porosity coupled with increase in membrane hydrophobicity (i.e. higher water contact angle value). As PPSF is naturally more hydrophobic than PSF, increasing the content of PPSF in the PSF membrane matrix would therefore reduce membrane hydrophilicity (Liu et al., 2012). In this study, it was found that the influence of PPSF on PSF membrane hydrophilicity became more obvious when more than 75% of the total polymer used was PPSF. Interestingly, the membrane prepared from pure PPSF exhibited zero flux (tested at 0.2 MPa), even though the dope solution containing same polymer weight as the PSF dope solution was used in preparing pure PSF membrane. The zero-flux membrane is mainly due to the formation of relatively dense skin layer which is the result of the high viscosity of PPSF dope solution and

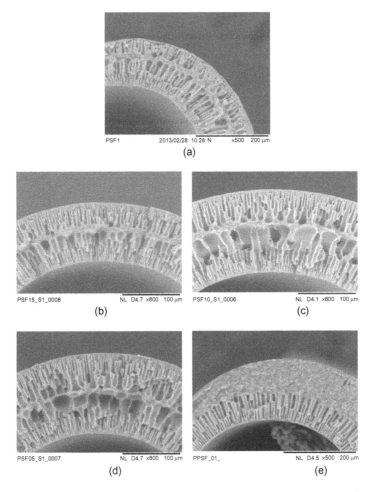

Figure 1.1.   Scanning electron microscope (SEM) images of membrane cross-section: (a) PSF, (b) 75PSF/25PPSF, (c) 50PSF/50PPSF, (d) 25PSF/75PPSF and (e) PPSF membrane.

partly caused by the increase in membrane hydrophobicity. With respect to surface roughness, no significant change was observed on all the membranes prepared (except PPSF membrane) as the R_a value of membrane was in the range of 2.13–6.89 nm.

With respect to BSA removal, it is found that the separation efficiency of membrane increased with increasing the PPSF concentration from PSF:PPSF weight ratio of 1:0 to 0.25:0.75 followed by very sharp drop in the pure PPSF membrane. Rejection rate of blended PSF/PPSF membranes are reported to be higher than that of pure PSF and PPSF membrane. Increasing BSA removal from PSF to 25PSF/75PPSF membrane can be explained by the fact that the membrane pore size tended to decrease with increasing PPSF content in the PSF membrane. Nevertheless, no pores were formed on the membrane prepared from PPSF as this membrane exhibited zero flux when tested at 0.2 MPa. Based on the results obtained, it is found that 50PSF/50PPSF membrane was the best performing membrane, owing to the good combination of water flux and BSA rejection rate achieved. In view of this, this 50PSF/50PPSF blend membrane was selected for further study by subjecting the membrane to POME treatment process. The detailed investigation together with the result discussion is provided in the following section.

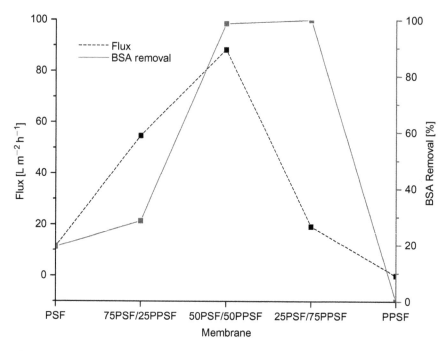

Figure 1.2.   Pure water flux and BSA removal of membranes prepared from different PSF:PPSF weight ratio.

Table 1.2.   Surface roughness, root mean square roughness, porosity and contact angle of the membranes prepared from different PSF:PPSF weight ratios.

Membrane	Surface roughness, R_a [nm]	Root mean square roughness, RMS [nm]	Porosity [%]	Contact angle [°]
PSF	3.26	4.52	56.93	77.9
75PSF/25PPSF	2.13	3.29	67.21	74.5
50PSF/50PPSF	2.44	3.13	79.33	70.5
25PSF/75PPSF	6.89	9.84	67.14	80.7
PPSF	11.44	13.84	5.35	86.5

1.3.2   *POME treatment using blended PSF/PPSF UF membrane*

Table 1.3 shows the performances of 50PSF/50PPSF membrane in treating the raw POME sample discharged from a local palm oil mill. As can be seen, the membrane demonstrated a very promising result in reducing the pollutant levels of several important parameters. BOD_3 and COD values of the raw POME were remarkably reduced from around 1995 mg L^{-1} and 8457 mg L^{-1} to around 250 mg L^{-1} and 780 mg L^{-1}, achieving close to 88 and 91% reduction, respectively. Compared to BOD_3 and COD reduction rate, almost complete elimination of sample turbidity could be achieved using in-house made UF membrane. This excellent reduction rate of turbidity can be explained by the presence of large particulates in POME sample which in general were much larger in terms of size in comparison to the pore size of UF membranes, making the separation of them very efficient and effective in UF membrane process.

Figure 1.3 compares the physical color of the feed POME sample and permeate samples treated by 50PSF/50PSSF membranes. Clearly, there was a significant reduction in color intensity of permeate sample in comparison to the feed POME sample. It must be pointed out that complete

(a)   (b)

Figure 1.3.   Palm oil mill effluent sample, (a) before membrane treatment (raw) and (b) after membrane treatment using 50PSF/50PPSF membrane.

Table 1.3.   The performance of 50PSF/50PPSF membrane in removing important pollutants from raw POME.

Parameter	Raw POME	Permeate	Removal
BOD_3 [mg L^{-1}]	1994.58	251.31	87.40%
COD [mg L^{-1}]	8456.67	780.55	90.77%
Turbidity [NTU]	8207.77	2.46	99.97%

elimination of color is almost impossible to achieve using UF membrane (even with commercial UF membrane), mainly because of the presence of nano-sized color components in the POME sample. The use of NF and RO membranes has potential to produce better quality of permeate, but they require much higher operating pressure during treatment and flux productivity are also significantly lower than that of UF membranes.

1.4 CONCLUSION

Blended PSF/PPSF membranes made of different PSF:PPSF weight ratio were successfully fabricated via dry-jet/wet spinning method. In addition to pure water flux and BSA rejection tests, the membranes were also characterized using SEM, AFM and contact angle goniometer. Results showed that the blended membrane made of equal amount of PSF and PPSF (designated as 50PSF/50PPSF membrane) was the best performing membrane after taking into account its highest water flux and excellent BSA rejection. Further investigations revealed that the 50PSF/50PPSF membrane was able to reduce significantly the values of several important pollutants in raw POME, achieving close to 88, 91 and 100% reduction for BOD_3, COD and turbidity, respectively. However, it was found that UF membrane was not efficient enough to completely eliminate the color of the effluent, mainly due to the existence of nano-sized color components in POME sample.

ACKNOWLEDGEMENT

The authors would like to express sincere gratitude to Ministry of Higher Education (MOHE) for the sponsorship and Universiti Teknologi Malaysia for the financial aid under Research University Grant (Tier 1) (Grant no. Q.J.130000.2542.04H86).

REFERENCES

Abdurahman, N.H., Rosli, Y.M. & Azhari, N.H. (2011) Development of a membrane anaerobic system (MAS) for palm oil mill effluent (POME). *Desalination*, 266 (1–3), 208–212.

Ahmad, A.L., Ismail, S. & Bhatia, S. (2005) Ultrafiltration behavior in the treatment of agro-industry effluent: pilot scale studies. *Chemical Engineering Science*, 60, 5385–5394.

Ahmad, A.L., Chong, M.F. & Bhatia, S. (2006) Drinking water reclamation from palm oil mill effluent (POME) using membrane technology. *Journal of Hazardous Materials*, 171 (1–3), 166–174.

Borja, R. & Banks, C.J. (1994) Anaerobic digestion of palm oil mill effluent using up-flow anaerobic sludge blanket reactor. *Biomass Bioenergy*, 6, 381–389.

Hwang, T.K., Ong, S.M., Seow, C.C. & Tan, H.K. (1978) Chemical composition of palm oil mill effluents. *Planter*, 54, 749–756.

Idris, A., Ahmad, I. & Jye, H.W. (2007) Performance cellulose acetate-polyethersulfone (CA-PS) blend membranes prepared using microwave heating for palm oil mill effluent treatment. *Water of Science Technology*, 56 (8), 167–177.

Ismail, N. (2014) *Preparation and characterization of polysulfone/polyphenylsulfone/titanium dioxide ultra-filtrationmembrane for palm oil mill effluent treatment.* Master Thesis, Universiti Teknologi Malaysia, Johor, Malaysia.

Liu, Y., Yue, X., Zhang, S., Ren, J., Yang, L., Wang, Q. & Wang, G. (2012) Synthesis of sulfonatedpolyphenyl-sulfone ascandidates for antifouling ultrafiltration membrane. *Separation and Purification Technology*, 98, 298–307.

Lo, Y.M., Cao, D., Argin-Soysal, S., Wang, J. & Hahm, T.-S. (2005) Recovery of protein from poultry processing wastewater using membrane ultrafiltration. *Bioresource Technology*, 96, 687–698.

Mameri, N., Halet, F., Drouiche, M., Grib, H., Lounici, H., Pauss, A., Piron, D. & Belhocine, D. (2000) Treatment of olive mill washing water by ultrafiltration. *Canadian Journal of Chemical Engineering*, 78, 590–595.

Onyia, C.O., Uyub, A.M., Akunna, J.C., Norulaini, N.A. & Omar, A.K.M. (2001) Increasing the fertilizer value of palm oil mill sludge: bioaugmentation in nitrification. *Sludge Management Entering the Third Millennium – Industrial, Combined, Water and Wastewater Residues*, 44, 157–162.

Perez, M., Romero, L.J. & Sales, D. (2001) Organic matter degradation kinetics in an anaerobic thermophilic fluidized bed bioreactor. *Anaerobe*, 7, 25–35.

Quah, S.K., Lim, K.H., Gillies, D., Wood, B.J. & Kanagaratnam, J. (1982) POME treatment and land application system. *Proceeding of Regional Workshop On Palm Oil Mill Technology and Effluent Treatment, PORIM, Kuala Lumpur, Malaysia.* pp. 193–200.

Tansel, B., Bao, Y.W. & Tansel, I.N. (2000) Characterization of fouling kinetics in ultrafiltration systems by resistance in series model. *Desalination*, 129 (1), 7–14.

Wu, T.Y., Mohammad, A.W., Md. Jahim, J. & Anuar, N. (2007) Palm oil mill effluent (POME) treatment and bioresourcesrecovery using ultrafiltration membrane: effect of pressure on membrane fouling. *Biochemical Engineering Journal*, 35, 309–317.

CHAPTER 2

Development and characterization of sago/PVA blend membrane for recovery of ethyl acetate from water

Abdulhakim M. Alamaria & M.M. Nawawia Ghazali

2.1 INTRODUCTION

Pervaporation (PV) is a membrane separation process that has been studied intensively to separate alcohol/water mixtures such as ethanol/water, iso-propanol/water and ethyl acetate/water (Ghazali *et al.*, 1997; Xia *et al.*, 2011; Zhang *et al.*, 2009). A good advantage of pervaporation is that no pollution occurs and high efficiency production is expected (Feng *et al.*, 1997; Lipnizki *et al.*, 1999). All chemical and pharmaceutical industries are presently showing incredible interest in this new technology that is low energy consuming, has high separation efficiency and is of an eco-friendly nature (Hyder *et al.*, 2006; Rachipudi *et al.*, 2011). Pervaporation has several applications like recovery of organic mixture from water, separation of mixtures of similar boiling points and dehydration of organic solvents (Feng *et al.*, 1997; Hasanoğlu *et al.*, 2005; Liua, Y. *et al.* 2011). Ethyl acetate is one of the important solvents, and wildly used in various manufacturing industries such as pharmaceutical industries and chemical industries for cleaning fluids, inks, coated papers and perfume (Hai-Kuan Yuana *et al.*, 2011; Yongquan *et al.*, 2012). Recovery of ethyl acetate from water is however very difficult because ethyl acetate forms azeotrope with the water and the separation becomes expensive by distillation which requires entrainers that must be removed after distillation is completed (Wua *et al.*, 2012). Hence PV is considered most appropriate for the recovery of ethyl acetate from water. Many different natural and synthetic polymers have been used to develop membranes for PV. As well, some polymers such as chitosan, polyvinyl alcohol, and cellulose were blended to fabricate membranes for dehydration of the alcohol/water mixture. It is also known that separation factor and permeation flux depend not only on membrane material but also on the operating conditions such as feed temperature and liquid composition (Huang *et al.*, 1999). Table 2.1 summarizes the effect of different materials for recovery of ethyl acetate from its aqueous solution. The hydrophilic polymers that have O–H group are usually preferred to prepare membranes for ethyl acetate recovery by dehydration. In addition to those listed in Table 2.1, much work has been done to develop starch-based membranes for producing renewable energy, reducing environmental effect, and finding for more applications (Lu *et al.*, 2009). Sago starch is evidently a hydrophilic polymer, which makes it an interesting material to be developed into membranes for dehydration of ethyl acetate. There are several chemical and physical modification methods to improve the properties of sago starch and one of those is blending with other polymers. Sago starch has not yet been studied as a membrane material, especially for PV. In this study, an attempt is made to separate azeotrope forming mixtures of ethyl acetate-water using sago starch/PVA blend membranes. The effects of operating conditions are also discussed in the work.

2.2 EXPERIMENTAL

2.2.1 *Materials*

Sago starch was supplied by Ng Kia Heng Sago Industries Sdn Bhd, Batu Pahat, Johor Malaysia. Polyvinyl alcohol (86,000 Mwt, 99–100% hydrolyzed) from New Jersey USA, ethyl acetate (99%)

Table 2.1.   Comparison between different materials on the separation of ethyl acetate/water mixture.

Water content in the feed [wt%]	Temperature [°C]	Membrane material	Crosslinking	Flux J [$\mathrm{kg\,m^{-2}\,h^{-1}}$]	Selectivity α	References
1–2.5	30–60	Polyvinyl alcohol	Yes	(0.0001–0.065)	(10^3–10^4)	Salt *et al.* (2005)
5.1	60	PVA/ceramic composite	No	1.45×10^{-4}	129	Xia *et al.* (2011)
9	35	Chitosan/poly(vinyl pyrrolidone)	No	0.953	746	Zhang *et al.* (2009)
97	30–50	Poly(vinylidene fluoride-co-hexafluoropropene)	No	0.2–2	100	Tian and Jiang (2008)
95	60	Commercial NaA zeolite	No	2.65	200–500	Shaha *et al.* (2000)

from USA, sulfuric acid from Thailand were obtained from QReC Chemicals (Asia) Sdn Bhd. Acetone was from Taman industry Rawang Selangor Malaysia, glutaraldehyde was purchased from Fisher Scientific Company, and water was deionized before use.

2.2.2   *Preparation of sago starch/PVA membranes*

Sago starch (3 wt%) and (10 wt%) of PVA were prepared separately by adding to and keeping in hot water of 90°C for 3 and 6 h, respectively, to make the solution homogeneous. The sago starch and PVA solutions were mixed (50/50 by weight) and kept stirred at 70°C for 24 h. The solution was then kept in an oven for 24 h before casting onto a glass plate and the film was dried in the ambient air for 72 h. The membrane so prepared was further cross-linked either by chemical reaction or by thermal treatment or both. For the chemical cross-linking the membrane was immersed in a solution containing 0.5 wt% sulfuric acid (H_2SO_4), 2.5 wt% glutaraldehyde, 48 wt% acetone and deionized water, the balance, for 30 min at room temperature. Then the membrane was washed many times by distilled water before being kept in distilled water for 7 h at room temperature to remove residual H_2SO_4. The membrane was then dried at room temperature for 24 h. For the heat treatment, the cast polymer film was kept in an oven together with the glass plate at 70°C for 4 h before it was removed from the glass plate. For the combined chemical/thermal cross-linking, the dried chemically cross-linked membrane was further heat treated. The membrane without cross-linking, cross-linked by chemical reaction, cross-linked by heat treatment and cross-linked by chemical reaction and heat treatment are called hereafter N, CH, TH and CH + TH, respectively. All membranes were subjected to PV for dehydration of ethyl acetate/water mixtures.

2.3   CHARACTERIZATION

2.3.1   *Differential scanning calorimetry (DSC) and thermogravimetric analysis (TGA)*

All membrane samples were characterized by DSC and TGA. DSC experiments were done with TA instruments model DSC 2920 differential scanning calorimeter, from 0 to 200°C at a heating rate of 10°C min^{-1}. Liquid nitrogen was used to decrease the temperature from 200 to -30°C before every run. DSC was run first with an empty aluminum pans and the weight of the samples in the cell, and obtained DSC curve was used as the baseline for DSC measurements of all

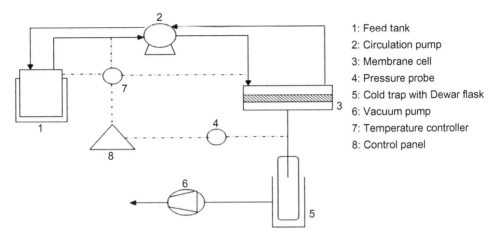

1: Feed tank
2: Circulation pump
3: Membrane cell
4: Pressure probe
5: Cold trap with Dewar flask
6: Vacuum pump
7: Temperature controller
8: Control panel

Figure 2.1. Schematic diagram of pervaporation process.

samples, whose weight was between 6 and 8 mg. TA instruments model SDT 2960 Simultaneous DSC-TGA was used for thermogravimetric analysis (TGA). The temperature range for TGA was 30–200°C with a heating rate of 10°C min^{-1}. DSC and TGA were both conducted in a helium atmosphere.

2.3.2 FTIR spectroscopy

Fourier transform infrared spectroscopy (FTIR) spectroscopy was used to quantify the chemical composition of a membrane film before and after cross-linking, and was recorded within the range of 4000–650 cm^{-1}. There were four samples named as pristine N, CH, TH and CH + TH.

2.3.3 Pervaporation

The pervaporation of ethyl acetate-water mixture was carried out using the apparatus which is described in Figure 2.1. The effective membrane area for the pervaporation was 78 cm^2. The vacuum pressure was kept under 3 mmHg and the experiments were conducted in the temperature range between 30–60°C. The concentration of water in the feed was 1–4 wt%. The permeate was collected in a collector cooled by ice. The ethyl acetate concentration was measured by refractometry using a calibration curve. The errors inherent in the pervaporation measurements were not more than 1.0 for both flux and separation factor. The total permeates flux (J) was determined as:

$$J = \frac{w}{A\Delta t} \qquad (2.1)$$

where w refers to the amount of the permeate [g], A refers to the effective area [m^2] of the membrane used in the pervaporation and (Δt) is the time [h].

The separation factor α was calculated by Equation (2.2) using the weight fractions of water and ethanol in the permeate, y_w and y_{eth}, and in the feed, x_w and x_{eth}:

$$\alpha = \frac{y_w / y_{eth}}{x_w / x_{eth}} \qquad (2.2)$$

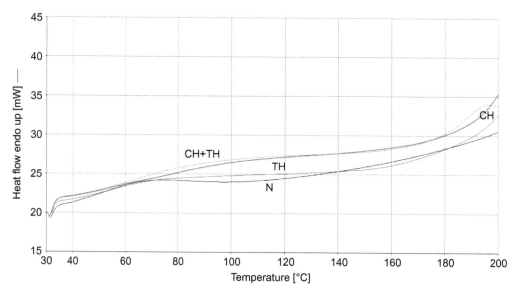

Figure 2.2.   Differential scanning calorimetry (DSC) curve of effect of different crosslinking on sago
starch/PVA blend membrane (N: non crosslinking, CH: chemical crosslinking, TH: thermal
crosslinking, CH + TH: chemical + thermal crosslinking).

2.4   RESULTS AND DISCUSSION

2.4.1   *Membrane characterization*

2.4.1.1   *DSC*
Sago starch has a helical structure which can be crystallized and polyvinyl alcohol is semi-crystalline. For both of them, the structure may change by cross-linking. From the DSC curves N and TH samples showed glass transition temperature (T_g) at 60°C, CH showed T_g at 96°C and (CH + TH) showed T_g and melting point (T_m) at 90°C and 208°C, respectively. Thus the structure of the membranes changed considerably by cross-linking as presented in Figure 2.2.

2.4.1.2   *Thermal stability TGA*
The thermal stability of sago starch/PVA membrane and its cross-linked derivatives were investigated by TGA analysis under nitrogen flow. The resulting thermograms are shown in Figure 2.3. From the thermograms, it is clear that all membranes exhibited three consecutive steps for weight loss. In N (Fig. 2.3a) the first weight loss of 20% occurred in a range of temperature between 60 and 250°C. The second weight loss of 6% occurred between 250 and 300°C, corresponding to physically absorbed water molecules. The majority of these absorbed water molecules exist in a bound state rather than in a free molecular state (Guana *et al.*, 2006; Shaoa *et al.*, 2003) and looks like to be bound directly to the polymer chain through hydrogen bonds. For CH membrane shown in Figure 2.3b, the first weight loss of 22% occurred at temperature between 50 and 300°C. Compared to N membrane the weight loss increased in a wider temperature range up to 300°C. This also is attributed to desorption of physically absorbed water molecules in the membrane. The second weight loss for CH membrane was 80% in a temperature range between 300 and 450°C, which corresponds to decomposition of sago starch and polyvinyl alcohol. For TH membrane shown in Figure 2.3c the first weight loss of 19.87% occurred between 60 and 200°C, which is due to a loss of absorbed water. The second weight loss of 83% occurred at around 200–400°C, which may correspond to the structural decomposition of sago starch/PVA. For CE + TH membrane the first weight loss of 12.17% occurred between 50 and 280°C, due to its physical change

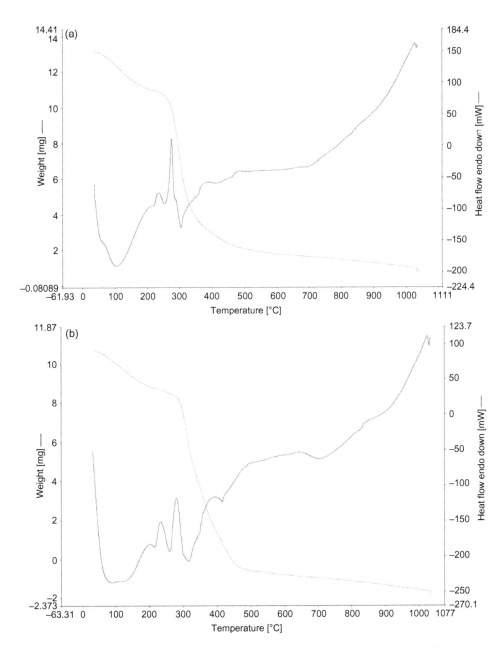

Figure 2.3.   (a) TGA for non cross-linking sago/PVA membrane, (b) TGA for chemical cross-linking membrane, (c) TGA for thermal cross-linking membrane and (d) TGA for chemical + thermal cross-linking membrane (N: non crosslinking, CH: chemical crosslinking, TH: thermal crosslinking, CH + TH: chemical + thermal crosslinking).

in the polymer such as dehydration. The second weight loss of 87% occurred between 300 and 450°C as shown in Figure 2.3d. It can be thus concluded that CH + TH membrane was the most stable thermally among all the tested membranes. Generally in pervaporation operating temperature is kept below 100°C. Hence, all the tested membranes seem to be thermally stable enough (Xiao *et al.*, 2006).

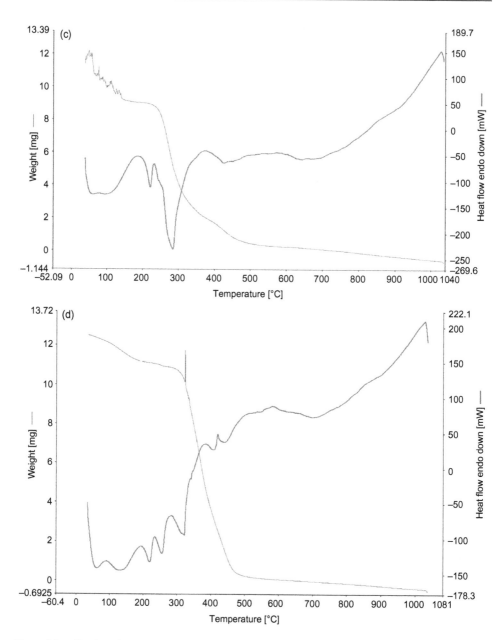

Figure 2.3.   *Continued.*

2.4.1.3   FTIR

Figure 2.4 shows the FTIR spectra for all tested membranes (N, TH, CH, CH + TH). The absorption peak of IR spectrum for N membrane appears at 3266.88 cm^{-1} corresponding to the hydroxyl (−OH) group. For chemical cross-linking (CH) as we can see the hydroxyl group peak decreased slightly due to the reaction with aldehyde group during the cross-linking reaction. Chemically cross-linked membrane showed a new peak at 1714.31 cm^{-1} for C−O−C and also another peak at 1256.40 cm^{-1} is likely due to C−C from acetal linkages. Sago starch and PVA have a very large number of hydroxyl groups, and not all of them were reacted with aldehyde during the limited

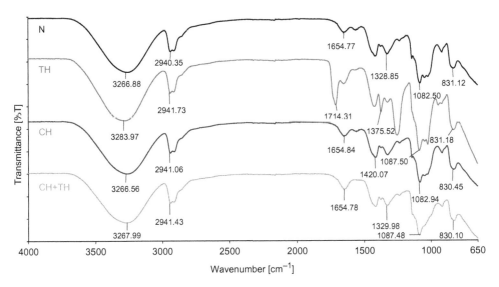

Figure 2.4.   Fourier transform infrared (FTIR) for sago/PVA membranes (N: non crosslinking, CH: chemical crosslinking, TH: thermal crosslinking, CH + TH: chemical + thermal crosslinking).

time. On the contrary, the thermally cross-linking showed increase in the absorption peak of 3283.97 cm^{-1} after heat treatment. Chemical + thermal cross-linked membrane showed the same result as thermally cross-linked membrane,which can be explained by completion of cross-linking by the thermal treatment.

2.4.2   *Pervaporation*

2.4.2.1   *Effect of feed temperature*

Operating temperature is an important parameter in pervaporation due to its effect on the sorption and diffusion rates (Moheb Shahrestani *et al.*, 2013). Figure 2.5a shows total flux in a range of 30–60°C when the water content in the feed ethyl acetate/water mixture is fixed to 2 wt%. From the figure, the total flux increases as the temperature increases. From Figure 2.5a N membrane's flux increased from 0.03 kg m^{-2} h^{-1} at 30°C to 0.3 kg m^{-2} h^{-1} at 60°C. This can be explained by the increase in mobility of polymer chains and the expansion of the free volume. CH membrane showed the flux of 0.162 kg m^{-2} h^{-1} at 30°C and 0.549 kg m^{-2} h^{-1} at 60°C, which were higher than the N membrane. For the TH membrane, the fluxes were 0.217 kg m^{-2} h^{-1} at 30°C and 1.192 kg m^{-2} h^{-1} at 60°C, which were even higher than both of N and CH membranes. Interestingly, the flux was increasing continuously with the temperature, while for the other membranes a flux drop at 50°C was observed. The CH + TH membrane achieved the highest permeation flux among all the membranes, i.e. the flux was 0.148 kg m^{-2} h^{-1} at 30°C, which was lower than the CH and TH membranes, but the highest flux of 4.967 kg m^{-2} h^{-1} was reached at 60°C. Figure 2.5b shows the complicated nature of the feed temperature effect on the separation factor. The separation factor at 60°C was 9000 for N membrane and 1960, 1760 and 400 for TH, CH and CH + TH membrane, respectively, demonstrating the highest selectivity for the N membrane which is without cross-linking.

2.4.2.2   *Effect of feed concentration*

The effect of water content on the permeation flux is shown in Figure 2.6. All PV experiments were carried out at 30°C. As seen from Figure 2.6, generally, flux increases with an increase in water content from 1 to 4 wt% for all the membranes. This can be explained by two reasons.

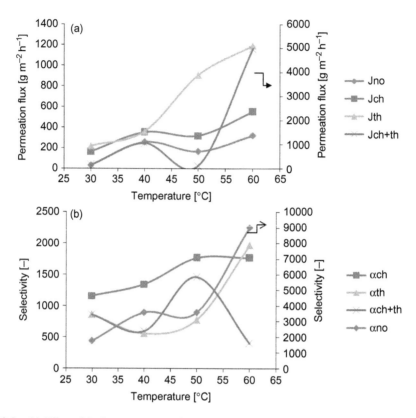

Figure 2.5.   (a) Effect of feed temperature on the permeation flux for all membranes and (b) effect of feed temperature on the separation factor for all membranes.

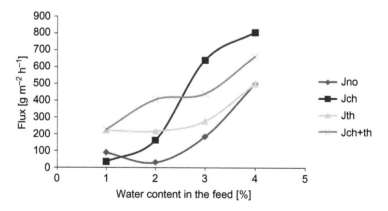

Figure 2.6.   Effect of feed concentration on the permeation flux for all membranes.

First, as the water content increases in the feed the water transport increases and hence the total permeation flux increases, and the second, at higher water content the membrane swells more, which leads to increased flexibility of the polymer network and both ethyl acetate and water cross the membrane more easily. Hence, the flux increases but the separation factor may decrease (Hyder *et al.*, 2006). The order in the flux changes with water content; i.e. at 1 wt% water content CH membrane showed the lowest flux while at 4 wt% water content CH membrane showed the highest flux. As mentioned above, when the water content increases in the feed, the degree of swelling of the membrane will increase. Moreover, when water content increases, the coupling of water and ethyl acetate transport is enhanced.

2.5   CONCLUSION

The intrinsic properties of sago starch/PVA blend membranes with and without cross-linking were characterized by FTIR, DSC and TGA. FTIR revealed that the hydroxyl groups of sago starch and PVA decreased by chemical cross-linking due to formation of acetal linkage. As a result, the membranes became less hydrophilic. On the other hand, CH+TH and TH membranes are more hydrophilic than CH membrane. DSC showed increase in T_g by cross-linking. PV of ethyl acetate/water mixture showed that the flux of the CH + TH membrane is the highest but the selectivity is the lowest due to the hydrophilic nature of the membrane. TH membrane with good permeation flux and high selectivity is considered to be the best.

ACKNOWLEDGEMENT

The authors acknowledge University of Technology Malaysia, and Chemical Engineering Department for their support (No. 03H73).

REFERENCES

Feng, X. & Huang, R.Y.M. (1997) Liquid separation by membrane pervaporation: a review. *Industrial & Engineering Chemistry Research*, 36 (4), 1048–1066.

Ghazali, M., Nawawi, M. & Huang, R.Y.M. (1997) Pervaporation dehydration of isopropanol with chitosan membranes. *Journal of Membrane Science*, 124 (1), 53–62.

Guan, H.-M. Chung T.-S., Huang Mei, Z., Chng, L. & Kulprathipanja, S. (2006) Poly(vinyl alcohol) multilayer mixed matrix membranes for the dehydration of ethanol-water mixture. *Journal of Membrane Science*, 268 (2), 113–122.

Hasanoğlu, A. Salt, Y., Keleşer, S., Özkan, D. & Dinçer, S. (2005) Pervaporation separation of ethyl acetate-ethanol binary mixtures using polydimethylsiloxane membranes. *Chemical Engineering and Processing: Process Intensification*, 44 (3), 375–381.

Huang, R.Y.M., Pal, R. & Moon, G.Y. (1999) Characteristics of sodium alginate membranes for the perva-poration dehydration of ethanol-water and isopropanol-water mixtures. *Journal of Membrane Science*, 160 (1), 101–113.

Hyder, M.N., Huang, R.Y.M. & Chen, P. (2006) Correlation of physicochemical characteristics with per-vaporation performance of poly(vinyl alcohol) membranes. *Journal of Membrane Science*, 283 (1–2), 281–290.

Lipnizki, F., Field, R.W. & Ten, P.-K. (1999) Pervaporation-based hybrid process: a review of process design, applications and economics. *Journal of Membrane Science*, 153 (2), 183–210.

Liu, Y., Zhu, M., Zhao, Q., An, Q., Qian, J., Lee, K. & Lai, J. (2011) The chemical crosslinking of poly-electrolyte complex colloidal particles and the pervaporation performance of their membranes. *Journal of Membrane Science*, 385–386, 132–140.

Lu, D.R., Xiao, C.M. & Xu, S.J. (2009) Starch-based completely biodegradable polymer materials. *EXPRESS Polymer Letters*, 6, 366–375.

Moheb Shahrestani, M., Moheb, A. & Ghiaci, M. (2013) High performance dehydration of ethyl acetate/water mixture by pervaporation using NaA zeolite membrane synthesized by vacuum seeding method. *Vacuum*, 92, 70–76.

Rachipudi, P.S., Kariduraganavara, M.Y., Kitturb, A.A. & Sajjana, A.M. (2011) Synthesis and character-ization of sulfonated-poly(vinyl alcohol) membranes for the pervaporation dehydration of isopropanol. *Journal of Membrane Science*, 383 (1–2), 224–234.

Salt, Y., Hasanoğlu A., Salt, I., Keleşer, S., Özkan, S. & Dinçer, S. (2005) Pervaporation separation of ethylacetate-water mixtures through a crosslinked poly(vinylalcohol) membrane. *Vacuum*, 79 (3–4), 215–220.

Shah, D., Kissick, K., Ghorpade A., Hannah, R. & Bhattacharyyaa, D. (2000) Pervaporation of alcohol-water and dimethylformamide-water mixtures using hydrophilic zeolite NaA membranes: mechanisms and experimental results. *Journal of Membrane Science*, 179 (1–2), 185–205.

Shaoa, C., Kim, H.-Y., Gong, J., Ding, B., Lee, D.-R. & Park, S.J. (2003) Fiber mats of poly(vinyl alcohol)/silica composite via electrospinning. *Materials Letters*, 57 (9–10), 1579–1584.

Tian, X. & Jiang, X. (2008) Poly(vinylidene fluoride-co-hexafluoropropene) (PVDFHFP) membranes for ethyl acetate removal from water. *Journal of Hazardous Materials*, 153 (1–2), 128–135.

Wu, S., Wang, J., Liu, G., Yang, Y. & Lu, J. (2012) Separation of ethyl acetate (EA)/water by tubular silylated MCM-48 membranes grafted with different alkyl chains. *Journal of Membrane Science*, 390–391, 175–181.

Xia, S., Dong, X., Zhu, Y., Wei, W., Xiangli, F. & Jin, W. (2011) Dehydration of ethyl acetate-water mixtures using PVA/ceramic composite pervaporationmembrane. *Separation and Purification Technology*, 77 (1), 53–59.

Xiao, S., Huang, R.Y.M. & Feng, X. (2006) Preparation and properties of trimesoyl chloride crosslinked-poly(vinyl alcohol) membranes for pervaporation dehydration of isopropanol. *Journal of Membrane Science*, 286 (1–2), 245–254.

Yongquan, D., Ming, W., Lin, C. & Mingjun, L. (2012) Preparation, characterization of P(VDF-HFP)/[bmim]BF4 ionic liquids hybrid membranes and their pervaporation performance for ethyl acetate recovery from water. *Desalination*, 295, 53–60.

Yuan, H.-K., Ren, J., Ma, X.-H. & Xu, Z.-L. (2011) Dehydration of ethyl acetate aqueous solution by pervaporation using PVA/PAN hollow fiber composite membrane. *Desalination*, 280 (1–3), 252–258.

Zhang, X.H., Liu, Q.L., Xiong, Y., Zhu, A.M., Chen, Y. & Zhang, Q.G. (2009) Pervaporation dehydration of ethyl acetate/ethanol/water azeotrope using chitosan/poly (vinyl pyrrolidone) blend membranes. *Journal of Membrane Science*, 327 (1–2), 274–280.

CHAPTER 3

Development of adsorbents based cellulose acetate mixed matrix membranes for removal of pollutants from textile industry effluent

Ramesh Kumar Saranya, Yogarathinam Lukka Thuyavan,
Gangasalam Arthanareeswaran & Ahmad Fauzi Ismail

3.1 INTRODUCTION

Polymeric modification using adsorptive inorganic particles is gaining prime importance for its significance in enhancing the filtration properties of membranes. Membranes can remove various pollutants through the separation mechanisms of impaction, diffusion, electrostatic interaction, hydrophobic property, and adsorption (Walther et al., 1988). Among several mechanisms, adsorption is the process most commonly used in to remove the dye constituents of textile industry effluent. The combination of different methods like adsorption and filtration for enhanced membrane separation is highly presumed to achieve significant reduction on pollutants due to textile industries. For a long time, application of activated carbon (AC) as dye adsorbent in textile and dye wastewater treatment has been studied (Choy et al., 1999; Kannan et al., 2001; Malik, 2003; Namasivayam and Kavitha, 2002) because of its large surface area, polymodal porous structure, high adsorption capacity and variable surface chemical composition (Marsh Rodríguez-Reinoso, 2006). However, the addition of AC in organic polymer matrix for the synthesis of mixed matrix membranes (MMMs) has been very rarely studied. Ballinas et al. (2004) studied the influence of characters of AC on polysulfone (PSF) composite membranes in a continuous operation system and concluded the high performance of a hybrid AC/PSF composite membrane better than non-hybrid ones. The effect of polyethersulfone concentration and AC loading on the performance of MMMs in terms of permeability and selectivity of O_2/N_2 gas separation has also been investigated (Kusworo et al., 2010). Yet, the studies on application of AC coupled with membrane separation in wastewater treatment show much less evidence.

Iron oxide (IO) particles due to their high surface reactivity as a result of their small size range of 1–100 nm (Zhang, 2003) can transform chemical pollutants such as PCBs, TCE, pesticides and chlorinated organic solvents in biosolids (Li et al., 2007). They have the ability to degrade pollutants effectively by reductive (Tratnyek and Johnson, 2006; Xu et al., 2005) and/or oxidative pathways depending on the state of iron (Laine and Cheng, 2007; Pignatello et al., 2006). Several methods to synthesize IO nanoparticles were available but the main challenges of these numerous and novel techniques lie in their capacity to obtain a narrow dispersion in particle size together with the desired compositional, structural and crystalline uniformity (Martínez-Mera et al., 2007). The interface between a polymer matrix and inorganic filler plays an important role in the performance of MMMs. Hence, in this study, the feasibility of synthesis of AC and IO based MMMs was examined. The recent development of MMMs makes it an ideal platform for incorporation of adsorptive particles for the removal of pollutants from textile industry effluent.

To fully exploit the use of AC and IO for an adsorbent enhanced membrane filtration, different wt% of AC and IO was blended in solution with cellulose acetate (CA). The influence of adsorption and reaction of incorporated nanoparticles owing to dispersion was confirmed by ultrasonication followed by which the CA/AC and CA/IO membranes have been successfully synthesized by phase inversion method. The characteristics of MMMs in terms of surface morphology, particle size distribution have been studied by field emission scanning electron microscopy (FESEM) and

X-ray diffraction (XRD), respectively. The performance of prepared MMMs was also evaluated for its permeability and rejection efficiency of chemical oxygen demand (COD), biological oxygen demand (BOD) and sulfates from textile dye effluent by means of polymer enhanced ultrafiltration.

3.2   EXPERIMENTAL

3.2.1   Materials

All chemicals and reagents used were of analytical grade and used without any further purification. Cellulose acetate (CA) was obtained from Sigma Aldrich, India Limited. N,N'-dimethyl formamide (DMF) was purchased from Qualigens fine chemicals, Glaxo India limited. Analytical grade AC prepared from charcoal with the surface area of 800 to 850 $m^2\,g^{-1}$ and methylene blue adsorption of 0.13 $g\,g^{-1}$ was purchased from Merck (India) Limited. IO particles were also procured from Merck India Limited. Polyethyleneimine (PEI) of 50 kDa MW was obtained from Alfa Aesar (India) Limited. Double distilled and Millipore water were produced in the laboratory.

3.2.2   Preparation of CA MMMs

The phase inversion method of membrane synthesis was carried out for the fabrication of CA/IO and CA/AC MMMs. The composition of various CA MMMs prepared is provided in Table 3.1. The homogenous casting solution for each composition was obtained by dissolving IO and AC initially in DMF solvent which after complete dissolution by means of ultrasonication was mixed with CA under mechanical stirring. The final dope solution was also sonicated for 30 min before casting to ensure homogeneity and casted on a cleaned glass plate. The thickness of the membrane was fixed as 400 μm using a film applicator (elicometer). After leaving for 30 s evaporation time, the membrane was immersed in distilled water maintained at 20°C.

3.2.3   Characterization of CA MMMs

The X-ray diffraction (XRD) pattern was recorded for both neat CA and CA MMMs using an X-ray diffractometer (Rigaku Corporation, Japan) constituting Ni-filtered Cu Kα as a monochromatic radiation source (40 kV, 30 mA) with scintillation counter (NaI) as a detector.

The top and cross-section morphology of synthesized CA MMMs were visualized with the help of field emission scanning electron microscopy (FESEM, Hitachi, S-4160, Japan) operated at an accelerating voltage of 20 kV. The dried membrane samples were pre-treated using Au sputtering to impart electrical conductivity.

Table 3.1.   Casting composition of prepared CA MMMs.

Membrane type	Casting solution composition			Membrane description
	CA [g]	AC/IO [g]	DMF solvent [mL]	
CA	4.375	–	21.7	Neat CA
CA/AC1	4.353	0.022	21.7	CA + 0.5 wt% AC
CA/AC2	4.310	0.065	21.7	CA + 1.5 wt% AC
CA/AC3	4.266	0.109	21.7	CA + 2.5 wt% AC
CA/IO1	4.354	0.022	21.7	CA + 0.5 wt% IO
CA/IO2	4.310	0.065	21.7	CA + 1.5 wt% IO
CA/IO3	4.266	0.109	21.7	CA + 2.5 wt% IO

3.2.4 *Performance of CA MMMs*

3.2.4.1 *Permeability studies*

The flat sheet CA MMMs were employed in stirred cell dead-end ultrafiltration (UF) unit (Model Cell-XFUF076, Millipore, USA) for studying the pure water permeability. The effective area of the membrane taken was $38.5\,cm^2$. The initial membrane compaction was performed for 1 h at 60 psi (≈ 0.414 MPa) trans-membrane pressure (TMP). Later, the test run for determining the pure water flux was done and the permeate volume was collected every 10 min. The pure water flux (J_w) was found using the following equation:

$$J_w = \frac{V}{A \times \Delta t} \tag{3.1}$$

where V, A and Δt are the permeate volume, membrane effective area and permeation time, respectively.

3.2.4.2 *Performance of CA MMMs in the treatment of textile industry effluent*

To evaluate the rejection efficiency of CA MMMs in reducing COD, BOD and sulfates from the textile dye effluent, polymer enhanced UF was performed. The feed solution of raw textile dye effluent was first complexed with the water-soluble polyelectrolyte PEI. The concentration of PEI was taken to be 1 wt% and completely mixed with PEI under stirring for 24 h (Arthanareeswaran *et al.*, 2007). The macroligands formed due to the PEI complexation with the organic pollutants in the effluent enables improved rejection using UF (Juang and Chen, 1996). The permeate volume was collected in replicates over distinct intervals at constant TMP of 60 psi (≈ 0.414 MPa) and which were then analyzed for COD, BOD and SO_4^{2-} concentration based on the standard methods (APHA, 2005).

Then, the rejection efficiency of each CA MMM was determined separately using the following equation:

$$\% \text{ Rejection } (R) = \frac{C_f - C_p}{C_f} \tag{3.2}$$

where C_f and C_p are the concentration of feed and permeate respectively for determining the rejection of COD, BOD and SO_4^{2-}. However, the actual rejection [%] can be evaluated using Equation (3.3) from the concentration of retentate obtained after filtration of specific volume at specific time:

$$\% \text{ Actual rejection } (R_{act}) = \frac{C_r - C_p}{C_r} \tag{3.3}$$

where C_r is the concentration of reject volume.

3.3 RESULTS AND DISCUSSION

3.3.1 *Morphological characteristics of CA MMMs*

The top surface morphology of CA/AC and CA/IO MMMs (Fig. 3.1) reveals that the ultrason-ication has its significance on dispersing the adsorbent particles in uniform manner in the CA matrix. The dense symmetric structure with no differentiation between top finger pores and bottom macrovoids has been observed for neat CA membrane. However, surface morphology of CA MMMs suggests the efficient dispersion of IO and AC has provided accessible active sites for pollutants to be adsorbed (Smuleac *et al.*, 2010). It can also be seen that most IO particles were distributed uniformly on CA as shown in CA/IO1 and CA/IO3 MMMs and the few clusters of IO particles created pores when they leached out of the CA matrix during phase inversion. This leads to increase in permeability of CA/IO1 MMM however the reason for the lesser permeability offered by CA/IO3 MMMs compared to CA/IO1 is attributed to the surface pore blockage on

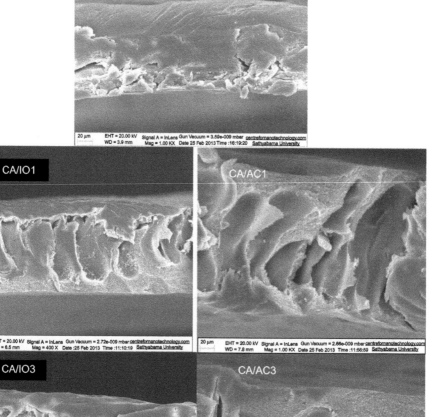

Figure 3.1.   Scanning electron microscope SEM micrographs of neat CA and CA MMMs.

highest addition of about 2.5 wt% of IO particles. With the highest IO loading, the particles completely lose structure and thus bringing deformation of the skin layer and pore structure (Jian *et al.*, 2006). AC that interacts with CA polymer can change dope solution properties significantly and thus enhancing the macrovoids formation in CA/AC MMMs. The cross-section FESEM images of CA/AC MMMs shows a number of elongated porous sub-structures when compared with neat CA membrane. From the results, it could be inferred that increase in AC concentration has caused the hydrophobic interactions between CA and carbon that subsequently helped to increase pore formation. These macrovoidal formation at the bottom layer helps for improved permeability of CA/AC MMMs.

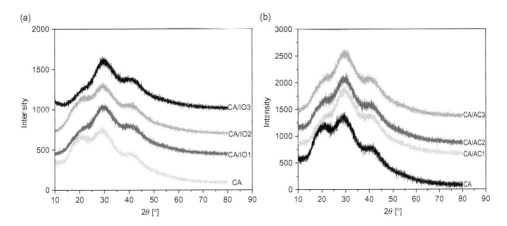

Figure 3.2.   XRD profile of CA/IO MMMs and CA/AC MMMs.

Table 3.2.   Crystallite size of CA/IO and CA/AC MMMs.

Membrane type	Crystallite size [nm]
CA	21.9
CA/IO1	4.9
CA/IO2	8.3
CA/IO3	10.6
CA/AC1	25.3
CA/AC2	29.2
CA/AC3	34.9

3.3.2   *XRD characterization of CA MMMs*

The effect of IO and AC addition on CA has also been studied with the help of XRD pattern as shown in Figure 3.2. The peaks corresponding to crystal lattice of (110) and (300) are characteristic of Fe (0) core and IO shell and are observed in all CA/IO MMMs. The broad peak at 2θ angle of $20.5°C$ corresponds to amorphous CA which changed into the crystalline nature of cubic lattice of IO as seen in CA/IO MMMs. The grain size of CA/IO and CA/AC MMMs of different compositions of IO and AC has been shown in Table 3.1. The crystallite size was observed to increase from 4.9 to 10.6 nm due to the increase of IO wt% in CA/IO3 MMMs. This confirms the presence of reactive IO nanoparticles and its crystalline behavior as the results were in agreement with the Braggs lattice of Magnetite: 01-111 (Daraei *et al.*, 2012). Similarly, the presence of AC in CA/AC MMMs was confirmed by the intensity peaks at 2θ angle of $40.6°$ and the crystallite size were lower compared to CA/IO3 MMMs.

3.3.3   *Effect of IO and AC on permeability of CA MMMs*

The effect of IO and AC on the pure water permeability of CA MMMs has been shown in Figure 3.3. The highest flux of $11.22 \, L \, m^{-2} \, h^{-1}$ was observed for CA/AC3 membrane, which suggests that the increase in AC content increases the permeability. The presence of AC in the CA matrix has changed the membrane pore morphology of CA (Hwang *et al.*, 2013) and hence the flux of CA/AC MMMs increased compared to pure CA. The macrovoidal pore formation owing to AC addition as seen in FESEM image of CA/AC3 MMM supports the flux enhancement. The mixing of AC and CA resulted in disintegration of CA polymeric chains that subsequently created

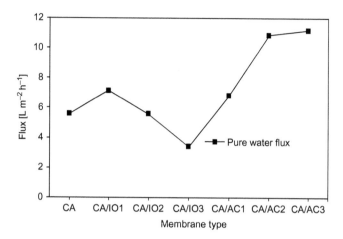

Figure 3.3.   Pure water flux studies on synthesized MMMs.

interfacial voids. This change in interfacial membrane morphology is mainly due to the adhesion of AC in CA matrix and thus contributes to the enhancement of permeability.

In case of CA/IO MMMs, the increase in wt% of IO reduced the water permeability from 7.17 to 3.63 L m^{-2} h^{-1} and the reason would be due to the obstruction of majority of holes and channels by the presence of IO particles at higher loading of 1.5 and 2.5 wt% (Gholami *et al.*, 2014). The permeation capacity also depends on the thinner top-layer of the membrane and hence the increase in thickness due to increased IO content tends to reduce the water permeability in case of CA/IO3 MMMs. Hence, the minimal loading of IO of about 0.5 wt% has only helped to retain the nanoparticle size of IO without aggregation that ultimately enhanced the permeability properties of CA.

3.3.4   *Effect of AC and IO on polymer assisted UF*

PEI complexation leads to the formation of macromolecules which resulted in decrease of permeate flux of the textile dye effluent (Fig. 3.4). The virgin CA showed 2.56 L m^{-2} h^{-1} of permeate flux for PEI complexed textile effluent which is 48.43% lower compared to the flux of effluent without PEI complexation. The binding capacity due to imine groups of PEI and its electrostatic interactions with molecules present in the effluent caused increase in liquid phase polymer-based retention (LPR) of PEI bound effluent that subsequently led to a decrease in permeate flux (Rivas *et al.*, 2003). However, the presence of IO and AC in the CA polymer matrix helped to decrease the reduction in permeate flux owing to PEI complexation.

Inspite of PEI complexation, for CA/AC3 MMM there was observed only a 25% flux decrease compared to the flux observed for effluent without PEI. The reason for decrease in reduction of flux difference would be due to the surface non-electrostatic interactions of AC (Moreno-Castilla, 2004). For CA/IO1 MMMs, there was observed a reduction in decrease of permeate flux of only about 35% compared to that of effluent having no PEI binding. The reactive and catalytic property of IO particles has been effectively utilized as they are situated at CA polymeric matrix. The reactivity of IO helped potentially for the inhibition of PEI complexed pollutants due to its surface acidity that tends to be involved in proton exchange and hence not much in reduction of effluent flux of CA MMMs compared to virgin CA.

3.3.5   *Effect of CA MMMs on COD removal*

Similar to BOD, the rejection efficiency of CA/IO in reducing COD was higher (by about 89.9%) with the reject concentration of 654 mg L^{-1}. The natural organic matter (NOM) and heavy metal

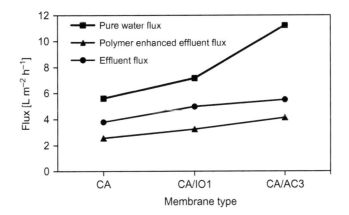

Figure 3.4.   Effect of PEI on flux of textile dye effluent.

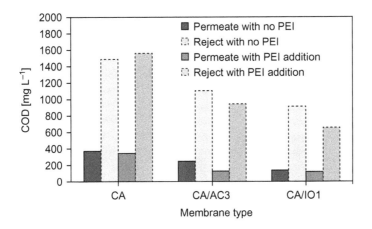

Figure 3.5.   Effect of CA MMMs on COD removal.

concentration were greatly reduced and there had been evidence in using iron nanoparticles coated membranes for such removal (Hosseini *et al.*, 2012; Ng *et al.*, 2010). Like IO nanoparticles, the presence of AC in CA/AC MMM has also influenced the COD rejection efficiency to about 89.2% owing to the addition of PEI. However, the reject COD concentration was high of about 944 mg L^{-1} as compared to 654 mg L^{-1}, which was the COD reject concentrate due to CA/IO MMMs (Fig. 3.5).

3.3.6   *Effect of CA MMMs on BOD removal*

The BOD removal is highly intense for CA/IO MMM operated with PEI bound effluent that showed about 96.7% reduction owing to fouling resistant ability of magnetite nanoparticles. The reject BOD concentration is 84 mg L^{-1} that ensures the adsorption capacity of IO impregnated in CA matrix. However, the reject BOD of neat CA and CA/AC MMMs as shown in Figure 3.6 indicated increase in concentration of BOD of about 284 and 372 mg L^{-1}, respectively. It can be concluded that superior hydrophilicity of CA/IO would also be the reason for reducing the irreversible fouling and resulted in higher flux with lesser % rejection.

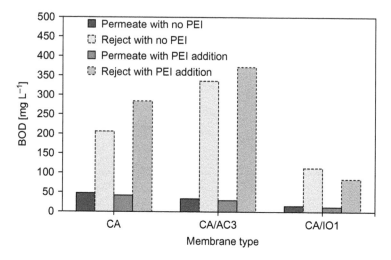

Figure 3.6.   Effect of CA MMMs on BOD removal.

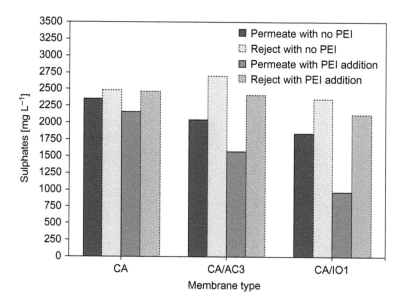

Figure 3.7.   Effect of CA MMMs on sulfate removal.

3.3.7   *Effect of CA MMMs on sulfate removal*

In contrast to BOD and COD rejection efficiency, the highest sulfate rejection was observed to be 65.8% for PEI complexed effluent treated using CA/IO MMM in Figure 3.7. The reason for lesser rejection is attributed to the fact that the dissolved salts pass through pores of UF membranes easily resulting in only less % rejection in comparison with those of COD and BOD. The CA/AC MMMs showed 44.72% sulfate rejection that was better than neat CA as it was observed to have only 12.8% sulfate rejection. These results confirm that the reducing ability of AC and IO nanoparticles present in CA matrix.

3.4 CONCLUSIONS

The effect of loading of IO and AC in CA matrix was widely investigated based on the permeation flux and pore morphology. Based on these investigations, some of the significant conclusions are as follows:

- Among various synthesized CA MMMs, CA/AC3 MMMs with the highest loading of 2.5 wt% of AC exhibited superior permeability and macrovoids formation in the sub-layer of its asymmetric structure.
- Compared to AC, the impregnation of IO offered high catalytic reduction of the pollutants thereby showing the rejection efficiency of about 96.7, 89.86 and 55.5% of BOD, COD and SO_4^{2-}, respectively, from the textile dye effluent.

The advantage of adsorption in decreasing the concentration of COD, BOD and sulfates in the retentate of textile effluent was the major breakthrough in this polymer enhanced UF using CA MMMs.

REFERENCES

APHA (2005) Standard methods for the examination of water and waste water. American Public Health Association, Washington, DC.

Arthanareeswaran, G., Thanikaivelan, P., Jaya, N., Mohan, D. & Raajenthiren, M. (2007) Removal of chromium from aqueous solution using cellulose acetate and sulfonatedpoly (ether ether ketone) blend ultrafiltration membranes. *Journal of Hazardous Materials* B, 139, 44–49.

Ballinas, L., Torras, C., Fierro V. & Garcia-Valls, R. (2004) Factors influencing activated carbon-polymeric composite membrane structure and performance. *Journal of Physics and Chemistry of Solids*, 65, 633–637.

Choy, K.K.H., McKay, G. & Porter, J.F. (1999) Sorption of acid dyes from effluents using activated carbon. *Resources Conservation Recycling*, 27 (1–2), 57–71.

Daraei, P., Madaeni, S.S., Ghaemi, N., Salehi, E., Khadivi, M.A., Moradian, R. & Astinchap, B. (2012) Novel polyethersulfonenanocomposite membrane prepared by PANI/Fe$_3$O$_4$ nanoparticles with enhanced performance for Cu(II) removal from water. *Journal of Membrane Science*, 415–416, 250–259.

Gholami, A., Moghadassi, A.R., Hosseini, S.M., Shabani, S. & Gholami, F. (2014) Preparation and characterization of polyvinyl chloride based nanocomposite nanofiltration membrane modified by iron oxide nanoparticles for lead removal from water. *Journal of Industrial Engineering Chemistry*, 20, 1517–1522.

Hosseini, S.M., Madaeni, S.S., Heidari, A.R. & Amirimehr, A. (2012) Preparation and characterization of ion-selective polyvinylchloride based heterogeneous cation exchange membrane modified by magnetic iron-nickel oxide nano-particles. *Desalination*, 284, 191–199.

Hwang, L., Chen, J. & Wey, M. (2013) The properties and filtration efficiency of activated carbon polymer composite membranes for the removal of humic acid. *Desalination*, 313, 166–175.

Jian, P., Yahui, H., Yang, W. & Linlin, L. (2006) Preparation of polysulfone-Fe$_3$O$_4$ composite ultrafiltration membrane and its behavior in magnetic field. *Journal of Membrane Science*, 284, 9–16.

Juang, R.S. & Chen, M.N. (1996) Measurement of binding constants of poly(ethylenimine) with metal ions and metal chelates in aqueous media by ultrafiltration. *Industrial and Engineering Chemistry Research*, 35, 1935–1943.

Kannan, N. & Sundaram, M.M. (2001) Kinetics and mechanism of removal of methylene blue by adsorption on various carbons – a comparative study. *Dyes and Pigments*, 51 (1), 25–40.

Kusworo, T.D., Ismail, A.F., Mustafa, A. & Budiyono, B. (2010) Application of activated carbon mixed matrix membrane for oxygen purification. *International Journal of Science and Engineering*, 1 (1), 21–24.

Laine, D.F. & Cheng, I.F. (2007) The destruction of organic pollutants under mild reaction conditions: a review. *Microchemical Journal*, 85 (2), 183–193.

Li, X.Q., Brown, D. & Zhang, W.X. (2007) Stabilization of biosolids with nanoscale zero-valent iron (nZVI). *Journal of Nanoparticle Research*, 9, 233–243.

Malik, P.K. (2003) Use of activated carbons prepared from sawdust and rice-husk for adsorption of acid dyes: a case study of Acid Yellow 36. *Dyes and Pigments*, 56 (3), 239–249.

Marsh, H. & Rodríguez-Reinoso, F. (2006) *Activated carbon.* Elsevier, Amsterdam, The Netherlands.

Martínez-Mera, I., Espinosa-Pesqueira, M.E., Perez-Hernandez, R. & Arenas-Alatorre, J. (2007) Synthesis of magnetite (Fe_3O_4) nanoparticles without surfactants at room temperature. *Material Letters*, 61, 4447–4451.

Moreno-Castilla, C. (2004) Adsorption of organic molecules from aqueous solutions on carbon materials. *Carbon*, 42, 83.

Namasivayam, C. & Kavitha, D. (2002) Removal of Congo Red from water by adsorption onto activated carbon prepared from coir pith, an agricultural solid waste. *Dyes and Pigments*, 54 (1), 47–58.

Ng, L.Y., Mohammad, A.W., Leo, C.P. & Hilal, N. (2010) Polymeric membranes incorporated with metal/metal oxide nanoparticles: a comprehensive review. *Desalination*, 261 (3), 313–320.

Pignatello, J.J., Oliveros, E. & MacKay, A. (2006) Advanced oxidation processes for organic contaminant destruction based on the Fenton reaction and related chemistry. *Critical Reviews on Environmental Science and Technology*, 36 (1), 1–84.

Rivas, B.L., Pereira, E.D. & Moreno-Villoslada, I. (2003) Water-soluble polymer-metal ion interactions. *Progress in Polymer Science*, 28, 173–208.

Smuleac, V., Bachas, L. & Bhattacharyya, D. (2010) Aqueous-phase synthesis of PAA in PVDF membrane pores for nanoparticles synthesis and dichlorobiphenyl degradation. *Journal of Membrane Science*, 346 (2), 310–317.

Tratnyek, P.G. & Johnson, R.L. (2006) Nanotechnologies for environmental cleanup. *Nanotechnology Today*, 1 (2), 44–48.

Walther, H., Faust, S.D. & Aly, O.M. (1988) Adsorption processes for water treatment. *Acta Hydrochimica et Hydrobiologica*, 16, 572–591.

Xu, J., Dozier, A. & Bhattacharyya, D. (2005) Synthesis of nanoscale bimetallic particles in polyelectrolyte membrane matrix for reductive transformation of halogenated organic compounds. *Journal of Nanoparticle Research*, 7 (4–5), 449–467.

Zhang, W.X. (2003) Nanoscale iron particles for environmental remediation: an overview. *Journal of Nanoparticle Research*, 5, 323–332.

CHAPTER 4

Studying the role of magnetite (Fe$_3$O$_4$) colloids functionality on PES membrane in removing of humic acid foulant using QCM-D

Qi-Hwa Ng, Jit-Kang Lim, Abdul Latif Ahmad & Siew-Chun Low

4.1   INTRODUCTION

Over the past few decades, the usage of membranes for potable water production has progressed remarkably due to the compact design of the membrane module, lower energy consumption, and reliable effluent quality. However, membrane fouling still remains as a critical issue in many applications of water filtration processes and it restricts the widespread application of membranes. In water treatment, humic acid (HA) has been considered to be one of the most significant foulants among many potential natural organic matters (NOMs) in both surface and ground water (Lahoussine-Turcaud *et al.*, 1990; Wang *et al.*, 2005). Previous studies of HA fouling were focused on the role of several important factors (e.g., pH and ionic strength) in humic acid solutions (Jones and O'Melia, 2000; Wang *et al.*, 2005). For example, Jones and O'Melia (2000) had demonstrated the effects of solution chemistry (pH and ionic strength) on the adsorption of HA to a membrane surface. The HA adsorption was found to increase with the increased of ionic strength and pH of the HA solution, as also observed in many studies dealing with various types of membranes. However, this kind of research approach is labor intensive, and normally requires tedious analysis of the feed stream before it enters the filtration process.

In general, hydrophilic membranes exhibit lower level of fouling (Jones and O'Melia, 2000; 2001). This is due to the formation of a water boundary between the hydrophilic membrane and the surrounding water molecules that resists the approach of hydrophobic foulants near the membrane surface. In turns out to make the membrane less vulnerable to fouling. Many approaches have been taken to improve the hydrophilicity of membranes, including the blending of the hydrophilic polymers and the surface modification of membranes (Ma *et al.*, 2001; Malaisamy and Bruening 2005; Peeva *et al.*, 2010). However, the polymer chain undergoing chemical modification sometimes becomes highly swollen, leading to the low mechanical properties of the membrane (Mbareck *et al.*, 2009). As a solution, the integration of nanoparticles into the membrane matrix allows both control of the membrane fouling and the ability to produce specific functionalities to the membrane (Li *et al.*, 2009; McLachlan, 2010). Generally, this type of composite membrane can be produced through polyelectrolyte multilayer modification (PEM) method, phase inversion techniques, self-assembly through covalent attachment or blending with a polymer casting solution (Liu *et al.*, 1997; Schlemmer *et al.*, 2009; Taurozzi *et al.*, 2008).

The quartz crystal microbalance with dissipation monitoring (QCM-D) technology is a surface sensitive technique which gives real time information about the deposition of the thin surface layers by measuring the changes of mass and viscoelastic properties at a surface. It has been used to evaluate the surface adsorption kinetics and to analyze the nanoparticle-surface interactions in real time (Chen and Elimelech, 2006; Xu *et al.*, 2010). For example, Xu *et al.* (2010) had investigated the adsorption behavior of Laponite and Ludox silica nanoparticles on surface of poly(diallyldimethylammonium chloride) through QCM-D. The adsorption rates of both silica nanoparticles were found to be concentration dependent and signifying that the Ludox adsorbed layer was more softly bound than the Laponite adsorbed layer. A series of reports have also been presented on the study of the membrane fouling using QCM-D technique and proved its effectiveness in assessing the adsorption and viscoelastic properties of the foulants (organic and inorganic) onto the membrane surfaces (Arkhangelsky *et al.*, 2012; Contreras *et al.*, 2011; Hashino *et al.*, 2011).

In the present study, we aimed to quantify the adsorption of functionalized magnetic nanoparticles (F-MNPs) onto the PES membrane surfaces in different dispersants (DI water and PSS) at conditions with or without PSS and PDDA as precursors through QCM-D measurement. First, the MNPs were functionalized using PSS polymer and flocculation studies for the nano-colloids were carried out. Then, different membrane surfaces functionalities in adsorptive fouling by common organic foulants (HA) were elucidated.

4.2 EXPERIMENTAL

4.2.1 *Materials*

The magnetite Fe_3O_4 nanoparticles (MNPs) were supplied by NanoAmor (USA). PSS and PDDA polyelectrolyte were purchased from Sigma (St. Louis, USA). Ultrafiltration PES membrane with molecular weight cut-off of 20,000 was obtained from GE Osmonics (USA). In this study, PES polymer supplied by BASF was used as the model membrane used in QCM-D measurement. The solvent *N*-Methyl-2-pyrrolidone (NMP) and humic acid used as the membrane foulant were purchased from Merck (Germany) and Aldrich (Switzerland), respectively. NaOH (Merck, Germany) and HCl were used for pH adjustment.

4.2.2 *Functionalization of nanoparticles*

A suspension with 2500 mg L^{-1} of MNPs was prepared using deionized water and ultrasonicated to break the existing aggregates. Similarly, PSS solution (0.00412 g mL^{-1}) was prepared and ultrasonicated to promote good dispersity of polymeric solution. Both colloids suspension and PSS solution were adjusted to pH 3.5 before the former was added drop wise into the latter solution, where the physisorption process was allowed to occur for 1 day. The PSS-coated nanoparticles (F-MNPs) were then separated using a permanent magnet and pre-washed before final dispersion in deionized water or PSS. Dynamic light scattering (Zetasized) was used to determine the size distribution and colloidal stability of naked Fe_3O_4 and functionalized Fe_3O_4 in suspension. The measurement was performed at an angle of 90° under a light source of 650 nm, and the calculation was based on the assumption of spherical particles in Brownian motion and single scattering.

4.2.3 *Quartz crystal microbalance with dissipation (QCM-D)*

The QCM-D measurements were performed with AT-cut quartz crystals mounted in an E1 system (Q-Sense, Goteborg, Sweden). 1% w/v of PES polymer dissolved in NMP was spin coated onto the gold quartz crystal (Q-Sense, Sweden) and evaporated at room temperature for 24 hours. All QCM-D measurements were conducted under flow-through condition via a digital peristaltic pump operating in sucking mode at a constant flow rate 50 μL min^{-1} and at 25°C. The ATR-FTIR spectra of the commercial PES membrane and the model PES membrane coated on the quartz crystal were analyzed using Thermo Scientific Fourier transform infrared spectrometer (NICOLET iS10, USA) to confirm the similarly of the functional groups. Each spectrum was obtained from 32 scans with 4 cm^{-1} resolutions at a 45° incident angle using a diamond crystal over the wavenumber range of 4000–600 cm^{-1}.

4.2.3.1 *Monitoring F-MNPs adsorption*
First, the solutions of PSS/PDDA precursor were alternately pumped through the QCM-D chamber at 50 μL min^{-1}. The flow-through precursor solution will be absorbed to the surface of the PES coated gold crystal cell. Negatively charged PSS solution will be first interacted with the

Table 4.1.   List of the different samples combination.

No. sample	Precursor	Dispersant of F-MNPs
1	–	Ultrapure water
2	–	PSS
3	PDDA	Ultrapure water
4	PDDA	PSS
5	PSS then PDDA	Ultrapure water
6	PSS then PDDA	PSS

PES polymer, followed by the positively charged PDDA solution. Throughout the layer-by-layer assembly, the composite precursor-PES surface was formed where outermost layer was ready to adsorb the functional F-MNPs. In this work, the dynamic adsorptions of F-MNPs onto the PES surface in condition with or without precursors (PSS/PDDA) as well as the effects of dispersants (DI water/PSS) were monitored using QCM-D, as the experimental protocols were shown in Table 4.1. Adsorbed mass of F-MNPs on different combination surfaces was modeled using the viscoelastic modeling in QTools provided by Q-Sense.

4.2.3.2   *Adsorption/cleaning of HA on the membrane surface*
The foulant (HA) analysis of the modified (PES-Fe₃O₄) and unmodified (PES) membranes were investigated using QCM-D through the adsorption study and cleaning of the humic acid from the membranes. In preparation of modified PES-Fe₃O₄ membrane, the precursor-PES composite layer was first generated according to the method described in section (Monitoring F-MNPs adsorption). Sequentially, $2500 \, mg \, L^{-1}$ of F-MNPs solution flowed-through the QCM-D chamber which crossed the precursor-PES surface at a constant flow rate of $50 \, \mu L \, min^{-1}$. The cationic (PDDA) precursor at the outermost layer will induce cooperative binding with the anionic F-MNPs and formed the modified PES-Fe₃O₄ membrane. After a stable baseline of the QCM-D signal was achieved while flowing with ultrapure water, HA foulants ($25 \, mg \, L^{-1}$) were allowed to flow across the quartz crystal surface for two hours, followed by washing using ultrapure water for another 30 min. The adsorbed mass Δm as well as the removed cleaning mass per unit surface of HA from the different quartz crystal surfaces (modified PES-Fe₃O₄ and unmodified PES model membranes) were evaluated based on the Sauerbrey equation (Sauerbrey 1959):

$$\Delta f = -\frac{n}{c}\Delta m \qquad (4.1)$$

where Δf refers to the changes of frequency [Hz], Δm refers to the changes of mass adsorbed per surface area of quartz crystal cell [$ng \, cm^{-2}$], n is the overtone number ($n = 3$ in this study) and C is the mass sensitivity constant of the QCM-D ($C = 17.7 \, ng \, cm^{-2} \, Hz^{-1}$ at $f = 4.95 \, MHz$).

4.3   RESULTS AND DISCUSSION

4.3.1   *Stability of the functionalized MNPs and characteristics of the model PES membrane*

Dynamic light scattering (DLS) was used to measure size distribution as well as to study the flocculation kinetics of MNPs and F-MNPs, where the colloidal suspension was diluted to $6 \, mg \, L^{-1}$ in $3.0 \, mL$ of deionized water. As shown in Figure 4.1, the bare MNPs were started to flocculate from its initial hydrodynamic diameter at around 306 nm to form the larger clusters up to ~2000 nm in size. The flocculation of MNPs was probably because the small particles tend to agglomerate to reduce the energy associated with high surface area to volume ratio of the nano-size particles (Lu, *et al.*, 2007). Moreover, formation of the large clusters may also be possible because of the

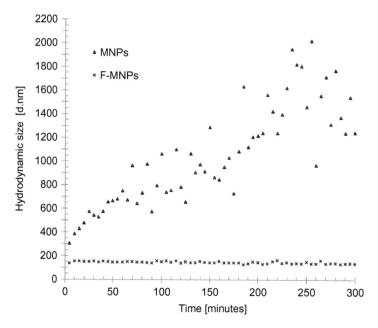

Figure 4.1.   Flocculation kinetic profile presenting the average hydrodynamic size [d.nm] of naked MNPs and PSS coated F-MNPs vs. time (minutes).

attractive van der Waals and magnetostatic forces working within the intermolecular MNPs (Lim *et al.*, 2012). Hence, surface modification/functionalization of MNPs is necessary to preserve good colloidal stability and provide a better layer dispersion of MNPs on the membrane surface.

As shown in Figure 4.1, MNPs functionalized using PSS (F-MNPs) have smaller hydrodynamic sizes compared to the bare MNPs at its initial stage. Moreover, the F-MNPs have not shown any obvious clustering behavior throughout the similar measuring time scale and remained almost constant in size (145.49 ± 7.60 nm). This is a result of the steric (osmotic and elastic) and electrostatic repulsion existing between the F-MNPs because of the adsorbed polyelectrolyte layer, rendering the restricted intermolecular interactions among the polyelectrolyte coated particles (Lim *et al.*, 2012). Hence, the MNPs stability is enhanced. ATR-FTIR was used to characterize the model PES coated on the gold crystal to confirm its similar functionalities to those of the commercial PES membrane. As shown in Figure 4.2 (spectra), and in Table 4.2 (peak assignments; Coates 2000), both commercial PES membrane and PES coated onto gold quartz crystal demonstrated the same spectra, with the revealed characteristic vibrations at the peaks for the aromatic bands at 1576 and 1484 cm^{-1}, the symmetric vibration of the SO$_2$ groups at 1146 cm^{-1}, the Ar–O–Ar groups at 1236 cm^{-1}, and the C–S groups at 1484 cm^{-1} (Liu and Bai, 2005).

4.3.2   *The F-MNPs adsorption onto PES model membrane*

After functionalized MNPs with PSS, the F-MNPs were dispersed in either PSS or DI solution. The higher charge density of the PSS polyelectrolyte present in the solution is expected to demonstrate higher adsorption of the F-MNPs onto the PES polymer. However, at all conditions stated in Table 4.1, the adsorbed mass of the F-MNPs dispersed in DI (conditions 1, 3 and 5) was relatively higher compared to the F-MNPs that were dispersed in PSS (conditions 2, 4 and 6). It was likely due to the adsorption competition between the F-MNPs and free PSS onto the coated surface of the quartz crystal. Thus, the average zeta potentials for the polyelectrolytes and F-MNPs were evaluated, as shown in Table 4.3. From Table 4.3, the free PSS polymer exhibited higher negative charge density compared to the F-MNPs. This is because the neutralization of charge occurs

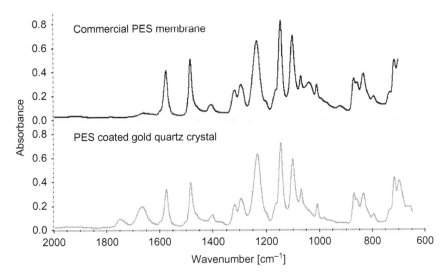

Figure 4.2.   ATR-FTIR spectra for commercial PES membrane and PES coated on crystal.

Table 4.2.   Possible assignments of the FTIR spectra of PES membrane.

Peak [cm^{-1}]	Range given in the literature (Coates, 2000)	Assignment
1576	About 1580	Aromatic stretching vibration
1484	1460–1550	C–S stretch
1236	1275–1200	Ar–O–Ar stretching vibration
1146	1160–1120	SO$_2$ stretching

during the functionalization of MNPs (positive charge) with PSS (negative charge). Hence, free PSS polymer has the high affinity to adsorb onto the surface compared to the F-MNPs. In addition, the components of lower charge density of the F-MNPs were removed easily from the surface during the washing process (Rojas *et al.*, 2002). For this reason, the F-MNPs that were dispersed in DI will be further used for the membrane fouling study by humic acid.

As for the effects of the presence of precursor on the adsorption of F-MNPs, the results (Fig. 4.3) show that F-MNPs adsorbtion on the PES surface is lowest when no precursor is applied. This might be because only weak hydrogen bonding is working between F-MNPs and the PES membrane (Diagne *et al.*, 2012). Comparing between the PDDA and PSS/PDDA precursors, the combination of PSS/PDDA precursor shows higher F-MNPs adsorbed mass (125.61 mg m^{-2}) than PDDA precursor (77.18 mg m^{-2}). This is due to the weak adsorption of PDDA on the PES polymer which further caused the detachment of the PDDA together with the adsorbed F-MNPs from the PES surface. However, when the PES surface is covered by coating PSS precursor prior to the assembly of PDDA onto the crystal cell, the combination (PES → PSS → PDDA) will form a thicker "polycation blanket" with a strongly charged outermost layer compared to the presence of only a layer of PDDA on the PES coated crystal cell (PES → PDDA) (Hua and Lvov, 2008). Consequently, increase in the adsorption of the F-MNPs on the polyelectrolyte multilayer (PES → PSS → PDDA) was observed. Furthermore, the addition of the polyelectrolyte layers would be expected to produce a more continuous film with increased surface charge (positive charge in this study) and more ion-exchange sites for the adsorption of F-MNPs (Liu *et al.*, 2010); nevertheless, the number of bilayer of PSS/PDDA was limited to 1 to reduce the membrane thickness to avoid flux decline during the membrane application later.

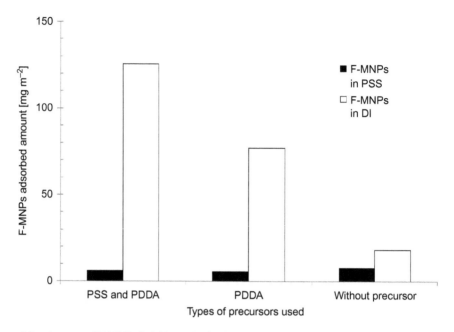

Figure 4.3.   Amount of F-MNPs in PSS or DI adsorbed on PES spin-coated quartz crystal sensor with or without precursor.

Table 4.3.   Details on average zeta potential for polyelectrolytes and F-MNPs.

Polyelectrolyte	Average zeta potential [mV]
PSS	-63.40 ± 3.41
PDDA	$+63.83 \pm 3.76$
F-MNPs	-40.53 ± 0.81

4.3.3   Adsorption of HA foulant on the membrane surface

QCM-D was used to study the interactions between foulants and membrane surface, and the results are shown in Table 4.4. The amount of HA adsorbed on the pure PES membrane was much higher than that of the modified PES-Fe$_3$O$_4$ membrane. This is due to the increase of negative charge density and hydrophilicity on the surface of modified membrane that makes them not easy to foul. Almost all foulants were washed out (95.27%) from the modified PES-Fe$_3$O$_4$ membrane surface, whereas only 33.63% of HA was removed or 56.15 ng cm^{-2} was retained on the pure PES membrane surface after the washing step. There is only weak interaction between the HA foulant and the modified membrane due to the impregnation of F-MNPs with negative charge and of high hydrophilicity to the PES membrane. Thus, the attachment of HA foulant which has the similar negative charge is reduced. In contrast, the stronger attachment of HA foulant occurred when strong hydrophobic-hydrophobic interactions occurred between the HA and the unmodified PES membrane.

However, some minor HA foulants were still deposited on the surface of the modified membrane (3 ng cm^{-2}) due to the higher membrane roughness by the presence of the nanoparticles. However, the deposited amount of HA foulant (3 ng cm^{-2}) was negligible compared to the unmodified PES membrane (56.15 ng cm^{-2}), showing that the lower irreversible fouling occurred for the modified PES-Fe$_3$O$_4$ during fouling experiments.

Table 4.4.   Adsorption level for HA on different membrane surfaces and percentual mass of adsorbed layer (HA) removed by pure water.

Membrane	Adsorbed mass [ng cm^{-2}]	HA mass removed after washing [%]
PES	84.60	33.63
PES-Fe$_3$O$_4$	63.51	95.27

4.4 CONCLUSION

Surface modified/functionalized MNPs were successfully end-capped to the surface of PES membrane due to improved colloidal stability. In this study, membrane modified with F-MNPs in the presence of PSS/PDDA precursors exhibited lower membrane fouling by the HA foulant, due to the formation of highly negatively charged and hydrophilic surface. The F-MNPs modified membrane developed in this work achieved high efficiency for the removal of humic substances up to 97.25%, which make the membrane processes more economical (longer life span of the membrane) and more sustainable.

ACKNOWLEDGEMENT

The authors wish to thank financial support granted by Universiti Sains Malaysia Research University (RU) Grant (1001/PJKIMIA/814230). All authors are affiliated to the Membrane Science and Technology Cluster of USM. Q.H. Ng is financially assisted by Ministry of Higher Education (MOHE) and Universiti Malaysia Perlis (UniMAP).

REFERENCES

Arkhangelsky, E., Wicaksana, F., Tang, C., Al-Rabiah, A.A., Al-Zahrani, S.M. & Wang, R. (2012) Combined organic-inorganic fouling of forward osmosis hollow fiber membranes. *Water Research*, 46, 6329–6338.

Chen, K.L. & Elimelech, M. (2006) Aggregation and deposition kinetics of fullerene (C60) nanoparticles. *Langmuir*, 22, 10,994–11,001.

Coates, J. (2000) Interpretation of infrared spectra, a practical approach. In: Meyers, R.A. (ed.) *Encyclopedia of analytical chemistry*. John Wiley & Sons Ltd, Chichester, UK. pp. 10,815–10,837.

Contreras, A.E., Steiner, Z., Miao, J., Kasher, R. & Li., Q. (2011) Studying the role of common membrane surface functionalities on adsorption and cleaning of organic foulants using QCM-D. *Environmental Science & Technology*, 45, 6309–6315.

Diagne, F., Malaisamy, R., Boddie, V., Holbrook, R.D., Eribo, B. & Jones, K.L. (2012) Polyelectrolyte and silver nanoparticle modification of microfiltration membranes to mitigate organic and bacterial fouling. *Environmental Science & Technology*, 46, 4025–4033.

Hashino, M., Hirami, K., Katagiri, T., Kubota, N., Ohmukai, Y., Ishigami, T., Maruyama, T. & Matsuyama, H. (2011) Effects of three natural organic matter types on cellulose acetate butyrate microfiltration membrane fouling. *Journal of Membrane Science*, 379, 233–238.

Hua, F. & Lvov, Y.M. (2008) *Layer-by-layer assembly*. Chapter 1 in Erokhin, V., Ram, M.K. & Yavuz, O. (eds.) *The new frontiers of organic and composite nanotechnology*. Elsevier, Amsterdam, The Netherlands. pp. 1–44.

Jones, K.L. & O'Melia, C.R. (2000) Protein and humic acid adsorption onto hydrophilic membrane surfaces: effects of pH and ionic strength. *Journal of Membrane Science*, 165, 31–46.

Jones, K.L. & O'Melia, C.R. (2001) Ultrafiltration of protein and humic substances: effect of solution chemistry on fouling and flux decline. *Journal of Membrane Science*, 193, 163–173.

Lahoussine-Turcaud, V., Wiesner, M.R. & Bottero, J.Y. (1990) Fouling in tangential-flow ultrafiltration: the effect of colloid size and coagulation pretreatment. *Journal of Membrane Science*, 52, 173–190.

Li, J.H., Xu, Y.Y., Zhu, L.P., Wang, J.H. & Du, C.H. (2009) Fabrication and characterization of a novel TiO₂ nanoparticle self-assembly membrane with improved fouling resistance. *Journal of Membrane Science*, 326, 659–666.

Lim, J.K., Chan, D.J.C., Jalak, S.A., Toh, P.Y., Yasin, N.H.M., Ng, B.W. & Ahmad, A.K. (2012) Rapid magnetophoretic separation of microalgae. *Small*, 8, 1683–1692.

Liu, C. & Bai, R. (2005) Preparation of chitosan/cellulose acetate blend hollow fibers for adsorptive performance. *Journal of Membrane Science*, 267, 68–77.

Liu, G., Dotzauer, D.M. & Bruening, M.L. (2010) Ion-exchange membranes prepared using layer-by-layer polyelectrolyte deposition. *Journal of Membrane Science*, 354, 198–205.

Liu, Y., Wang, A. & Claus, R.O. (1997) Layer-by-layer electrostatic self-assembly of nanoscale Fe_3O_4 particles and polyimide precursor on silicon and silica surfaces. *Applied Physics Letters*, 71, 2265–2267.

Lu, A.H., Salabas, E.L. & Schüth, F. (2007) Magnetic nanoparticles: synthesis, protection, functionalization, and application. *Angewandte Chemie International Edition*, 46, 1222–1244.

Ma, H., Hakim, L.F., Bowman, C.M. & Davis, R.H. (2001) Factors affecting membrane fouling reduction by surface modification and backpulsing. *Journal of Membrane Science*, 189, 255–270.

Malaisamy, R. & Bruening, M.L. (2005) High-flux nanofiltration membranes prepared by adsorption of multilayer polyelectrolyte membranes on polymeric supports. *Langmuir*, 21, 10,587–10,592.

Mbareck, C., Nguyen, Q.T., Alaoui, O.T. & Barillier, D. (2009) Elaboration, characterization and application of polysulfone and polyacrylic acid blends as ultrafiltration membranes for removal of some heavy metals from water. *Journal of Hazardous Materials*, 171, 93–101.

McLachlan, D. (2010) The defouling of membranes using polymer beads containing magnetic micro particles. *Water SA*, 36, 641–650.

Peeva, P.D., Pieper, T. & Ulbricht, M. (2010) Tuning the ultrafiltration properties of anti-fouling thin-layer hydrogel polyethersulfone composite membranes by suited crosslinker monomers and photo-grafting conditions. *Journal of Membrane Science*, 362, 560–568.

Rojas, O.J., Ernstsson, M., Neuman, R.D. & Claesson, P.M. (2002) Effect of polyelectrolyte charge density on the adsorption and desorption behavior on mica. *Langmuir*, 18, 1604–1612.

Sauerbrey, G. (1959) Verwendung von Schwingquarzen zur Wägung dünner Schichten und zur Mikrowägung. *Zeitschrift für Physik*, 155, 206–222.

Schlemmer, C., Betz, W., Berchtold, B., Rühe, J. & Santer, S. (2009) The design of thin polymer membranes filled with magnetic particles on a microstructured silicon surface. *Nanotechnology*, 20, 255–301.

Taurozzi, J.S., Arul, H., Bosak, V.Z., Burban, A.F., Voice, T.C., Bruening, M.L. & Tarabara, V.V. (2008) Effect of filler incorporation route on the properties of polysulfone-silver nanocomposite membranes of different porosities. *Journal of Membrane Science*, 325, 58–68.

Wang, Z., Zhao, Y., Wang, J. & Wang, S. (2005) Studies on nanofiltration membrane fouling in the treatment of water solutions containing humic acids. *Desalination*, 178, 171–178.

Xu, D., Hodges, C., Ding, Y., Biggs, S., Brooker, A. & York, D. (2010) Adsorption kinetics of laponite and ludox silica nanoparticles onto a deposited poly(diallyldimethylammonium chloride) layer measured by a quartz crystal microbalance and optical reflectometry. *Langmuir*, 26, 18,105–18,112.

CHAPTER 5

The effect of spinning parameters on PLA hollow fiber membrane formation

Arisa Jaiyu, Kanungnuch Keawsupsak, Julaluk Pannoi,
Passakorn Sueprasit & Chutima Eamchotchawalit

5.1 INTRODUCTION

Development of renewable and decomposable material from the agriculture feed stocks has become one of the most popular topics for research interest due to environmental ecological awareness and a shortage of petroleum resource (Berkesch, 2005). Poly(lactic acid) (PLA) is the one of the most attractive decomposable and renewable polymers derived from corn and sugar cane (Datta and Henry, 2006; Lim *et al.*, 2008; Liu *et al.*, 2000). Nowadays, the manufacture of PLA successfully developed into the commercial scale and the price of PLA has become much cheaper compared to the past.

Membranes play a central role of our daily life (Mulder, 1996) and have gained an important place in chemical technology because they can be used in broad range of applications (Baker, 2004). Most of the polymeric membranes were made from petroleum-based polymers. If these polymer membranes are discarded into the environment, ending up as wastes that do not degrade in a hundred years, it causes a serious global problem in the management of the increasing amount of solid waste. In this study, PLA was chosen to use as the membrane material for hollow fiber membrane fabrication.

The fabrication of hollow fiber is still a complex process especially when the new material is used. Nonsolvent induced phase separation (NIPS) has been widely used for fabrication of hollow fiber membrane on a commercial scale (Moriya *et al.*, 2009; Witte *et al.*, 1996) since NIPS is a simple and convenient method to prepare porous membranes. There are few publications about preparation of PLA membrane in flat sheet and hollow fiber (Moriya *et al.*, 2009; Witte *et al.*, 1996; Zereshki *et al.*, 2010a; 2010b). Moriya *et al.* (2009) studied the effect of solvent, nonsolvent and PEG on membrane morphologies and filtration performance of PLA hollow fiber membranes prepared via NIPS. In the NIPS method, factors involved in the hollow fiber membrane formation include (i) polymer dope formulation such as polymer and additive concentrations and (ii) spinning parameters such as air gap, take-up speed, coagulation and bore fluid temperature (Chakrabarty *et al.*, 2008; Garca-Payo *et al.*, 2009; Loh *et al.*, 2011). In this chapter, PLA hollow fiber with good ultrafiltration performance was fabricated via NIPS using only one formulation of dope solution containing 20% PLA, 2% glycerin 0.5% Polyethylene glycol (PEG1500) and *N*-methyl-2-pyrollidone (NMP). The effects of spinning parameters on the structure, morphologies and filtration performance of hollow fiber membrane have been studied.

5.2 EXPERIMENTAL

5.2.1 *Materials*

Poly(lactic acid) (Ingeo 2003D) was purchased from NatureWorks LLC and dried in a conventional oven at 50°C for 24 hours before use. NMP was purchased from POSH. S.A. (Labscan). Glycerin was purchased from Iltalmar company. Bovine serum albumin (BSA) and PEG1500 were

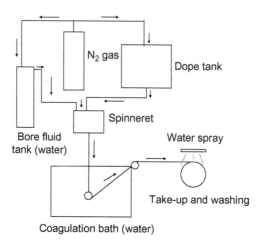

Figure 5.1.   Schematic of spinning process.

purchased from S.M. Chemical Supplied Co. Ltd. All materials were used as received without further purification.

5.2.2   Fabrication of PLA hollow fiber membranes

PLA resin (20 wt%), glycerin (2 wt%), PEG1500 (0.5%) were dissolved in NMP at approximately 80°C for about 7 hours to form an homogeneous dope solution. Then, the dope solution was transferred into the polymer dope tank and kept overnight at 70°C for eliminating the air bubbles formed during the stages of stirring and pouring. The degassed dope solution was used to fabricate PLA hollow fiber membranes using the dry-wet spinning process shown schematically in Figure 5.1. Distilled water and tap water at room temperature were used as bore fluid and coagulant, respectively. The dope solution and bore fluid (water) were pressurized by nitrogen gas through a spinneret, with an outer tube diameter of 1.06 mm and inner tube diameter of 0.66 mm, to a coagulation bath of water. The preparation parameters such as feed pressure, air gap, and take-up speed have been varied, while the dope solution temperature and bore fluid rate were fixed at 70°C and 10 mL min^{-1}, respectively. After phase separation and solidification of membrane, the membrane was collected by a roller. The spinning parameters are summarized in Table 5.1. The obtained membranes were immersed in water over 3 days to completely remove the NMP used in membranes fabrication. Next, the membranes were immersed in 10% aqueous glycerin solution for 10 min to preserve the pore structure during drying. The membranes were dried in the air at ambient for further tests.

5.2.3   Characterization of the PLA hollow fiber membranes

5.2.3.1   Characterization of the structures and morphologies
of PLA hollow fiber membranes by FESEM

A field scanning electron microscope (FESEM, JEOL model JSM-6340F) was used to characterize the cross-section, outer and inner surface morphologies of asymmetric PLA hollow fiber membranes. Specimens of the cross-section hollow fiber membranes for FESEM were prepared by fracturing the dried membrane sample in liquid nitrogen. The specimens were covered with a thin layer of gold using a sputter coater before the FESEM analysis. The diameter of hollow fiber membranes was measured from FESEM image using Image J 1.44 P program.

Table 5.1.   Spinning parameters and structure of hollow fiber membranes.

Code	Feed pressure [MPa]	Air gap [cm]	Take-up speed [m min^{-1}]	i.d. [μm]*	o.d. [μm]*	Wall thickness [μm]
HF-PLA1	0.045	25	220	916	1237	161
HF-PLA2	0.015	5	220	955	1206	125
HF-PLA3	0.045	5	270	808	1017	105
HF-PLA4	0.065	5	220	938	1223	142
HF-PLA5	0.1	5	220	909	1207	149
HF-PLA6	0.1	5	270	746	991	123

*i.d.: inner diameter; o.d.: outer diameter (of hollow fiber membrane).

5.2.3.2   Measurement of hollow fiber pure water flux

The hollow fiber membranes were prepared for measurement of pure water flux using a lab-scale permeation system. Prior to the test, the hollow fiber membranes were immersed in pure water to remove glycerin for overnight. Pure water was circulated through the inside of membrane under pressure (0.1 MPa) and then, the pure water permeate from the membrane was collected. The pure water flux of the membrane (J) was calculated by the following equation:

$$J = \frac{Q}{At} = \frac{Q}{n\pi dlt} \tag{5.1}$$

where Q is the total filtrate volume [m^3] within the operation time t [h], A the membrane area [m^2], n the number of fibers in one testing, d the inner diameter of the testing fibers [m], and l the effective length of the fiber [m].

5.2.3.3   BSA rejection

The BSA (MW = 67,000) solution at 1000 mg L^{-1} was used as the feed solution. The BSA concentration of feed and permeate was evaluated using UV-visible spectroscopy (Shimadzu model UV-1700) at 280 nm.

The % BSA rejection can be calculated using the following equation:

$$R\,[\%] = \left(1 - \frac{C_p}{C_f}\right) \times 100 \tag{5.2}$$

where C_p and C_f are the BSA concentration [mg L^{-1}] of permeate and feed, respectively.

5.2.3.4   Mechanical properties

Tensile strength and elongation at break of each sample were determined using a tensometer (Monsanto model T10) at room temperature with a crosshead rate of 50 mm min^{-1} and a 25 mm gauge length. At least 5 replicates were tested, and average values were reported.

5.3   RESULTS AND DISCUSSION

The fabrication of hollow fiber is still a complex process especially when the new material is used. The spinning parameters play an important role in formation of membrane hollow fiber via NIPS. In this study, a number of PLA hollow fiber membranes were prepared by dry-wet solution spinning process via NIPS. The spinning parameters such as feed pressure, air gap and take-up speed were varied while the formulation of polymer dope solution was fixed.

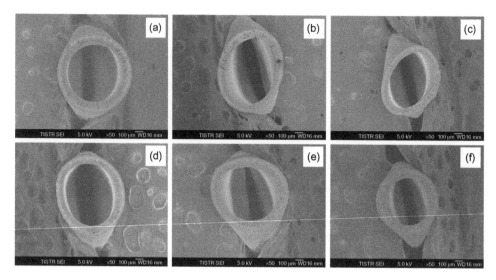

Figure 5.2.   FESEM image of cross-section PLA hollow fiber: (a) HF-PLA1, (b) HF-PLA2, (c) HF-PLA3, (d) HF-PLA4, (e) HF-PLA5 and (f) HF-PLA6.

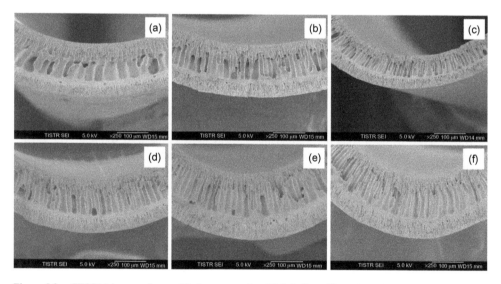

Figure 5.3.   FESEM image of magnified cross-section PLA hollow fiber: (a) HF-PLA1, (b) HF-PLA2, (c) HF-PLA3, (d) HF-PLA4, (e) HF-PLA5 and (f) HF-PLA6.

5.3.1   *The effect of spinning parameters on the hollow fiber membrane structure and morphologies*

The PLA hollow fiber spinning parameters and structure are summarized in Table 5.1. The results showed that the diameter and wall thickness of membranes decreased with the increase of take-up speed of membranes. Moreover, the wall thickness of membranes increased with the increase of feed pressure. The air gap between spinneret and coagulant bath also affected the structure of hollow fiber membrane.

The asymmetric structure of the hollow fiber membranes was examined by FESEM. The FESEM image of cross-sections in Figures 5.2 and 5.3 shows that the PLA hollow fiber membranes

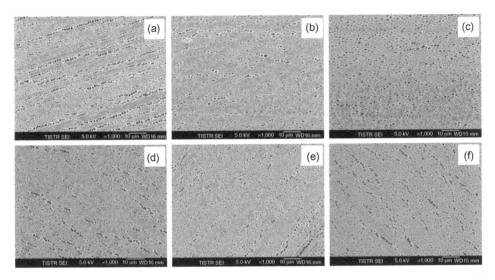

Figure 5.4.   FESEM image of inner surface PLA hollow fiber: (a) HF-PLA1, (b) HF-PLA2, (c) HF-PLA3, (d) HF-PLA4, (e) HF-PLA5 and (f) HF-PLA6.

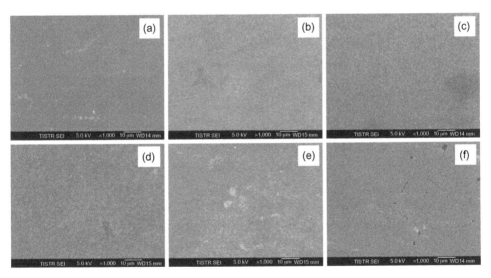

Figure 5.5.   FESEM image of outer surface PLA hollow fiber: (a) HF-PLA1, (b) HF-PLA2, (c) HF-PLA3, (d) HF-PLA4, (e) HF-PLA5 and (f) HF-PLA6.

have double finger-like structure (macrovoid) which is the normal structure formed in NIP method. These results indicate that phase separation was induced by nonsolvent (water) that has penetrated from both inner and outer surfaces of hollow fiber membranes (Moriya *et al.*, 2009). There is a sponge-like layer between 2 macrovoids in the middle of the membranes. FESEM image in Figure 5.3 clearly shows the 3 layers of membrane (the finger-like structure near the inner side of the membrane, sponge-like in the middle of membrane and the finger-like near the outer side of membrane). The finger-like structure layers near inner side of membrane are thicker than the layer near outer side of membrane, which may be due to the velocity of the bore fluid. The porous structure was formed at the inner surface and outer surface (see Figs. 5.4 and 5.5) of the hollow fiber membrane maybe because the solvent diffused out slowly or nonsolvent diffused into the

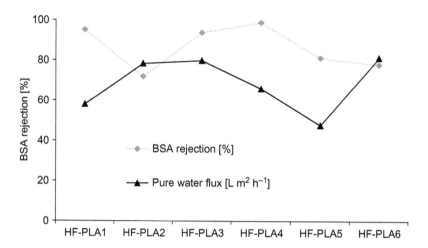

Figure 5.6.   BSA rejection [%] and pure water flux of PLA hollow fiber membrane.

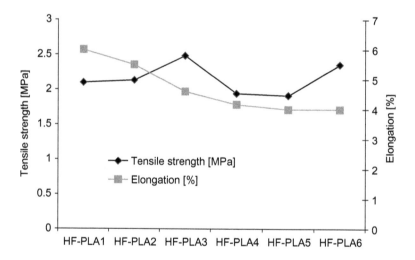

Figure 5.7.   Tensile strength and elongation of PLA hollow fiber membrane.

dope solution rapidly. The inner surface showed larger pores compared to the outer surface due to the velocity of the bore fluid. The larger finger-like structure was formed when the air gap increased from 5 cm to 25 cm.

5.3.2   *Pure water flux and BSA rejection*

The pure water flux and BSA rejection of PLA hollow fiber membranes are shown in Figure 5.6. The PLA hollow fiber membranes HF-PLA-2, HF-PLA3 and HF-PLA6 showed high pure water flux because the wall thickness of these membranes were thinner compared to the other membranes.

The PLA hollow fiber membranes exhibited 71–99% in BSA rejection. The BSA rejection of HF-PLA1, HF-PLA3 and HF-PLA4 were higher than 90%. These results indicated that these membranes exhibited excellent ultrafiltration performance.

5.3.3 *Tensile strength and elongation*

Results related to the tensile strength and the elongation at break of PLA hollow fiber membranes are presented in Figure 5.7. The PLA hollow fiber membranes exhibited 1.9–2.5 MPa in tensile strength and 4–6% in elongation. HF-PLA3 showed the highest tensile strength, while HF-PLA1 showed the highest elongation.

5.4 CONCLUSION

PLA hollow fiber membranes were successfully prepared via nonsolvent induced phase separation by dry-wet solution spinning method. The solution containing 20% PLA, 2% glycerin, 0.5% PEG1500 and NMP at 70°C and water at RT were used, respectively, as the spinning dope and the bore fluid as well as the external coagulant. The spinning parameters, such as feed pressure, air gap and take-up speed, affected the structure, morphologies and the filtration performance of the membrane. PLA hollow fiber membrane obtained from this work exhibited excellent ultrafiltration performance with 99% in BSA rejection. These results indicated that PLA hollow fiber membrane can be a good candidate for hollow fiber membrane made of nonpetroleum-based polymer.

ACKNOWLEDGEMENT

The authors gratefully acknowledge Thailand Institution of Scientific and Technological Research for financial, material and instrument support.

REFERENCES

Baker, R.W. (2004) *Membrane technology and applications.* 2nd edition. John Wiley & Sons Ltd., Chichester, UK.

Berkesch, S. (2005) *Biodegradable polymer: a rebirth of plastic.* Michigan State University, East Lansing, MI.

Chakrabarty, B., Ghoshal, A.K. & Purkait, M.K. (2008) Effect of molecular weight of PEG on membrane morphology and transport properties. *Journal of Membrane Science*, 309, 209–221.

Datta, R. & Henry, M. (2006) Lactic acid: recent advances in products, processes and technologies – a review. *Journal of Chemical Technology & Biotechnology*, 81, 1119–1129.

Garca-Payo, M.C., Essalhi, M. & Khayet, M. (2009) Preparation and characterization of PVDF-HFP copolymer hollow fiber membranes for membrane distillation. *Desalination*, 245, 469–473.

Lim, L.T., Auras, R. & Rubino, M. (2008) Processing technologies for poly(lactic acid). *Progress in Polymer Science*, 33, 820–852.

Liu, L., Li, S., Garreau, H. & Vert, M. (2000) Selective enzymatic degradations of poly(l-lactide) and poly(ε-caprolactone) blend films. *Biomacromolecules*, 1, 350–359.

Loh, C.H., Wang, R., Shi, L. & Fane, A.G. (2011) Fabrication of high performance polyethersulfone UF hollow fiber membranes using amphiphilic pluronic block copolymers as pore-forming additives. *Journal of Membrane Science*, 380, 114–123.

Moriya, A., Maruyama, T., Ohmukai, Y., Sotani, T. & Matsuyama, H. (2009) Preparation of poly(lactic acid) hollow fiber membranes via phase separation methods. *Journal of Membrane Science*, 342, 307–312.

Mulder, M. (1996) *Basic principles of membrane technology.* Kluwer Academic Publishers, Dordrecht, The Netherlands.

Witte, P., Esselbrugge, H., Dijkstra, P.J., Berg, J.W.A. & Feijen, J. (1996) Phase transitions during membrane formation of polylactides. I. A morphological study of membranes obtained from the system polylactide-chloroform-methanol. *Journal of Membrane Science*, 113, 223–236.

Zereshki, S., Figoli, A., Madaeni, S.S., Simone, S., Jansen, J.C., Esmailinezhad, M. & Drioli, E. (2010a) Poly(lactic acid)/poly(vinyl pyrrolidone) blend membranes: effect of membrane composition on pervaporation separation of ethanol/cyclohexane mixture. *Journal of Membrane Science*, 362, 105–112.

Zereshki, S., Figoli, A., Madaeni, S.S., Galiano, F. & Drioli, E (2010b) Pervaporation separation of ethanol/ETBE mixture using poly(lactic acid)/poly(vinyl pyrrolidone) blend membranes. *Journal of Membrane Science*, 373, 29–35.

CHAPTER 6

Fabrication of microporous polypropylene membrane via thermally induced phase separation for support media

Norlisa Harruddin, Norasikin Othman, Ani Idris,
Norul Fatiha Mohamed Noah & Zing-Yi Ooi

6.1 INTRODUCTION

In recent years, microporous membranes have been widely used in a variety of industrial applications involving ultrafiltration, microfiltration, as support material for manufacturing thin film composite membranes for nanofiltration. There are many techniques that have been used to fabricate microporous membranes such as modified microwave casting solution, melt casting and stretching, feasible freeze (Mu *et al.*, 2010), liquid-liquid phase separation, vapor induced phase separation and thermally induced phase separation (Fu *et al.*, 2003; Matsuyama *et al.*, 2002).

Among all these methods, thermally induced phase separation (TIPS) is found to be the most versatile, simplest and an effective way of producing microporous membrane (Fu *et al.*, 2003). TIPS offers several advantages such as: it is applicable in wide range of polymers, it can be used to prepare membranes from semi-crystalline polymers, it is highly capable of producing a variety of microstructure membranes and fewer variables need to be controlled (Khayet and Matsuura, 2011). TIPS process includes two mechanisms; solid-liquid (S-L) and liquid-liquid (L-L) phase separation. L-L phase separation occurs when the polymer diluent solution is cooled from one phase region to the temperature below cloud point curve whereas in S-L phase separation, the polymer diluent solution is cooled from one phased region to the temperature below crystal point curve and the polymer crystallizes prior to L-L phase separation.

The behavior of phase separation and thermodynamic interaction between polymer and diluent in TIPS process are greatly influenced by several parameters such as type of polymer, polymer concentration, type of diluents and quenching condition. Among these parameters, polymer concentration and quenching condition were selected as more effective parameters since they directly affect the phase separation mechanism and membrane morphology (Lloyd *et al.*, 1990). In most cases, polymer concentration is responsible for determining the strength and durability of membrane structure whereas quenching conditions are useful for fixing the structure of the polymer solution at a certain stage of solidification during liquid-liquid phase separation.

In this work, microporous isotactic polypropylene membranes were fabricated by using TIPS method. The effect of polymer concentration and quenching condition on membrane morphology and porosity were investigated by SEM analysis.

6.2 EXPERIMENTAL

6.2.1 *Reagents and solutions*

To fabricate a microporous membrane, isotactic polypropylene (iPP) pellets (Mw: 250,000) and diphenylether (DPE) were used as polymeric support and diluents. Both chemicals were obtained from Sigma Aldrich. In order to extract the diluents from the membrane, methanol was chosen as an extractant since methanol does not swell iPP to any appreciable extent and is miscible with

Table 6.1. Properties of isotactic polypropylene (iPP), diphenyether (DPE) and methanol.

Properties	Isotactic polypropylene	Diphenylether	Methanol
Physical appearance	White pellet	Yellow liquid	Colorless liquid
Melting point [°C]	171	27.5	−98
Boiling point [°C]	Not applicable	258	65

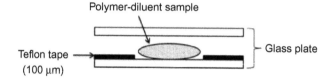

Figure 6.1. Apparatus for membrane preparation.

DPE at high concentration (McGuire *et al.*, 1995). The properties of isotactic polypropylene, diphenylether and methanol are tabulated in Table 6.1.

6.2.2 *Fabrication procedures via TIPS method*

TIPS method is almost similar to the previous study and certain modification steps have been proposed for producing the desired morphology of membrane (Lin *et al.*, 2009; McGuire *et al.*, 1995). Homogeneous polymer-diluent sample was prepared by dissolving iPP and DPE in a beaker at different polymer concentrations (10, 15 and 20%) and the beaker was purged with nitrogen gas and sealed with aluminum foil to prevent oxidation. The mixture was then heated at a temperature of 200°C for almost 3 hours and continuously stirred until the solution became homogeneous. After that, the polymer-diluent solution was cooled at room temperature until a white solid polymer was formed. The white solid polymer was sliced into desired pieces and placed in between two glass plates with a thickness 0.3 cm and width 10 cm to avoid evaporation of diluent. The thickness of the polymer solution was adjusted to 100 μm by inserting a teflon film in the square opening of the glass plate. (Fig. 6.1 exhibits the apparatus of membrane preparation). As shown in Figure 6.1, a teflon tape and vacuum grease were applied to the edges of the bottom glass plate to avoid diluent loss during the evaporation process and prevent the polymer-diluent sample from being compressed by the top glass plate. The thickness of the polymer-diluent sample should be constant in the whole experiment.

After that, the glass plate containing the solid polymer was re-melted in an oven at 473K for 5 minutes to cause melt-blending. Then, the sample was quenched at two different quenching conditions, either by air cooling or ice cooling in a horizontal position (Fu *et al.*, 2003; Kim *et al.*, 1991). Different quenching conditions considerably affect the time for coarsening the droplets in liquid-liquid separation (Fu *et al.*, 2003). After the liquid-liquid separation is induced, the membrane was immersed in methanol for one day to extract the diluent from the microporous structure of the membrane that was produced by evaporation of methanol.

6.2.3 *Fabrication by varying polymer concentration*

Polymer concentration is the most important factor that affects membrane pore structure since it directly affects phase separation and coarsening on the morphology of the membrane. The influence of polymer concentrations on fabricated membrane was investigated by varying polymer-diluent compositions in the range of 10% polymer – 90% diluent, 15% polymer – 85%

diluent and 20% polymer – 80% diluent. The fabrication method was similar to those described in Section 6.2.2.

6.2.4 Fabrication by varying quenching condition

Quenching temperature has significantly affected the polymer crystallization and structure of the resulting membrane. Hence, the effects of quenching conditions on the final morphology of the membrane were investigated. As stated in Section 6.2.2, after solid polymer was re-melted in an oven at 473K, the sample was quenched at two different quenching conditions either at air cooling, 29°C or ice cooling, 7°C. Each condition considerably affects the crystallinity and formation of porous membrane. The fabrication method was similar to Section 6.2.2 but different in solidification process.

6.2.5 Porosity measurement

The fabricated membrane was dried in a vacuum oven at 80°C for 24 hour to ensure that no water is present in the pores of the membrane and weighed as W_1. After that, the membrane was fully immersed in oil to ensure all pores were filled with the oil. The wet surface of the membrane was absorbed by filter paper and the membrane was weighed as W_0. The membrane porosity was calculated using the equation:

$$P = \frac{W_0 - W_1}{Ah} = \frac{V_{\text{pores}}}{V_{\text{total}}} \times 100 \qquad (6.1)$$

where different weight of oil in the pores [g] = volume of pores in membrane [cm³], P is the membrane porosity [%], W_0 is the weight of the wet membrane [g], W_1 is the weight of the dry membrane [g], A is the membrane area [cm²], and h is the membrane thickness [cm], V_{pores} is the volume of membrane pores [cm³] and V_{total} is the total volume of the membrane [cm³].

6.3 RESULTS AND DISCUSSION

6.3.1 Membrane formation via thermally induced phase separation (TIPS) – morphology of fabricated membrane support

The morphology of the membrane resulting from the thermal separation process reflects the thermodynamics and phase separation kinetic of the polymer-diluent solution. In this research, membrane was fabricated using TIPS method induced by liquid-liquid phase separation which resulted in different porous structures of the membrane. Figure 6.2 exhibits the SEM micrograph of membrane support prepared by various polymer-diluent concentrations and quenching conditions.

Scanning the membrane at low magnification, 50 μm revealed that all iPP membranes exhibit identical morphologies and similar pore shapes. Table 6.2 presents the pore size and porosity of the membrane support at different polymer concentrations and quenching conditions. The pore size was identified from the SEM image and the porosity was obtained from calculation using Equation (6.1). It can be observed that the resulting membranes have an average pore size of 17 μm for 10 wt%, 11 μm for 15 wt% and 9 μm for 20 wt% polymer concentration for air cooling condition. Briefly, the pore size of the membrane decreases by increasing polymer concentration and it seems that membrane with air cooling condition is likely to produce membrane with larger pore size than ice cooling condition. A similar trend was observed in porosity value, where increasing polymer concentration decreased porosity of the membrane. In addition, membrane quench at the air cooling condition produced membrane with higher porosity compared to that of the ice cooling condition. Generally, porosity of membrane can be related to pore size of membrane. Large pore sizes indicate that the membrane contains a lot of empty space inside and around pores, hence results in increased membrane porosity (Akbari and Yegani, 2012; Khayet and Matsuura, 2011).

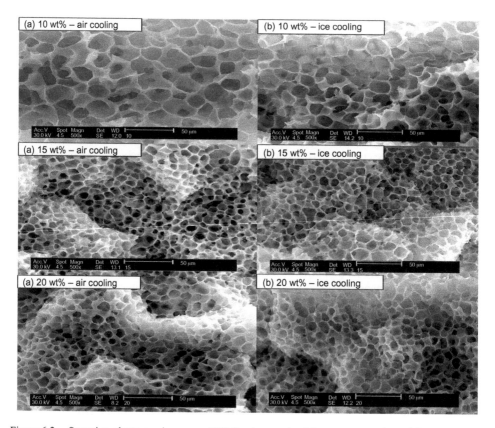

Figure 6.2.   Scanning electron microscope (SEM) micrograph of inner cross section of iPP membrane: (a) air cooling: 10, 15, 20 wt% polymer concentration and (b) ice cooling: 10, 15, 20 wt% polymer concentration.

Table 6.2.   Pore size and porosity of membrane support at different polymer concentrations and quenching conditions.

Polymer concentration [%]	Quenching condition	Pore size [μm]	Porosity [%]
10	Air cooling	17.4	93
	Ice cooling	15.7	90
15	Air cooling	11.5	87
	Ice cooling	10.3	85
20	Air cooling	9.0	85
	Ice cooling	7.8	84

The influence of polymer concentration and quenching condition on the morphology of membrane will be discussed in detail in next sections.

Furthermore, the membrane possessed closed pores, beady morphology with good interconnectivity throughout the membrane, which indicated that the membrane was formed in the metastable region during liquid-liquid (L-L) phase separation (Budyanto *et al.*, 2009; Martinez-Perez *et al.*, 2011; Matsuyama *et al.*, 2000; Tao *et al.*, 2006; Vanegas *et al.*, 2009). When a homogeneous polymer solution is cooled from the one-phase region of the phase diagram to a temperature below the binodal curve, liquid-liquid phase separation occurs and droplets of one phase form within

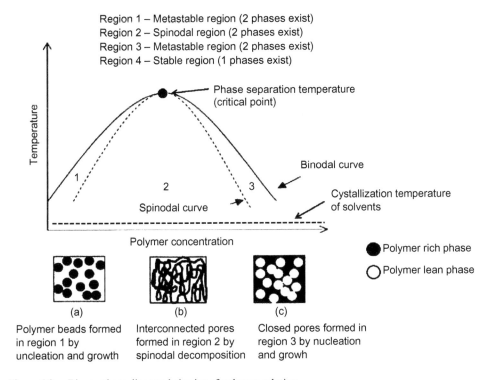

Figure 6.3.   Binary phase diagram behavior of polymer solution.

a continuous matrix of a second phase. Phase separation can occur either in spinodal region or metastable region. The resulting morphology of the membrane is dependent on the position of the polymer solution in the binary phase diagram of polymer solution as illustrated in Figure 6.3. The figure shows that phase separation can occur at four different positions which are position 1: metastable region, position, 2: spinodal region, position 3: metastable region and position 4: stable region (Khayet and Matsuura, 2011).

Related to the SEM micrograph in Figure 6.2, it was observed that all resulting membranes appear similar to Figure 6.3c, which indicates that the phase separation occurs in the metastable region by the nucleation and growth mechanism. The microstructure formed reflects that the crystallization of solvent can induce the polymer to separate into polymer rich and polymer lean phases when the temperature decreases (Ramaswamy *et al.*, 2002) and the cooling temperature is located under the metastable region. Two major factors that control the size of diluent droplet and shape of diluent rich phase in L-L phase separation are polymer concentration and the quenching condition (Caleb, 2008; Gu *et al.*, 2006; Martinez-Perez *et al.*, 2011). Hence, noticeable differences in membrane morphology and pore structural were identified between three different polymer concentrations and two quenching conditions.

Figure 6.4 illustrates the SEM micrograph of the whole cross-section of iPP membranes prepared at different polymer concentrations and quenching conditions. It can be observed that all membranes exhibit a symmetric structure from top to bottom parts of the membrane. A symmetric structure is defined as a membrane with uniform pore size throughout the membrane. Commonly, a symmetric membrane is suitable for the supported liquid membrane (SLM) process because it has higher stability compared to that of an asymmetric membrane as explained previously by several researchers (Lv *et al.*, 2007; Scott and Hughes, 1996; Yang *et al.*, 2007). According to Lv *et al.* (2007), a force exerted on both sides of the symmetric membranes is likely to be almost the same thus there is a possibility of improving the SLM process. According to the experiment done

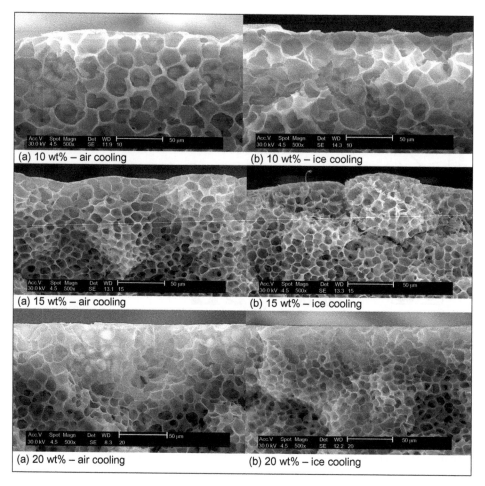

Figure 6.4.   SEM micrograph of whole cross section of iPP membranes: (a) air cooling for 10, 15 and 20 wt% polymer concentration and (b) ice cooling for 10, 15 and 20 wt% polymer concentration.

by Fu *et al.* (2004), almost 100% of metal ions were removed by using a microporous membrane with a symmetric structure as a support material.

From the figure, it is noticed that membrane with low concentration of polymer has numerous defect pores mainly due to the solidification during cooling and entrapment of air bubbles (Ferrer, 2007). Beside, small defects can be observed at high concentration of polymer due to high viscosity of the polymer solution caused by entrapment of air bubbles in the polymer solution. Therefore, a compromise must be attained between low and high concentration in order to reduce the cooling effect and entrapment of air bubbles. It has also been observed that membrane quenched by the ice cooling condition is likely to produce membrane with imperfected structure and defect pores throughout the membrane. This phenomenon occurs since high cooling rate causes incomplete pore formation, hence leads to defective pore structure after the extracting out of the diluent.

In addition, several studies done by previous researchers have also demonstrated similar morphology of membrane by inducing L-L phase separation (Budyanto *et al.*, 2009; Martinez-Perez *et al.*, 2011; Matsuyama *et al.*, 2000; Ramaswamy *et al.*, 2002; Tao *et al.*, 2006). A study done by Budyanto *et al.* (2009) showed that poly L-lactic acid (PLLA) scaffolds had a bicontinuous structure, where both the polymer-rich phase and polymer-lean phase were completely interconnected when induced by L-L phase separation.

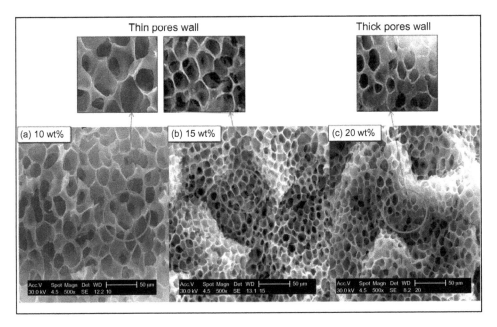

Figure 6.5.   SEM micrograph of inner cross section of iPP membrane as a function of polymer concentration: (a) 10 wt%, (b) 15 wt% and (c) 20 wt% (quench at air cooling condition).

6.3.2   *Effect of polymer concentration on the morphology and porosity of membrane*

In order to investigate the effect of polymer concentration, the quenching condition was fixed to air cooling condition to avoid any disruption affect. Figure 6.5 shows the SEM micrograph of the cross-section of iPP membranes prepared at different polymer concentrations (10, 15 and 20 wt%). It can be observed that by increasing the polymer concentration from 10 to 20 wt%, the interconnectivity between cellular pores and the pore size are decreased throughout the membrane. As can be seen, membrane (c) has a thicker wall and smaller pore diameter compared to membrane (a) and (b). Polymer concentration affects considerably the final morphology of membrane where the interconnectivity of pores increases with decrease of polymer concentration, leading to high tensile strength in the polymer-diluent system (Khayet and Matsuura, 2011; Lin *et al.*, 2009).

Different polymer concentrations resulted in different pore sizes of the membrane, as shown in Table 6.3. It shows that the average pore size decreases with increasing polymer concentration from 10 to 20 wt%. Membrane (a) had larger cellular pores with average pore size of 17.4 μm, membrane (b) had a slightly smaller pore size of 11.5 μm whereas membrane (c) exhibited closed pore structure with smaller average pore size of 9 μm. The decrease in pore size upon increasing polymer concentration could be attributed to the growth rate and viscosity of the polymer solution. A solution of the high polymer concentration has the higher viscosity, hence leading to reduction in the droplet growth rate of membrane pores. This finding is in agreement with the previous research done by Wu *et al.* (2012) and Sun *et al.* (2012). Besides, increasing the polymer concentration results in a larger number of nuclei, which in turn reduces the area of the polymer domain in the final membrane morphology. Reduction of the polymer domain leads to smaller pores and membrane of greater integrity.

Polymer concentrations influence the pore density and therefore the overall porosity of the membrane (Ferrer, 2007). Table 6.4 exhibits the porosity value of the fabricated membranes at different polymer concentrations. It was identified that increase the polymer concentration decreases the porosity of membrane. By increasing the concentration of polymer from 10 to 20 wt%, the porosity of membrane decreased from 93 to 85%. This can be explained by the fact that

Table 6.3.   Pore size of membrane at different polymer concentrations (quench at air cooling condition).

Membrane	Polymer concentration [wt%]	Pore size [μm]
(a)	10	17.4
(b)	15	11.5
(c)	20	9.01

Table 6.4.   Porosity of membrane at different polymer concentrations (quench at air cooling condition).

Membrane	Polymer concentration [wt%]	Porosity [%]
(a)	10	93
(b)	15	87
(c)	20	85

an increase in the polymer concentration will increase the nucleation density. High nucleation density induces the formation and a number of small pores, which leads to the lower porosity. However, membranes with higher nucleation density are stronger and have high structural integrity because they contain a large number of pores with compact arrangement (Khayet and Matsuura, 2011), hence it indicates that membrane (c) and (b) are stronger compared to membrane (a). The effects of polymer concentration on nucleation density were studied by other researchers and similar findings were obtained by Akbari and Yegani (2012) and by Ferrer (2007).

The relationship between polymer concentration and morphology of membrane has been widely investigated by previous researchers (Akbari and Yegani, 2012; Budyanto et al., 2009; Pavia et al., 2012; Tao et al., 2006). A previous study by Lin et al. (2009) demonstrated a similar result, in which increasing polymer concentration of iPP from 10 to 60 wt% decreased the mean pore size of membrane. In addition, the tensile strength of membrane increased also with increasing polymer concentration. Another study done by Budyanto et al. (2009), showed that at low and medium concentration of PLLA polymer, membrane with powder like structure and interconnected pores were produced meanwhile high concentration of PLLA produced membrane with closed pore structure. Overall, all fabricated membranes exhibited symmetric porous network with good interconnectivity and it was highly depending on the polymer concentration.

6.3.3   *Effect of quenching condition on the morphology and porosity of membrane*

In membrane fabrication via TIPS, the quenching condition is responsible for inducing the phase separation and crystallization of polymer (Ferrer, 2007; Fu et al., 2003; Martinez-Perez et al., 2011). Hence, quenching condition has a strong influence on the porous structure of the membrane finally obtained. In order to investigate the effect of quenching condition, concentration of polymer was fixed to 15 wt%. Figure 6.6 exhibits the effect of quenching condition on final morphology of membrane at 15 wt% polymer concentrations. The figure shows that all membranes exhibited similar morphology with pores throughout the membrane. Membrane (a) has well-structured pores with perfect shape throughout the membrane due to adequate time for the formation of pore membrane. However, slight defect structure is detected on membrane (b) due to the entrapment of air bubbles within the matrix. This was supported by a previous study by Ferrer (2007), who stated that the presence of defect within the structure of membrane is due to solidification of the mixture and entrapped air bubbles. The blocked air bubbles in the pore of the membrane failed to diffuse out due to the high cooling rate during solidification. Beside, high cooling rate

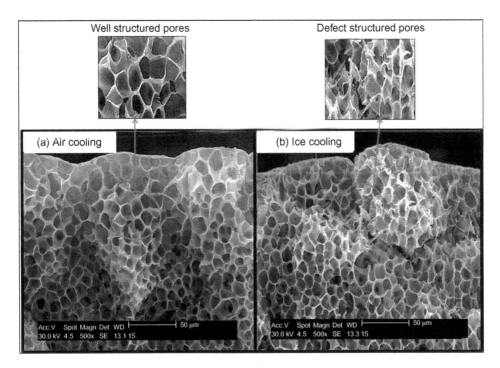

Figure 6.6.   SEM micrograph of inner cross section of iPP membrane as a function of quenching condition at 15 wt% polymer concentration: (a) air cooling condition and (b) ice cooling condition.

Table 6.5.   Pore size of membrane with 15 wt% polymer concentration at different quenching condition.

Membrane	Quenching condition	Pore size [μm]
(a)	Air cooling	11.5
(b)	Ice cooling	10.3

upon solidification led to imperfect pore shape due to insufficient time for pore formation of membrane.

Table 6.5 represents the pore size value of the membranes at different quenching conditions. Apparently, membrane (b) has smaller pore size value compared to that of membrane (a) due to a nucleation phenomenon. Quenching membrane at low temperature causes the polymer crystallization to occur immediately after phase separation, which results in reduction of time for droplet growth in liquid-liquid phase region (Ramaswamy *et al.*, 2002). As a result, membrane with smaller pore size is produced. However, quenching at high temperature enhances the mobility of the polymer chain in polymer diluent solution and also increases the time for polymer crystallization (Tao *et al.*, 2006). Ferrer (2007) also pointed out that quenching at low temperature reduces the pore size due to the nucleation phenomenon whereas quenching at high temperature tends to produce large pores due to the presence of less nuclei and the occurrence of the growth phenomenon.

Quenching condition also gives an impact on the porosity of membrane. Table 6.6 exhibits the porosity of membrane at different quenching conditions. It shows that increase in quenching temperature increases the membrane porosity. Apparently, porosity value of membrane quenched by air cooling is slightly higher than the membrane quenched by ice cooling. This result can be

Table 6.6.   Porosity of membrane with 15 wt% polymer concentration
at different quenching condition.

Quenching condition	Polymer concentration [wt%]	Porosity [%]
Air cooling	15	87
Ice cooling	15	85

explained by the fact that when the quenching temperature increases, the cooling rate decreases, hence pores have sufficient time to complete their growth and leafy structure of membrane is created (Akbari and Yegani, 2012). Hence, an increment on pore size and space between pores lead to higher porosity of membrane. However, the porosity values for both membranes are not much different.

The effect of the quenching condition on the resulted membrane structure has been discussed by previous researchers (Akbari and Yegani, 2012; Ferrer, 2007; Lin *et al.*, 2009; Ramaswamy *et al.*, 2002). Similar finding was obtained by Tao *et al.* (2006), who found that pore size of membrane poly(4-methyl-1-pentene) was greatly influenced by quenching temperature, in which smaller pore size was produced by quenching at lower temperature and vice versa. Fu *et al.* (2003) also demonstrated a similar observation in which iPP membrane quenched by ice water has smaller pore size due to high cooling rate. High cooling rate corresponds to shorter time for coarsening the droplets generated by liquid-liquid phase separation process.

6.4   CONCLUSION

Microporous polypropylene membranes with bicontinuous and symmetric structure was prepared from the iPP-DPE system via TIPS technique by varying polymer concentration and quenching condition. The mechanism of phase separation and final morphology was greatly influenced by polymer concentration and quenching condition. Increasing the polymer concentration caused a decrease in pore size and membrane porosity due to high viscosity of polymer solution and reduced droplet growth rate. Meanwhile, at different quenching conditions, smaller pore sizes were obtained by quenching in ice cooling, and the larger and well-structured pores were formed when quenched in air cooling condition. The growth rate of pores was higher in air cooling than ice cooling, as the polymer-rich phase had more time for polymer crystallization in the L-L phase separation region.

ACKNOWLEDGEMENT

The authors would like to acknowledge Ministry of Higher Education (FRGS GRANT: 4F048) and Universiti Teknologi Malaysia (UTM) for financial and technical support of this research.

REFERENCES

Akbari, A. & Yegani, R. (2012) Study on the impact of polymer of poly-ethylene membranes fabricated via TIPS method. *Journal of Membrane and Separation Technology*, 1, 100–107.

Budyanto, L., Gohand Y.Q. & Ooi, C.P. (2009) Fabrication of porous poly(L-Lactide) (PLLA) scaffolds for tissue engineering using liquid-liquid phase separation and freeze extraction. *Journal of Materials Science*, 20, 105–111.

Caleb, V.F. (2008) *The microporous mixed matrix (ZeoTIPS) membranes*. PhD Thesis, University of Texas at Austin, Austin, TX.

Ferrer M.C.-H. (2007) *Development and characterisation of completely degradable composite tissue engineering scaffolds.* PhD Thesis, Polytechnic University of Catalonia, Barcelona, Spain.

Fu, S.S., Matsuyama, H., Teramoto, M. & Lloyd, D.R. (2003) Preparation of microporous polypropylene membrane via thermally induced phase separation as support of liquid membrane used for metal ion recovery. *Journal of Chemical Engineering of Japan*, 36, 1397–1404.

Fu, S.S., Mastuyama, H. & Teramoto, M. (2004) Ce(III) recovery by supported liquid membrane using polyethylene hollow fiber prepared via thermally induced phase separation. *Separation and Purification Technology*, 36, 17–22.

Gu, M., Zhang, J., Wang, X., Tao, H. & Ge, L. (2006) Formation of poly(vinylidene fluoride) (PVDF) membranes via thermally induced phase separation. *Desalination*, 192, 160–167.

Khayet, M. & Matsuura, T. (2011) *Membrane distillation: principles and applications.* Elsevier, Amsterdam, The Netherlands.

Kim, S.S., Lim, G.B.A., Alwattari, A.A., Wang, Y.F. & Lloyd, D.R. (1991) Microporous membrane formation via thermally induced phase separation. V. Effect of diluent mobility and crystallization kinetics on membrane structuresof isotactic polypropylene membranes. *Journal of Membrane Science*, 64, 41–53.

Lin, Y.K., Chen, G., Yang, J. & Wang, X.L. (2009) Formation of isotactic polypropylene membranes with bicontinuous structure and good strength via thermally induced phase separation method. *Desalination*, 236, 8–15.

Lloyd, D.R., Kinzer, K.E. & Tseng, H.S. (1990) Microporous membrane formation via thermally induced phase separation. I. Solid-liquid phase separation. *Journal of Membrane Science*, 52, 239–261.

Lv, J., Yang, Q., Jiang, J. & Chung, T.S. (2007) Exploration of heavy metal ions transmembrane flux enhancement across a supported liquid membrane by appropriate carrier selection.*Chemical Engineering Science*, 62, 6032–6039.

Martinez-Perez, C.A., Olivas-Armendariz, I., Castro-Carmona, J.S. & Garcia-Casillas, P.E. (2011) *Scaffolds for tissue engineering via thermally induced phase separation.* Intech Open Science Publisher, Rijeka, Croatia.

Matsuyama, H., Kudari, S., Kiyofuji, H. & Kitamura, Y., (2000). Kinetic studies of thermally induced phase separation in polymer-diluent system. *Journal of Applied Polymer Science*, 76, 1028–1036.

Matsuyama, H., Man Kim, M. & Lloyd, R.D. (2002) Effect of extraction and drying on the structure of microporous polyethylene membranes prepared via thermally induced phase separation. *Journal of Membrane Science*, 204, 413–419.

McGuire, K.S., Laxminarayan, A. & Lloyd, D.R. (1995) Kinetics of droplet growth in liquid-liquid phase separation of polymer-diluent systems: experimental results. *Polymer*, 36, 4951–4960.

Mu, C., Su, Y., Sun, M., Chen, W. & Jiang, Z. (2010) Fabrication of microporous membranes by a feasible freeze method. *Journal of Membrane Science*, 361, 15–21.

Pavia, F.C., Carrubba, V.L. & Brucato, V. (2012) Morphology and thermal properties of foams prepared via thermally induced phase separation based on polylacticacid blends. *Journal of Cellular Plastics*, 48, 399–407.

Ramaswamy, S., Greenberg, A.R. & Krantz, W.B. (2002) Fabrication of poly (ECTFE) membranes via thermally induced phase separation. *Journal of Membrane Science*, 210, 175–180.

Scott, K. & Hughes, R. (1996) *Industrial membrane separation technology.* 1st edition. Blackie Academic & Professional, an imprint of Chapman & Hall, London, UK.

Sun, Y., Salehand, L. & Bai, B. (2012) *Measurement and impact factors of polymer rheology in porous media.* In: Vicente, J.D. (ed.) *Rheology.* Intech Open Science, Rijeka, Croatia.

Tao, H., Zhang, J., Wang, X. & Gao, J. (2006) Phase separation and polymer crystallization in a poly (4-methyl-1-pentene)-dioctylsebacate-dimetylphtalate system via thermally induced phase separation. *Journal of Polymer Science*, 45, 153–161.

Vanegas, M.E., Quijada, R. & Serafini, D. (2009) Microporous membranes prepared via thermally induced phase separation from metallocenic syndiotactic polypropylenes. *Polymer*, 50, 2081–2086.

Wu, Q.Y., Wan, L.S. & Xu, Z.K. (2012) Structure and performance of polyacrylonitrile membranes prepared via thermally induced phase separation. *Journal of Membrane Science*, 409, 355–364.

Yang, Q., Chung, T.S., Xiao, Y. & Wang, K. (2007) The development of chemically modified P84 co-polyimide membranes as supported liquid membrane matrix for Cu(II) removal with prolonged stability. *Chemical Engineering Science*, 62, 1721–1729.

CHAPTER 7

Synthesis and characterization of superhydrophobic alumina membrane with different types of silane

Nor Aini Ahmad, Thinesh Rachandran, Choe Peng Leo & Abdul Latif Ahmad

7.1 INTRODUCTION

Hydrophobic membrane permits the passage of nonpolar compounds selectively in a feed stream. This is because the hydrophobic surface with "water repellent" feature has less bonds with polar molecules such as water.

Hydrophobic membranes with large liquid entry pressure, fast permeation of nonpolar molecules, excellent rejection of polar compounds, great mechanical strength and outstanding chemical stability are nowadays required in the advanced membrane separation process. Membranes with superhydrophobic features (water contact angle >150°) are most desirable in membrane distillation, osmotic evaporation, membrane gas absorption and filtration. Hydrophobic membranes have been successfully applied in membrane distillation for desalination, with salt rejection reported to be higher than 90% at a feed temperature more than 90°C (Cerneaux et al., 2009; Gazagnes et al., 2007; Hendren et al., 2009; Khemakhem and Amar, 2011; Krajewski et al., 2006; Larbot et al., 2004). Hydrophobic membrane with a water contact angle of 139° has also been effectively used to concentrate fruit juices (Vargas-Garcia et al., 2011). Hydrophobic polymeric membranes such as polytetrafluoroethylene (PTFE), polypropylene (PP), polyvinylidenefluoride (PVDF), poly(amide-ethylene oxide) copolymer (Pebax), poly(1-trimethylsilyl-1-propyne) (PTMSP) and polymers of intrinsic microporosity (PIMs) are commonly used due to their low cost, ease of preparation and wide range of commercial availability (Pierre and Christian, 2008).

However, polymeric membranes are morphologically unstable in corrosive media, under high operating temperature and/or pressure. Membrane deformation, membrane swelling and plasticization impair the membrane performance which leads to higher operating cost and maintenance expenditure. (Brodard et al., 2003). Due to these reasons, ceramic membranes with superior mechanical strength, excellent chemical resistance and great thermal stability are of great interest. Unlike polymeric membranes, ceramic membranes naturally possess hydrophilic characteristic because their metal oxides can form hydroxyl (OH) groups encouraging adsorption of water (Mansur et al., 2008). Thus, ceramic membranes must be modified into superhydrophobic ceramic membranes for the mentioned applications.

Ceramic membranes are naturally hydrophilic; hence they show low rejection of polar components and permeate flux of non-polar species. The conversion of a hydrophilic ceramic membrane into a hydrophobic ceramic membrane can be easily achieved by chemical grafting of silane (Cot et al., 2000; Han et al., 2012; Jun et al., 2008; Lu et al., 2009; Mansur et al., 2008; Picard et al., 2001; Zhang et al., 2008), n-dodecanethiol (Pan et al., 2008) and lipid solution (Romero et al., 2006) onto the membrane surface. The preparation of hydrophobic ceramic membranes via silane grafting has been widely reported (Gazagnes et al., 2007; Krajewski et al., 2004; 2006; Larbot et al., 2004; Krajewski et al., 2004). Membrane grafting via immersion of the ceramic membrane in silane solution is a simple procedure which requires a short time to produce hydrophobic ceramic membrane. The reactive silane solution is prepared by hydrolyzing silane in solvents like hexane (Lu et al., 2009) and alcohol (Sugimura, 2012; Zhang et al., 2008).

Membrane hydrophobicity formed after membrane grafting is significantly influenced by the chemistry of silane. Fluoroalkylsilanes (FAS) is the most common silane used for the synthesis of hydrophobic ceramic surface due to its ease of handling. More importantly, superhydrophobic alumina membrane with contact angle more than 150° could be produced by grafting of FAS (Khemakhem and Amar, 2011; Lu et $al.$, 2009). The n-octadecyltrimethoxysilane ($C_{18}H_{37}Si(OCH_3)_3$, ODS), n-(6-aminohexyl) aminopropyltrimethoxysilane ($NH_2(CH_2)_6NH$ $(CH_2)_3Si(OCH_3)_3$, AHAPS) and fluoroalkysilane ($C_8F_{17}(CH_2)_2Si(OC_2H_5)_3$, FAS) were used to prepare hydrophobic surface on ceramic surfaces (Sugimura, 2012). The maximum water contact angles on ceramic surfaces using AHAPS, ODS and FAS were 62, 105 and 112°, respectively. Grafting with FAS generates the highest water contact angle since it consists of hydrophobic groups; Si–C, C–C, C–O, $-CF_2-CH_2-$, $-CF_2-CF_2-$ and $-CF_2-CF_3$ (Sugimura et $al.$, 2002). The presence of $-CF_2-CF_2-$ and $-CF_2-CF_3$ groups provide FAS with lower surface tension, because the surface tension of the typical substituent groups decreases in the following order: CH_2 (36 dyn cm^{-1}) > CH_3 (30 dyn cm^{-1}) > CF_2 (23 dyn cm^{-1}) > CF_3 (15 dyn cm^{-1}) [1 dyn = 1 g cm s^{-2}] (Bernett and Zisman, 1960; Dettre and Johnson Jr, 1966). Besides that the hydrophobicity on the ceramic surface is influenced by the length of fluorocarbon chain (Yim et $al.$, 2013). Researchers frequently used silane with long-CF_2-backbone such as $C_8F_{17}C_2H_4Si(OCH_2CH_3)_3$, $C_6F_{13}C_2H_4Si(OCH_3)_3$, $C_{18}H_{37}Si(OCH_3)_3$ and $NH_2(CH_2)_6NH(CH_2)_3Si(OCH_3)$ to produce hydrophobic ceramic membranes (Cerneaux et $al.$, 2009; Gazagnes et $al.$, 2007; Khemakhem and Amar, 2011; Koonaphapdeelert and Li, 2007; Krajewski et $al.$, 2006; Larbot et $al.$, 2004; Picard et $al.$, 2001; Sugimura, 2012).

The membranes grafted with silane exhibited additional mass transfer resistance due to the growth of a needle-like structure which blocks the membrane pores (Ahmad et $al.$, 2013; Koonaphapdeelert and Li, 2007; Lu et $al.$, 2009; Wei and Li, 2009). The needle-like structure originated from the long molecule chain of silane which substitutes the OH groups on the membrane surface during the grafting process. The additional resistance in the FAS layer on the membrane surface caused the decline of the gas permeability (Ahmad et $al.$, 2013; Koonaphapdeelert and Li, 2007). Hence, the major aim of this work is to study the effects of different silane solvents, grafting times and organosilane chains on mass transfer resistance of hydrophobic alumina membrane.

7.2   EXPERIMENTAL

7.2.1   *Membrane preparation*

Alumina powder with an average particle size of 0.53 μm was used to prepare alumina membrane support. The alumina powder was mixed with polyethylene glycol (Fluka 10000, Sigma Aldrich, Malaysia), methyl cellulose (Sigma Aldrich, Malaysia) and deionized water to prepare membrane support of 2 cm diameter as described elsewhere (Passalacqua et $al.$, 1998). Three different types of silane were used (as shown in Fig. 7.1); (heptadecafluoro-1,1,2,2-tetrahydrodecyl) trimethoxysilane ($CF_3(CF_2)_7C_2H_4Si(OCH_3)_3$ (LC, Sigma Aldrich, Malaysia), (tridecafluoro-1,1,2,2-tetrahydrodecyl) trimethoxysilane ($CF_3(CF_2)_5C_2H_4Si(OCH_3)_3$) (SC, Gulf Chemical, Singapore) and 3-aminopropyl-triethoxysilane ($H_2N(CH_2)_3Si(OC_2H_5)_3$) (RC, Gulf Chemical, Singapore). Three parameters were studied in this research, namely solvent selection, grafting duration and silane structures. Either ethanol or n-hexane was used to prepare grafting solutions. LC was mixed with the solvents with the volume ratio of 1 mL (LC): 50 mL solvents. The alumina support was immersed into the different grafting solutions for 20 min and rinsed with the solvents to remove the excess silane. The membrane grafting times were varied by immersing the alumina support for 5, 10 and 30 min. Different silane chains (SC and RC) were mixed with the ethanol with a similar volume ratio. Then, the alumina support was immersed into the grafting solutions for 20 min and rinsed with ethanol. All membrane samples were later dried in the oven (Memmert) at 60°C for 24 h. The designation on the membrane samples was shown in Table 7.1.

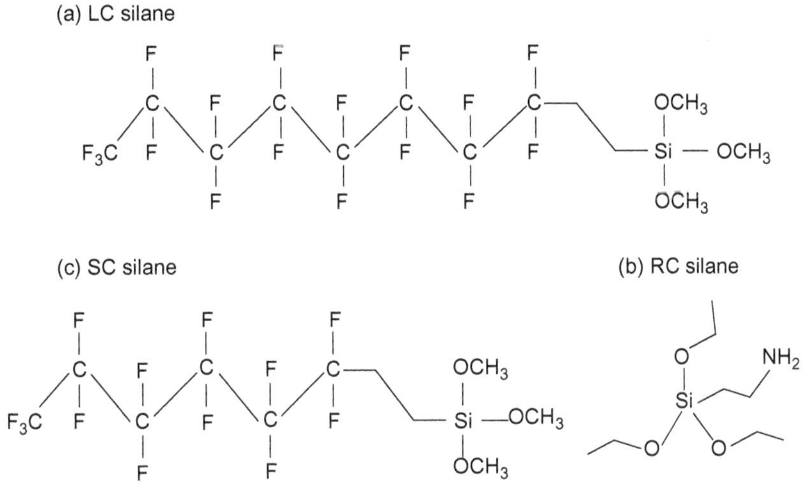

(a) LC silane

(c) SC silane

(b) RC silane

Figure 7.1.   Molecular structures of silanes.

Table 7.1.   Designation and the water contact angles of the membrane samples.

Sample	Description	Contact angle [°]	Grafting time [min]
S	Alumina membrane support	0	0
ETLC-5	Immersion of membrane sample in 1 mL LC: 50 mL ethanol for 5 min	129.34 ± 7.24	5
ETLC-10	Immersion of membrane sample in 1 mL LC: 50 mL ethanol for 10 min	134.32 ± 8.12	10
ETLC-20	Immersion of membrane sample in 1 mL LC: 50 mL ethanol for 20 min	152.26 ± 3.92	20
ETLC-30	Immersion of membrane sample in 1mL LC: 50 mL ethanol for 30 min	147.73 ± 4.27	30
HXLC-5	Immersion of membrane sample in 1 mL LC: 50 mL *n*-hexane for 5 min	148.73 ± 1.14	5
HXLC-10	Immersion of membrane sample in 1 mL LC: 50 mL *n*-hexane for 10 min	144.70 ± 3.28	10
HXLC-20	Immersion of membrane sample in 1 mL LC: 50 mL *n*-hexane for 20 min	148.24 ± 1.53	20
HXLC-30	Immersion of membrane sample in 1 mL LC: 25 mL *n*-hexane for 30 min	149.51 ± 8.38	30
ETSC-20	Immersion of membrane sample in 1 mL SC: 50 mL ethanol for 20 min	128.06 ± 3.92	20
ETRC-20	Immersion of membrane sample in 1 mL RC: 50 mL ethanol for 20 min	0	20

7.2.2   *Membrane characterization*

A goniometer (Ramé-Hart Instruments Co.) was used to determine the water contact angle on alumina (Al_2O_3) membranes. A few drops of deionized water were placed on each of the membrane samples to get an average value of contact angle. A scanning electron microscopy (SEM, Crest System (M) Sdn. Bhd, Quanta Feg 450) with energy dispersive X-ray (EDX) was used to observe the morphology and chemical elements of the membranes before and after grafting. Further analysis with porometer (Porolux1000) was made to investigate membrane resistance and pore reduction after membrane modification. The membrane resistance was determined from dry measurement using equation:

$$J = \frac{\Delta P}{\eta R_m}$$

(7.1)

where J is nitrogen permeance, ΔP is pressure difference, η is nitrogen viscosity and R_m is membrane resistance.

7.3   RESULTS AND DISCUSSION

7.3.1   Contact angle

The wetting property of a membrane surface is frequently examined using water contact angle. A membrane surface is classified as hydrophilic when its static water contact angle (CA_w) is less than 90° or hydrophobic if CA_w is larger than 90°. Further enhancement on the membrane hydrophobicity and roughness can result a superhydrophobic surface with CA_w higher than 150°. In contrast, a porous membrane tends to adsorb water droplets into its pores, resulting in lower CA_w compared to dense membrane. Figure 7.2 shows the time dependent hydrophilic property of grafted alumina membranes and alumina support (S).

The unmodified alumina membrane support (S) showed an early CA_w of 17° which then decreased to 0°. The water droplet was drastically adsorbed into the porous membrane barrier after 5 s. Based on Figure 7.2, superhydrophobicities on an alumina support were successfully developed after grafting with LC solution prepared using different solvents (n-hexane and ethanol). CA_w of more than 140° was observed for sample HXLC-20 and ETLC-20, and it remained stable in the ambient air. CA_w of the other grafted alumina membrane with LC in varied grafting durations were recorded after 3 min as represented in Table 7.1. The hydrophilic nature of the S sample was changed into a hydrophobic feature once grafted with long chains of silane. Membrane grafting with long silane chains (LC) by different solvents (ethanol and n-hexane) had an ability to produce hydrophobic membranes with hydrophobic features. Short grafting time in 5 min with LC in different solvents (ethanol or n-hexane) produced hydrophobic membrane with water contact angle more than 120°. LC-hexane grafting solution produced nearly superhydrophobic alumina membranes in a short time compared to LC-ethanol membranes. However, membrane

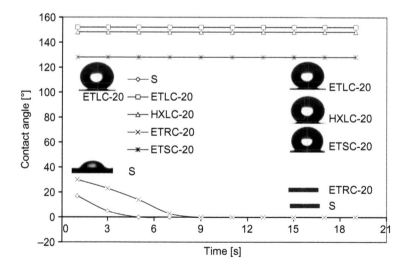

Figure 7.2.   Changes of the contact angle over time after the first measurement at 1 s.

grafting with LC-ethanol produced more superhydrophobic ETLC-20 with the water contact of $152.26 \pm 3.92°$ than membrane grafting with LC-hexane. The chemical structure of the solvent could affect the membrane hydrophobicity. Meanwhile, membrane grafting with different silane chains showed a different scenario as shown in Figure 7.2. Membrane grafting with SC produced ETSC-20 membrane with lower $128.06 \pm 3.92°$. About 24° reduction in CA_w occurred compared to ETLC-20 membrane even though grafted with similar grafting duration (20 min). This might have been caused by the different length of $-CF_2-$ chain. The grafted ETRC-20 membrane showed CA_w of 34° when water was dropped to the membrane surface, however the water contact angle surprisingly decreased to 0° within 10 s. This situation might be due to the presence of amino group ($R-NH_2$) in the silane structure.

As reported in some works (Ahmad *et al.*, 2013; Gao *et al.*, 2011; Krajewski *et al.*, 2006), membrane grafting with a long silane chain (LC), (heptadecafluoro-1,1,2,2-tetrahydrodecyl) triethoxysilane ($CF_3(CF_2)_7C_2H_4Si(OCH_3)_3$), commonly recognized as FAS, changed the hydrophilic nature into hydrophobic. A smooth surface modified with FAS rendered the surface hydrophobic with the CA_w of about 120°. In addition, CA_w values rose proportionally with the grafting time as more hydrolysable groups from LC reacted with OH groups of the alumina membrane surface. The chemical reaction between the hydrolysable and OH groups forms a hydrophobic layer. Nevertheless, extended grafting duration until 30 min changed the water contact angle only to $147.73 \pm 4.27°$ of ETLC-30 and to $149.51 \pm 8.38°$ of HXLC-30 membranes. It may be due to the saturation of bonding between the OH and the hydrolysable groups of LC. Membrane grafting with long $-CF_2-$ silane chain introduces mass transfer hindrance as reported in some earlier researches (Ahmad *et al.*, 2013; Lu *et al.*, 2009). In order to form a superhydrophobic membrane with low membrane resistance, the alumina membrane supports were grafted with different silane chains. SC silane has two less $-CF_2-$ chains compared to LC. The presence of two extra $-CF_2-$ chains with lower surface tension (23 dyn cm^{-1}) (Bernett and Zisman, 1960; Dettre and Johnson Jr, 1966; 1969) in LC enhanced the hydrophobic strength on the grafted membrane. In contrast, the introduction of functional amino groups in the silane structure resulted in highly water permeable hydrophilic features as reported in other studies (Herder *et al.*, 1988; Paradis *et al.*, 2012; Wang *et al.*, 2010). The protonated amino groups are bound to the alumina surface, thereby orienting the hydrophilic silanol end groups. Thus, silane with amino groups could not be used as the grafting agent to produce hydropobic membrane.

7.3.2 *Morphological characterization and elements analysis on the membrane surface*

A scanning electron microscope (SEM) with energy-dispersive X-ray spectroscopy (EDX) was used to study the membrane's morphological difference and the element variation of FAS grafted alumina membranes. Figure 7.3a–f shows the surface structure and cross-sectional morphology of the alumina support (S) and the grafted membranes with LC and SC. There was no significant difference in the membrane structure after direct grafting with LC, SC or RC silane, concurring with the reports by other scholars (Gao *et al.*, 2011; Koonaphapdeelert and Li, 2007; Noack *et al.*, 2000; Vargas-Garcia *et al.*, 2011). Membrane grafting with varied grafting solvents, times and silane chains did not show any significant changes on the membrane morphology. Although there was no clear structural difference from the SEM images, the chemical element analysis by EDX (Fig. 7.4) showed the presence of additional elements of fluorine (F), carbon (C) and silicon (Si). The elements of hydrogen (H) and nitrogen (N) were not included in the EDX image of ETRC-20. This is because nitrogen was present in small abundance and it was unable to detect chemical element abundances with atomic numbers less than 11 (Na) (Swapp, 2012).

7.3.3 *Mass transfer in membrane barriers*

Direct membrane grafting with silane causes serious mass transfer resistance in the membrane barrier due to the existence of silane chains on the membrane surface and pore. A capillary flow porometer with nitrogen permeability measurement could be used to determine membrane

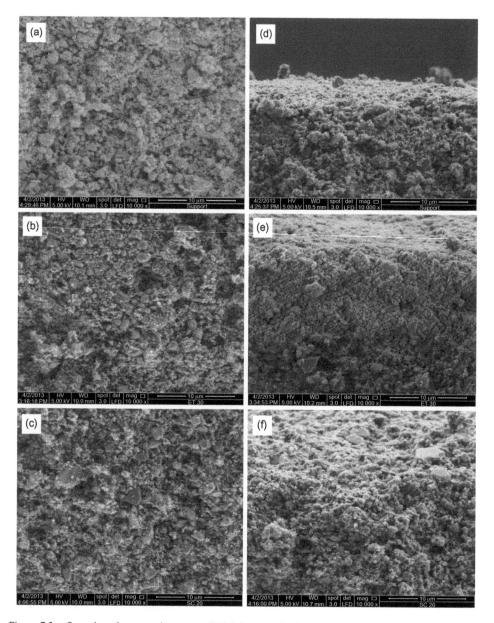

Figure 7.3.   Scanning electron microscope (SEM) images for the surface view of: (a) S, (b) ETLC-30, (c) ETSC-20 and cross-sectional view of (d) S, (e) ETRC-30 and (f) ETSC-20.

resistance. Figure 7.5 shows the nitrogen flow rate through membrane samples. Nitrogen flow rate through the hydrophobic membranes decreased when the membrane grafting duration and the length of the silane chain increased. Extended membrane grafting duration up to 30 min caused the coupling reaction between the OH groups of the alumina and the difuctional or trifunctional silane groups of silane. Consequently, chemically bonded layers of Si–O–Al were formed, which caused an additional membrane barrier and reduced the internal pore size by pore blockage (Lu *et al.*, 2009; Van Gestel *et al.*, 2003). Membrane grafting with short silane chain (SC and

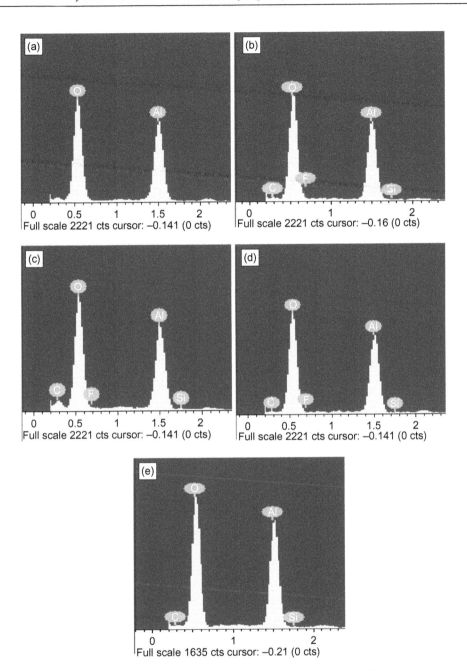

Figure 7.4. EDS results of (a) S, (b) ETLC-30, (c) HXLC-30, (d) ETSC-20 and (e) ETR-20 at the membrane surface.

RC) is expected to lower membrane resistance. However, SC silane can be used to produce superhydrophobic alumina membrane by enhancing the surface roughness. Nitrogen permeation flux reduced from $3.966 \, L \, min^{-1}$ (S) to $3.782 \, L \, min^{-1}$ (ETSC-20), $1.018 \, L \, min^{-1}$ (ETRC-20) and $0.6672 \, L \, min^{-1}$ (ETLC-20), respectively, after grafting with short, amino and long silane chain respectively for 20 min.

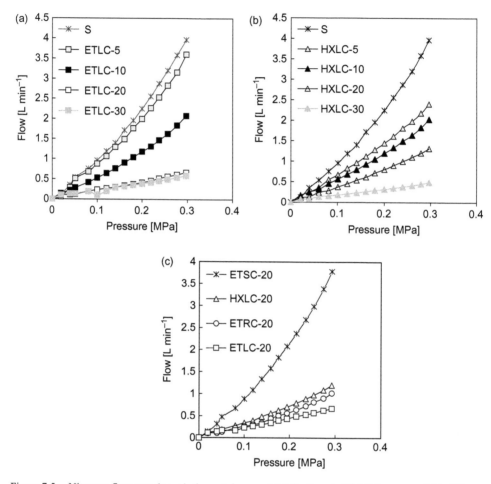

Figure 7.5.   Nitrogen flow rate through the membranes: (a) LC-ethanol, (b) LC-hexane and (c) different silane types.

7.4   CONCLUSION

Membrane grafting with silane is well known as a method to produce hydrophobic membrane. However, not all silane types are capable to modify the hydrophilic nature of ceramic membrane into hydrophobic membrane. Silane structure with long –CF$_2$– chain and low surface tension group could form a superhydrophobic alumina membrane. Membrane grafting with LC-ethanol and LC-hexane produced the superhydrophobic alumina membranes ETLC-20 ($152.26 \pm 3.92°$) and HXLC-20 ($148.24 \pm 1.53°$). However, LC silane with long –CF$_2$– chain showed higher membrane resistance due to pore blockage which caused reduction in membrane pore size. The amino group in the SC was protonated to form chemical bonding with OH on the membrane surface and water. Thus, the presence of amino groups makes the hydrophilic nature of the ceramic membrane remained unchanged when amino group was present in the saline compound. In conclusion, the selection of the silane chains is significantly important to control the membrane hydrophobicity and mass transfer resistances.

ACKNOWLEDGEMENTS

The authors would like to acknowledge Ministry of Higher Education Malaysia (MOHE) for providing LRGS (304/PJKIMIA/6050296). The authors would also like to thank Universiti Sains Malaysia for the facilities and equipment (Membrane Science and Technology Cluster 1001/PSF/8610013). Last but not least, the authors would like to thank Mr. Wan Muhammad Haniffah Wan Husin for proof-reading this manuscript.

REFERENCES

Ahmad, N.A., Leo, C.P. & Ahmad, A.L. (2013) Superhydrophobic alumina membrane by steam impingement: minimum resistance in microfiltration. *Separation and Purification Technology*, 107, 187–194.
Bernett, M.K. & Zisman, W.A. (1960) Wetting properties of tetrafluoroethylene and hexafluoropropylene copolymers 1. *The Journal of Physical Chemistry*, 64, 1292–1294.
Brodard, F., Romero, J., Belleville, M.P., Sanchez, J., Combe-James, C., Dornier, M. & Rios, G.M. (2003) New hydrophobic membranes for osmotic evaporation process. *Separation and Purification Technology*, 32, 3–7.
Cerneaux, S., Strużyńska, I., Kujawski, W.M., Persin, M. & Larbot, A. (2009). Comparison of various membrane distillation methods for desalination using hydrophobic ceramic membranes. *Journal of Membrane Science*, 337, 55–60.
Cot, L., Ayral, A., Durand, J., Guizard, C., Hovnanian, N., Julbe, A. & Larbot, A. (2000) Inorganic membranes and solid state sciences. *Solid State Sciences*, 2, 313–334.
Dettre, R.H. & Johnson, R.E., Jr. (1966) Surface properties of polymers. I. The surface tensions of some molten polyethylenes. *Journal of Colloid and Interface Science*, 21, 367–377.
Dettre, R.H. & Johnson, R.E., Jr. (1969) Surface tensions of perfluoroalkanes and polytetrafluoroethylene. *Journal of Colloid and Interface Science*, 31, 568–569.
Fang, X., Yu, Z., Sun, X., Liu, X. & Qin, F. (2009) Formation of superhydrophobic boehmite film on glass substrate by sol-gel method. *Frontiers of Chemical Engineering in China*, 3, 97–101.
Gao, N., Li, M., Jing, W., Fan, Y. & Xu, N. (2011) Improving the filtration performance of ZrO_2 membrane in non-polar organic solvents by surface hydrophobic modification. *Journal of Membrane Science*, 375, 276–283.
Gazagnes, L., Cerneaux, S., Persin, M., Prouzet, E. & Larbot, A. (2007) Desalination of sodium chloride solutions and seawater with hydrophobic ceramic membranes. *Desalination*, 217, 260–266.
Han, K.D., Leo, C.P. & Chai, S.P. (2012) Fabrication and characterization of superhydrophobic surface by using water vapor impingement method. *Applied Surface Science*, 258, 6739–6744.
Hendren, Z.D., Brant, J. & Wiesner, M.R. (2009) Surface modification of nanostructured ceramic membranes for direct contact membrane distillation. *Journal of Membrane Science*, 331, 1–10.
Herder, P., Vågberg, L. & Stenius, P. (1988) ESCA and contact angle studies of the adsorption of aminosilanes on mica. *Colloids and Surfaces*, 34, 117–132.
Jun, Y., Wei, D., Li-Ya, G. & Ya-Pu, Z. (2008) Surface forces, surface energy, and adhesion of SAMs and superhydrophobic films. In: Biresaw, G. & Mittal, K.L. (eds.) *Surfactants in tribology*, Volume 1. CRC Press, Boca Raton, FL.
Khemakhem, S. & Amar, R.B. (2011) Grafting of fluoroalkylsilanes on microfiltration Tunisian clay membrane. *Ceramics International*, 37, 3323–3328.
Koonaphapdeelert, S. & Li, K. (2007) Preparation and characterization of hydrophobic ceramic hollow fibre membrane. *Journal of Membrane Science*, 291, 70–76.
Krajewski, S.R., Kujawski, W., Dijoux, F., Picard, C. & Larbot, A. (2004) Grafting of ZrO_2 powder and ZrO_2 membrane by fluoroalkylsilanes. *Colloids and Surfaces* A: *Physicochemical and Engineering Aspects*, 243, 43–47.
Krajewski, S.R., Kujawski, W., Bukowska, M., Picard, C. & Larbot, A. (2006) Application of fluoroalkylsilanes (FAS) grafted ceramic membranes in membrane distillation process of NaCl solutions. *Journal of Membrane Science*, 281, 253–259.
Larbot, A., Gazagnes, L., Krajewski, S., Bukowska, M. & Wojciech, K. (2004) Water desalination using ceramic membrane distillation. *Desalination*, 168, 367–372.
Lu, J., Yu, Y., Zhou, J., Song, L., Hu, X. & Larbot, A. (2009) FAS grafted superhydrophobic ceramic membrane. *Applied Surface Science*, 255, 9092–9099.

Mansur, A.A.P., do Nascimento, O.L., Vasconcelos, W.L. & Mansur, H.S. (2008) Chemical functionaliza-tion of ceramic tile surfaces by silane coupling agents: polymer modified mortar adhesion mechanism implications. *Materials Research*, 11, 293–302.

Noack, M., Kölsch, P., Caro, J., Schneider, M., Toussaint, P. & Sieber, I. (2000) MFI membranes of different Si/Al ratios for pervaporation and steam permeation. *Microporous and Mesoporous Materials*, 35–36, 253–265.

Pan, Q., Wang, M. & Wang, H. (2008) Separating small amount of water and hydrophobic solvents by novel superhydrophobic copper meshes. *Applied Surface Science*, 254, 6002–6006.

Paradis, G.G., Kreiter, R., van Tuel, M.M.A., Nijmeijer, A. & Vente, J.F. (2012) Amino-functionalized microporous hybrid silica membranes. *Journal of Materials Chemistry*, 22, 7258–7264.

Passalacqua, E., Freni, S. & Barone, F. (1998) Alkali resistance of tape-cast SiC porous ceramic membranes. *Materials Letters*, 34, 257–262.

Picard, C., Larbot, A., Guida-Pietrasanta, F., Boutevin, B. & Ratsimihety, A. (2001) Grafting of ceramic mem-branes by fluorinated silanes: hydrophobic features. *Separation and Purification Technology*, 25, 65–69.

Picard, C., Larbot, A., Tronel-Peyroz, E. & Berjoan, R. (2004) Characterization of hydrophilic ceramic mem-branes modified by fluoroalkylsilanes into hydrophobic membranes. *Solid State Sciences*, 6, 605–612.

Pierre, A. & Christian, G. (2008) Current status and prospects for ceramic membrane applications. In: Pabby, A.K., Rizvi, S.S.H. & Sastre, A.M. (eds.) *Handbook of membrane separations*. CRC Press, Boca Raton, FL. pp. 139–179.

Romero, J., Draga, H., Belleville, M.P., Sanchez, J., Combe-James, C., Dornier, M. & Rios, G.M. (2006) New hydrophobic membranes for contactor processes – applications to isothermal concentration of solutions. *Desalination*, 193, 280–285.

Sugimura, H. *Self-assembled monolayer on silicon*. Kyoto University, Kyoto, Japan. pp. 1–22. Available from: www.mtl.kyoto-u.ac-jp/groups/sugimura-g/PDF/SAM-on-si.pdf [Accessed January, 18, 2012].

Sugimura, H., Hozumi, A., Kameyama, T. & Takai, O. (2002) Organosilane self-assembled monolayers formed at the vapour/solid interface. *Surface and Interface Analysis*, 34, 550–554.

Swapp, S. (2012) Scanning electron microscopy (SEM). Available from: http://serc.carleton.edu/research_education/geochemsheets/techniques/SEM.html [accessed April 2012].

Van Gestel, T., Van der Bruggen, B., Buekenhoudt, A., Dotremont, C., Luyten, J., Vandecasteele, C. & Maes, G. (2003) Surface modification of γ-Al$_2$O$_3$/TiO$_2$ multilayer membranes for applications in non-polar organic solvents. *Journal of Membrane Science*, 224, 3–10.

Vargas-Garcia, A., Torrestiana-Sanchez, B., Garcia-Borquez, A. & Aguilar-Uscanga, G. (2011) Effect of grafting on microstructure, composition and surface and transport properties of ceramic membranes for osmotic evaporation. *Separation and Purification Technology*, 80, 473–481.

Wang, J., Bratko, D., Luzar, A. & Widom, B. (2010) Probing surface tension additivity on chemically heterogeneous surfaces by a molecular approach. *Chemistry*, 108, 6374–6379.

Wei, C.C. & Li, K. (2009) Preparation and characterization of a robust and hydrophobic ceramic membrane via an improved surface grafting technique. *Industrial & Engineering Chemistry Research*, 48, 3446–3452.

Yim, J.H., Rodriguez-Santiago, V., Williams, A.A., Gougousi, T., Pappas, D.D. & Hirvonen, J.K. (2013) Atmo-spheric pressure plasma enhanced chemical vapor deposition of hydrophobic coatings using fluorine-based liquid precursors. *Surface and Coatings Technology*, 234, 21–32.

Zhang, X., Honkanen, M., Järn, M., Peltonen, J., Pore, V., Levänen, E. & Mäntylä, T. (2008) Thermal stability of the structural features in the superhydrophobic boehmite films on austenitic stainless steels. *Applied Surface Science*, 254, 5129–5133.

Part II
Water and wastewater treatment
(applications)

CHAPTER 8

Cross-flow microfiltration of synthetic oily wastewater using multi-channel ceramic membrane

Yousof H.D. Alanezi

8.1 INTRODUCTION

Produced water is water formed in underground formations and is brought up to the surface along with crude oil during production. It comprises mainly dispersed oil, organic compounds, and suspended solids. The most popular preference to deal with produced water is to re-inject it back into the formation. Produced water re-injection (PWRI) needs a modified treatment such as separation units to eliminate oil and suspended solids before re-injection for pressure build up. De-oiling treatment normally consists of gravity or corrugated plate separator and a gas flotation unit. However, gravity separation is not successful with emulsified oil droplets smaller than 20 μm. The reason is that as the oil droplets size reduces, the essential retention time to obtain acceptable separator efficiency increases considerably.

A promising membrane technology is cross-flow microfiltration (MF) for removal of suspended particles and emulsified oil droplets in the size range of 0.1–20 μm from their feed suspensions. For the technique to be industrially acceptable it must provide an increase of the filtrate volume with an oil concentration of less than 5 mg L^{-1} and also eliminate any solids in suspension. In contrast, the main drawback associated with MF relates to fouling, i.e. the membrane surfaces or pores become clogged.

8.1.1 *Case study*

Due to the increasing amounts of produced water during oil production in Kuwait, the establishment of wastewater treatment units for produced water re-injection purposes had become essential. It is estimated there that oil wells generate in quantity of 15 to 40% of produced water. The unit consists of surge tank, oily water treatment, and oil drum. The oily water treatment comprises parallel/corrugated plate separator and induced gas flotation. The main objective of this treatment train is to reduce the oil in water concentration from 2000 to 10 mg L^{-1}, the maximum allowable concentration for reinjection and disposal. The produced water characteristic are presented in Table 8.1.

8.2 EXPERIMENTAL

8.2.1 *Materials*

Dodecane and Sorbitan monooleate (Aldrich Chemical) were used to form oil in water emulsions. Emulsion particle size distributions were measured using a Malvern Zetasizer 3000HS; the average particle size of the distributions was 3–5 μm. The membrane used was a tubular ceramic (zirconia) microfiltration module obtained from Fairey Industrial Ceramics Ltd. The membrane average pore diameter was 0.2 μm and its effective length was 0.55 m. The membrane consists of 7 channels, each being circular with an inner diameter of 4.7 mm. The membrane was mounted horizontally in a stainless steel module.

Table 8.1.   Produced water characteristics.

Chemical component	Concentration [mg L^{-1}]
Cl$^-$	145900
HCO$_3^-$	196
Ca^{2+}	23250
Na$^+$	42191
SO$_4^{2-}$	256
Oil	2000
H$_2$S	150

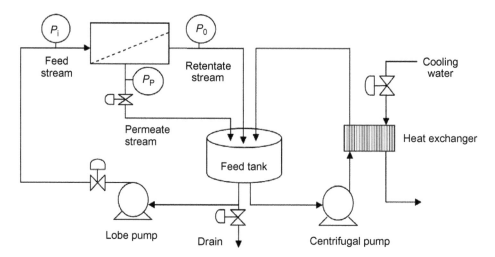

Figure 8.1.   Schematic diagram of the experimental setup.

8.2.2   Cross-flow microfiltration rig

The schematic diagram of the microfiltration rig used in this study is shown in Figure 8.1. The oil in water emulsions were pumped into the membrane module via a variable speed lobe pump. The transmembrane pressure was monitored using three pressure gauges, one at each of the feed and retentate ends of the membrane and one in the permeate stream. The temperature of the feed stream was regulated using a secondary circuit in which a plate type heat exchanger kept the feed temperature at 25 ± 2°C. This secondary circuit also provided most of the mixing effects, to keep particles in the suspension well dispersed. The ΔP for a given cross-flow velocity was therefore controlled by manipulating the permeate pressure, P_p, using the valve in the permeate stream.

8.2.3   Experimental procedure

Prior to the start of a filtration experiment, the oil-in-water emulsions were prepared by mixing n-dodecane (model oil) and sorbitan monooleate (surfactant) with deionized water for half an hour by using a high shear laboratory mixer at a speed of 4000 rpm. The emulsions were then flowed through the rig with the permeate valve closed for 10 min before the start of an experiment to stabilize the cross-flow velocity and to allow equilibrium to be achieved between the suspension and the surfaces in the rig (including the heat exchanger circuit). The clean water flux and total membrane resistance were determined before and after each experiment to ensure that the permeability of the membrane was approximately the same at the start of each experiment. This was necessary to enable analysis of the extent of irreversible fouling and the efficiency of the cleaning method.

J_{crit} is the critical flux; this was measured experimentally by successive increments/decrements of transmembrane pressure using a step by step technique. The technique consisted of systematic increases of ΔP, each step had minimum 15 min duration or until the equilibrium permeate flux had been reached. The first unstable permeation flux was determined when the flux declined over the course of time at a given ΔP step. After incrementing the pressure to a point beyond the critical ΔP, the pressure is then decremented. The flux points corresponding to the upwards and downwards steps are plotted against ΔP and the deviation point from the clean water flux or first step extrapolation line was obtained.

8.3   RESULTS AND DISCUSSION

8.3.1   *Effect of oil feed concentration on the critical flux*

With the 0.2 μm tubular multi-channel ceramic membrane, the critical flux was determined at four different cross-flow velocities (1.14, 1.52, 1.92, 2.28 m s^{-1}). At each cross-flow velocity, four oil feed concentrations were examined: 300, 600, 1200, and 2400 mg L^{-1}. The TMP stepping method was used to determine the critical flux in each case and Table 8.2 summarizes these results. From the results shown Table 8.2, it is seen that as the oil concentration increased in the feed, the critical flux value decreased at different operating conditions. This behavior is in agreement with several studies conducted before (Chen, 1998; Field *et al.*, 1995; Fradin and Field, 1999; Kwon *et al.*, 2000; Madaeni, 1997).

Gesan-Guiziou *et al.* (2000) observed that as the latex concentration in the feed increased from 400 to 2000 mg L^{-1}, the critical flux was decreased by two thirds. At higher concentrations (3000–8000 mg L^{-1}) the critical flux stayed approximately invariant. In the current study, when the oil concentration in the feed increased from 300 to 2400 mg L^{-1}, the critical flux values were decreased by almost one third while operating at cross-flow velocities of 1.14, 1.52, and 1.92 m s^{-1}. For the case of operating at cross-flow velocity of 2.28 m s^{-1}, the critical flux values decreased approximately by one quarter as the oil feed increased from 300 to 2400 mg L^{-1}, as shown in the Table 8.2.

As a result of increases in the oil feed concentration, the mass transfer rate of the oil droplets to the membrane surface increased, which led to more accumulation in the boundary layer near the membrane surface. Consequently, an increase in the membrane resistance is observed due to the expansion of concentration polarization and the formation of a thicker cake layer. During filtration of BSA at two different feed concentrations (at 0.1 and 1.0 wt%), the critical flux for the lower

Table 8.2.   Summary of estimated critical flux values from TMP stepping experiments at varying oil feed (dodecane) concentrations without adding NaCl.

Oil feed concentration [mg L^{-1}]	Cross-flow velocity [m s^{-1}]			
	1.14 Case A	1.52 Case B	1.92 Case C	2.28 Case D
	Apparent critical flux [L m^{-2} h^{-1}]			
300	65	118	143.36	238
600	73	115.67	137.13	217
1200	47	89	116.42	197.8
2400	44	79.74	99.33	179.9

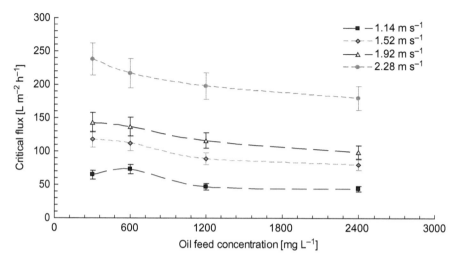

Figure 8.2.   Variation of critical flux with oil feed concentration at different cross-flow velocities (1.14, 1.52, 1.92, and 2.28 m s^{-1}).

concentration was measured to be approximately 66 L m^{-2} h^{-1} while the higher concentration was 200 L m^{-2} h^{-1} (Chen, 1998). Chen proposed that the most influential parameter that affected the critical flux measurement was local concentration.

It can also be seen from Figure 8.2 and Table 8.2 that the critical flux value decreased by 2–4% as the concentration increased from 300 to 600 mg L^{-1} for the cases B and C, and decreased by 9% for case D. With the exception for the case A, increasing oil feed concentration from 300 to 600 mg L^{-1} resulted in an increase in the critical flux value. Similar behavior was reported by Wakeman and Tarleton (1993) where they noticed that the membrane fouling rate was found higher at lower feed concentration, where a comparatively high rate of TMP increase was noticed. The accumulated particles tend to plug the membrane pores at low feed concentration, because the competition for settling over optimal positions (at the membrane pores) is not too high.

In contrast, the lower rate of TMP increase at higher feed concentration might be caused by the particles bridging over the pores of the membrane to compete for the optimal positions which lead to cake layer formation (Kwon et al., 2000).

Furthermore, by increasing the oil feed concentration from 600 to 1200 mg L^{-1}, the critical flux decreased by 35% for case A, by 23% for case B, by 15% for case C, and by 10% for case D. As the oil feed concentration increased from 1200 to 2400 mg L^{-1}, for case A the critical flux declined by 2%; for cases B and D the critical flux declined by 9–10%; and for case C the critical flux declined by 15%. Case D showed an inverse relationship between critical flux and feed concentration. As the oil feed concentration increased by 100% (doubled), the critical flux decreased by approximately 9–10% for each increment of oil concentration.

In general the variation trend of critical flux values was to decline as the oil feed increased up to an oil concentration of 1200 mg L^{-1}, after which the critical flux was almost independent of concentration, as presented in Figure 8.2. The high flux at lower oil concentration (300–600 mg L^{-1}) appears to point to the inadequate formation of an oil layer onto the membrane surface, which can be removed by the hydrodynamic effect (Mohhammadi et al., 2004; Ohya et al., 1998) while at higher oil concentrations (1200–2400 mg L^{-1}), the hydrodynamic effect cannot take away the formed oil layer. Kwon et al. (2000) observed a similar trend that both critical fluxes (based on mass balance and increase in TMP) decreased as the feed concentration increased. They suggested that this behavior was due to the higher accumulation of particles onto the membrane surface. Hence, at higher feed concentration more particles tend to accumulate onto the membrane surface for the similar flux step.

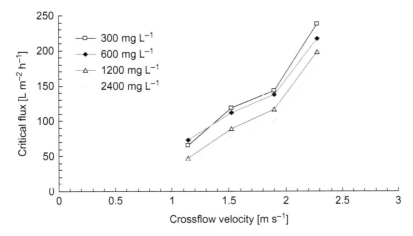

Figure 8.3.   Variation of critical flux with cross-flow velocity at different oil feed concentration (300, 600, 1200 and 2400 mg L^{-1}).

On the other hand, Bacchin *et al.* (2006) stated that the plotting the permeate flux as a function of the logarithm of the concentration was not a straight line relationship, which demonstrated incompliance with the film model. Therefore, the mechanism of particle mass accumulation could not be explained by such a model. This inconsistency could be attributed to assumptions made for such a model where the effects for the changes of diffusion coefficient and viscosity across the boundary layer were not considered. Furthermore, this disagreement could be credited to the existence of surface interactions between the particle and membrane or particles themselves.

8.3.2   *Effect of cross-flow velocity and wall shear stress on the critical flux*

Figures 8.3 and 8.4 demonstrate the variation of critical flux with increasing cross-flow velocity for *n*-dodecane emulsions.

The general observed trend was an increase in critical flux as the cross-flow velocity (shear) was increased as shown in Table 8.2 and Figure 8.3. At the higher cross-flow velocities, higher shear wall stress was generated and hence less particle deposition occurred at the membrane surfaces due to removal by erosion. For 300 mg L^{-1} emulsions, increasing the cross-flow velocity from 1.14 to 2.28 m s^{-1} led to an increase in critical flux by a factor of 3.7. Also, for 600 mg L^{-1} emulsions, increasing the cross-flow velocity from 1.14 to 2.28 m s^{-1} led to an increase in critical flux by a factor of 3. While for 1200 mg L^{-1} emulsions, increasing the cross-flow velocity from 1.14 to 2.28 m s^{-1} led to an increase in critical flux by a factor of 4.2. Similarly, for 2400 mg L^{-1} emulsions, increasing the cross-flow velocity from 1.14 to 2.28 m s^{-1} led to an increase in critical flux by a factor of 4.1.

In general, increasing the cross-flow velocity from 1.14 to 2.28 m s^{-1} resulted in an increase in critical flux by a factor approximately between 3 and 4. A comparable trend with a similar factor was reported by previous researchers such as Chen (1998) and Chiu and James (2005) when the cross-flow velocity increased. Chen (1998) accredited their results to the inception of turbulence at higher cross-flow velocities. Similarly in the present study, the Reynolds number at lowest cross-flow velocity (1.14 m s^{-1}) was estimated to 5358 and at the highest cross-flow velocity (2.28 m s^{-1}) it was approximated to 10,716. Hence, the flow regime could be described to lie in the turbulent regions.

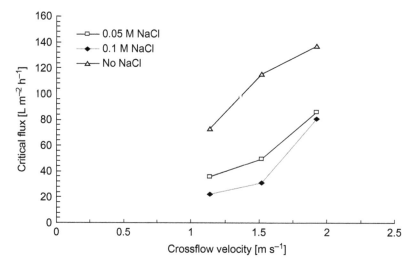

Figure 8.4.   Variation of critical flux with cross-flow velocity at different NaCl electrolyte concentration.

The critical flux values decreased when the NaCl salt amount increased in a 600 mg L^{-1} emulsion at different cross-flow velocities, as shown in Figure 8.4. When no salt was added, increasing the cross-flow velocity from 1.14 to 1.92 m s^{-1} led to an increase in critical flux by a factor of 1.9. Adding 0.05 M NaCl to the 600 mg L^{-1} emulsion and increasing the cross-flow velocity from 1.14 to 1.92 m s^{-1} led to an increase in critical flux by a factor of 2.4. By adding 0.1 M NaCl to the 600 mg L^{-1} emulsion and increasing the cross-flow velocity from 1.14 to 1.92 m s^{-1}, an increase in critical flux by a factor of 4 resulted.

The impact of increasing both ionic strength (0.05 M and 0.1 M NaCl) and cross-flow velocity on critical fluxes demonstrated similar curve behavior, as shown in Figure 8.4. When cross-flow velocity increased from 1.14 to 1.52 m s^{-1}, the critical fluxes increased for both ionic strength by a factor of almost 1.4, however, as the cross-flow velocity increased further from 1.52 to 1.92 m s^{-1}, the critical flux for the higher ionic strength (0.1 M) increased by a factor 2.6 while for the moderate ionic strength (0.05 M) by a factor of 1.7. In other words, the influence of operating at higher shear rate was more profound on critical flux for emulsions with higher ionic strength, where the flux was almost 60% higher compared to that for moderate ionic strength. By operating at higher shear rate, there is a tendency to remove relatively smaller droplets from the membrane surface by a scouring effect.

In contrast, for the 600 mg L^{-1} emulsion (without NaCl) when cross-flow velocity increased from 1.14 to 1.52 m s^{-1}, the critical fluxes increased by a factor of approximately 1.6, however, as the cross-flow velocity increased further from 1.52 to 1.92 m s^{-1}, the critical flux increased by a factor of roughly 1.2 (Fig. 8.4). This behavior suggested that there might be breakage of large oil droplets to finer particles which are not removed by shear rate and there may be internal fouling of the membrane. While for emulsion with added NaCl, the oil emulsion stability seemed strong enough to withstand these higher shear rates.

Generally, increasing cross-flow velocity led to an increase in the critical flux values as shown in Table 8.2. The best critical flux level was 238 L m^{-2} h^{-1} at experimental conditions where the cross-flow velocity was 2.28 m s^{-1}, 300 mg L^{-1} and no NaCl salt was added. Adding 0.1 M NaCl salt at the same experimental conditions, the critical flux value decreased to 88 L m^{-2} h^{-1}. On the other hand, the least critical flux value was 22 L m^{-2} h^{-1}, where experimental conditions were at the lowest cross-flow velocity (1.14 m s^{-1}) and the highest NaCl concentration (0.1 M) for 600 mg L^{-1} emulsion.

Table 8.3. The critical parameter (J_{crit}/τ_w) values at different cross-flow velocity for emulsion A, emulsion B, and emulsion C.

	Oil feed concentration [mg L^{-1}]		
	600	1200	2400
Cross-flow	Emulsion A	Emulsion B	Emulsion C
Velocity	J_{crit}/τ_w	J_{crit}/τ_w	J_{crit}/τ_w
[m s^{-1}]	[L m^{-2} h^{-1} Pa^{-1}]	[L m^{-2} h^{-1} Pa^{-1}]	[L m^{-2} h^{-1} Pa^{-1}]
1.14	12.15	7.82	7.32
1.52	11.51	8.85	7.93
1.92	9.34	7.93	6.76
2.28	10.74	9.79	8.90
Mean	10.93	8.60	7.73

Assuming cake filtration, the total particle accumulation onto the membrane surface occurs as a result of the convection flow dragging particles toward the membrane becoming higher than the removal rate of particles away to the bulk stream, which is believed to be proportional to shear stress (Huisman, 1999):

$$A(JC_p - a\tau_w) = \frac{dm}{dt} \tag{8.1}$$

where A is the surface area, J is the permeate flux, C_p is the mass concentration of suspension, a is an experimentally determined constant, τ is the shear stress and m is the cake mass. At higher wall shear stresses, fewer particles deposited onto membrane surface and hence higher critical flux values were attained (Fig. 8.4).

Gèsan-Guiziou et al. (2002) suggested a critical parameter (J_{crit}/τ_w), which stayed constant over the range investigated, where J_{crit} increased linearly with τ_w and J_{crit} was found to be independent of initial pore size of the membrane. Their argument was justified based on the work of Huisman (1999) who claimed that at the condition of critical flux there is no particle accumulation (dm/dt) and hence dm/dt becomes zero in Equation (8.1). Therefore, the convective particle deposition ($J_{crit}C_p$) is proportional to the shear stress (τ_w), that is $J_{crit}C_p \propto a\tau_w$. Gèsan-Guiziou et al. (2002) stated that the proof that "a" in Equation (8.1) is constant is inadequate; for this to be so implies that the ratio ($J_{crit}C_p$)/τ_w has to stay constant, which is not observed experimentally. Similarly, the experimental results obtained in this work demonstrate that the critical parameter (J_{crit}/τ_w) did not remain constant as shown in Table 8.3. Therefore, Gèsan-Guiziou et al. (2002) interpretation of such experimental data is that "a" is not constant, however a function of C_p, which implies that the hypothesis to establish Equation (8.1) was not applicable for such cases.

From Table 8.3 and Figure 8.5, it is apparent that no single linear correlation exists for all the concentrations. In addition, for every concentration there was some non-linearity in the variation of critical flux with shear stress and the proportional correlation between the convective particle deposition and particle back transport erosion (shear stress) exists over a limited range of values. These observations concerning the constancy of (J_{crit}/τ_w) do not entirely agree with the work of Gèsan-Guisiou et al. (2002) whose experimental results demonstrated a linear variation of critical flux and wall shear stress. The gradient of the linear relationship (J_{crit}/τ_w) for skimmed milk was 0.95 L m^{-2} h^{-1} Pa^{-1} and for latex particles was 18 L m^{-2} h^{-1} Pa^{-1}.

8.3.3 *Comparisons between experimental results and models*

Generally, it is known that permeate flux drags particles towards the membrane whilst diffusion induces particle back transport into the bulk. Various mechanisms have been proposed to account

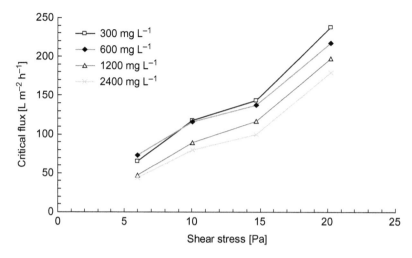

Figure 8.5.   Variation of critical flux with wall shear stress at different oil feed concentration (300, 600, 1200 and 2400 mg L⁻¹).

for this back transport such as shear-induced diffusion, Brownian diffusion, axial transport (particle rolling), inertial lift forces, and particle-particle interaction forces (Belfort et al., 1994). Presuming Brownian diffusion back transport is the dominant mechanism, predicted fluxes for micron-sized particles were found to be one or more of orders of magnitude less than those observed experimentally (Howell, 1995). This has been referred to as the 'flux paradox' for colloidal suspensions. Axial transport models are derived from solving fully-developed laminar flow equations and the flow in this work the turbulent regions.

A particle deposition could be initiated by the forces pushing it towards the membrane, such that the permeate drag force and attractive physico-chemical interactions, are bigger than the forces taking away the particle from the membrane (e.g. the shear drag force and repulsive physico-chemical interactions). Such conditions are better portrayed by torque-balance models. These models state that particles will roll along the membrane surface only if the positive torque produced by the cross-flow velocity, repulsive interactions, and lift forces is larger than the negative torque generated by the convective permeate flow. As the permeate flux increases, the negative torque increases until it balances the positive torque so that the net torque on the particle is nil, hence the particle will stop rolling and the critical flux is approached.

The critical flux is then expressed by (Belfort et al., 1994; Howell, 1995):

$$J_{crit} = \frac{2.36a\tau_w}{\mu_0 \tan \theta (a^2 \hat{R}_m)^{2/5}} + \frac{0.463F_i}{\pi\mu_0 \sin \theta (a^2 \hat{R}_m)^{2/5}} \tag{8.2}$$

where a is the particle radius, τ_w is the wall shear stress, μ_0 is the permeate viscosity, θ is the angle of repose (a measure for the surface roughness), $\hat{R}_m$ the specific membrane resistance, and F_i is the membrane-particle interaction force. F_i is positive if the surface charge sign of both the membrane and the particle are the same. F_i is negative if the surface charge sign of both the membrane and the particle are opposite. F_i diminishes if the membrane or the particle are uncharged (or at isoelectric point). Since most of critical flux values were measured at solution pH 5.5–6 and the reported iso-electric point for ceramic membrane was 5.8–6. Thus, the second term of Equation (8.2) will be cancelled and the equation is reduced to:

$$J_{crit} = \frac{2.36a\tau_w}{\mu_0 \tan \theta (a^2 \hat{R}_m)^{2/5}} \tag{8.3}$$

Table 8.4. Summary of prominent back-transport and lift models (Baruah and Belfort, 2003).

Flux model	Approaches	Flux equation	Equation number	Applicable range
Brownian diffusion	Use Leveque solution for laminar flow in a solid wall tube and Stokes-Einstein diffusion	$J = 0.114 \left(\dfrac{\gamma \kappa^2 T^2}{\mu_0^2 a^2 L} \right)^{1/3} \ln \left(\dfrac{\varphi_w}{\varphi_b} \right)$	(8.4)	Applicable for very small diameter particles ($<1\,\mu m$); under predicts flux by 1–2 orders of magnitude for large particles
Inertial lift	Include the inertial terms in solving the force balance around a single particle	$J = \left(\dfrac{0.036 \rho a^3 \gamma^2}{\mu_0} \right)$	(8.5)	Applicable for large diameter particles ($>20\,\mu m$) and consider only single particles
Shear-induced diffusion	Replace diffusion coefficient with a shear-dependent	$J = 0.078 \left(\dfrac{a^4}{L} \right)^{1/3} \gamma \ln \left(\dfrac{\varphi_w}{\varphi_b} \right)$	(8.6)	Applicable for intermediate diameter particles (1–$20\,\mu m$)

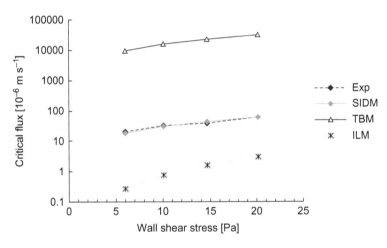

Figure 8.6. Comparison of predicted critical fluxes using SIM, TBM, ILM models with experimental measured critical fluxes for $600\,mg\,L^{-1}$ *n*-dodecane emulsion.

Here critical flux is directly proportional to particle radius (a) and wall shear stress (τ_w), since the term $((a^2 \hat{R}_m)^{2/5})$ is treated as a single unit.

A torque balance model (TBM), assuming axial transport as the major mechanism gave unrealistically high permeate fluxes. Several researchers, such as Huisman *et al.* (1999) and Belfort *et al.* (1994), noticed the same results experimentally and proposed that a promising remedy for the flux paradox will be to take into account the inertial lift consequence. Nonetheless, inertial lift effects were found to be insignificant for micron-sized particles in the current study. Normally, the inertial lift model (ILM) is applicable for particle sizes bigger than $20\,\mu m$ and consider only single particles (Table 8.4), whereas the particle size distribution in the present study is between 0.1–$10\,\mu m$. The results are plotted on Figures 8.6, 8.7 and 8.8, in which the TBM and ILM models were compared with the experimental data.

The back transport mechanism is more likely to govern the flux behavior, bearing in mind the range of the measured particle sizes is the shear-induced diffusion. Howell (1995) claimed that the shear-induced diffusion would be the main back transport mechanism for particle within the micron size range. Hence, the shear-induced diffusion model is fitted to the experimental data by

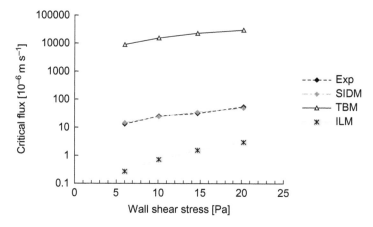

Figure 8.7.   Comparison of predicted critical fluxes using SIM, TBM, ILM models with experimental critical fluxes for 1200 mg L^{-1} n-dodecane emulsions.

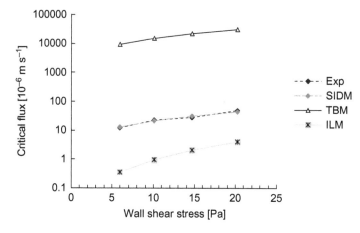

Figure 8.8.   Comparison of predicted critical fluxes using SIM, TBM, ILM models with experimental critical fluxes for 2400 mg L^{-1} n-dodecane emulsion.

using the particle size as the curve fitting parameter, providing physical illustrations of the particle sizes at critical flux value for different cross-flow velocities. The shear-induced model (SIDM) makes use of the shear-induced hydrodynamic diffusivity rather than the Brownian diffusivity, determined by using Stokes-Einstein correlation. The model employed here has been obtained from prior researchers (Belfort *et al.*, 1994; Howell, 1995) where detailed description of the model was provided. Comparisons with other models are shown in Figures 8.6, 8.7 and 8.8; the SID model showed a better prediction of experimental results of critical fluxes at different oil feed concentrations and various cross-flow velocities than did either the ILM or IBM models.

The distinction between employing Brownian diffusion (D_{BD}) and employing shear-induced diffusion (D_s) in evaluating particle depolarization is important. D_{BD} is a function of particle size and increases as the particle size decreases ($D_{BD} \propto a^{-1}$). The implication of Brownian diffusion becoming the predominant mechanism for particle back transport is that the back diffusion turns out to be more significant for finer particles and hence they tend to depolarize from the membrane surface. However, this implication is in disagreement with several experimental observations in which the smaller particles have a preference for deposition on the membrane surface during cross-flow filtration. In addition, because D_{BD} is invariable for specified particle size, the degree

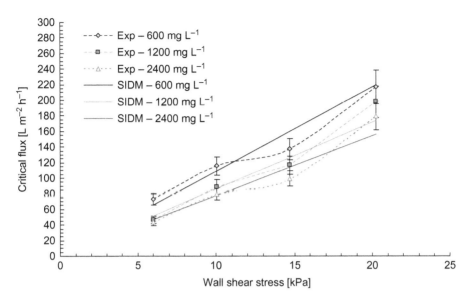

Figure 8.9.   Shear-induced diffusion model (SIDM) fitted to experimental data obtained for (600, 1200 and 2400 mg L^{-1}) emulsions at various cross-flow velocities.

of back transport founded on D_{BD} would be insensitive to variations of hydrodynamic conditions. This contradicts a number of experimental observations where the rate of particle back transport has increased when the shear rate was increased.

On the contrary, D_s values are revealed to be enhanced with increase in particle size and the shear rate ($D_S \propto a\gamma$). Hence, analyzing particle depolarization in cross-flow filtration by using the shear-induced diffusion theory has been more practical for a number of applications (Baruah and Belfort, 2003; Ripperger and Altmann, 2002; Tien, 2006). Baruah and Belfort (2003) summarized a few prominent models of cross-flow microfiltration with their fundamental assumption and applicability as shown in Table 8.4.

They suggested that the Brownian diffusion model is more appropriate for particle diameters below ~0.1 μm and at low wall shear rates. For the microfiltration case where $0.1 < a < 10$ μm, the permeation flux estimated by using Equation (8.4) was under-predicted by 1–2 orders of magnitude. As a response to such a flux paradox, Green and Belfort suggested that the inertial lift mechanism be evaluated using Equation (8.5), which showed that the sensitivity of flux was higher for particle size (a^3) and shear (γ^2).

8.3.4   *Critical flux prediction by shear-induced diffusion model (SIDM)*

The shear induced diffusion model (SIDM) has been fitted to experimental data by calculating the average critical particle size for the four cross-flow velocities by rearranging the following shear induced model equation:

$$J_{crit} = \frac{\tau_w}{\mu_p} \left(\frac{1 \times 10^{-4} a^4}{\varphi_b X} \right)^{1/3} \tag{8.7}$$

where φ_b is the volume fraction of particles in the bulk, and X is the length of membrane. The comparison between critical fluxes measured experimentally and those estimated using Equation (8.7) at different oil concentration and operating conditions is illustrated in Figure 8.9.

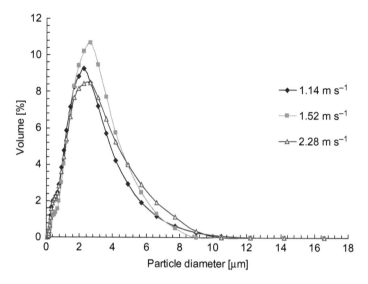

Figure 8.10.   Oil droplet size distribution at various cross-flow velocity for 600 mg L^{-1} n-dodecane emulsion with surfactant 60 mg L^{-1} after 60 minutes of recirculation.

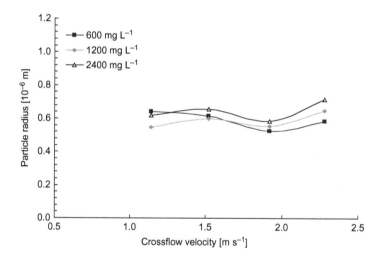

Figure 8.11.   Particle radii calculated from the shear-induced model at different cross-flow velocities.

The experimental critical flux values were inserted and the Equation (8.7) was rearranged to:

$$a_{\text{crit}} = \left(\frac{J_{\text{crit}} \mu_{\text{p}}}{\tau_{\text{w}}} \right)^{3/4} \left(\frac{\varphi_{\text{b}} X}{1 \times 10^{-4}} \right)^{1/4} \tag{8.8}$$

Then a_{crit} was calculated using Equation (8.8) and compared with the measured particle size shown in Figure 8.10.

Hence a_{crit} is considered to be a fitting parameter to give relatively smooth curves that would give 'best' fit the experimental data points. The particle sizes used in the SIDM to fit the model to experimental data at the four cross-flow velocities are plotted against cross-flow velocity in Figure 8.11.

Table 8.5. Estimated critical particle radii for *n*-dodecane emulsions with different oil feed concentration at various cross-flow velocities.

n-Dodecane concentration	600 [mg L^{-1}]	1200 [mg L^{-1}]	2400 [mg L^{-1}]
Cross flow velocity [m s^{-1}]	a_{crit} [μm]	a_{crit} [μm]	a_{crit} [μm]
1.14	0.641	0.548	0.620
1.52	0.616	0.601	0.658
1.92	0.526	0.554	0.584
2.28	0.585	0.648	0.718
Mean a_{crit}	0.592	0.588	0.645

Table 8.6. Estimated critical particle radii for 600 *n*-dodecane emulsions at various cross-flow velocities with different NaCl electrolyte concentrations.

Cross-flow velocity [m s^{-1}]	No NaCl a_{crit} [μm]	0.05 M NaCl a_{crit} [μm]	0.1 M NaCl a_{crit} [μm]
1.14	0.641	0.376	0.263
1.52	0.616	0.327	0.230
1.92	0.526	0.371	0.355
Mean	0.592	0.358	0.282

The 'fitted' mean particle size for the emulsion with 1200 mg L^{-1} *n*-dodecane was lower than the others (Fig. 8.11), where its mean particle size was about 0.588 μm. While for each individual cross-flow velocity the particle radii were approximately 0.548 at the cross-flow velocity of 1.14 m s^{-1}, 0.601 μm at the cross-flow velocity of 1.52 m s^{-1}, 0.554 μm at the cross-flow velocity of 1.92 m s^{-1}, and 0.648 μm at the cross-flow velocity of 2.28 m s^{-1}. The other 'fitted' mean particle sizes were 0.592 μm for the emulsion with 600 mg L^{-1} *n*-dodecane and 0.645 μm at the for the emulsion with 2400 mg L^{-1} *n*-dodecane. A summary table of particle radii calculated by using the rearranged Equation (8.8) is presented in Tables 8.5 and 8.6. While for estimated critical particle radii for 600 mg L^{-1} *n*-dodecane emulsions at various cross-flow velocities with different NaCl electrolyte concentrations, it could be observed that the smallest mean particle radius (0.282 μm) was found at the high ionic strength (0.1 M NaCl). For the moderate ionic strength (0.05 M NaCl), the mean particle radius was 0.358 μm, while for emulsion with no salt, the mean particle radius 0.592 μm.

8.4 CONCLUSIONS

The experimental results from this study have supported the concept that equilibrium flux is influenced by both hydrodynamics (i.e. cross-flow velocity) and particle interactions (emulsion stability through changes in ionic strength) during oily water filtration. An increase in cross-flow velocity for the oil emulsions from 1.14 to 1.92 m s^{-1} caused an increase in the equilibrium permeates flux. In contrast, as feed oil concentrations increased from 300 to 2400 mg L^{-1}, equilibrium permeate fluxes were decreased. Likewise, when the ionic strength for the feed emulsions increased, the permeate flux declined.

Particle back transport models such as torque balance model, inertial lift model, and shear-induced diffusion model, were compared to experimental data where particle size was used as a fitting parameter at different oil feed concentration and wall shear stresses. The torque balance over-predicted by three orders of magnitude the experimental fluxes, while the inertial lift model under-predicted by one order of magnitude. However, the shear-induced model showed a better prediction for the experimental data. The 'fitted' particle sizes were in the lower ranges of the measured feed particle size distributions, suggesting that these smaller particles were responsible for the initial permanent particle fouling on the membrane surface at J_{crit}. It is difficult to provide proof directly by experiment because of the lack of ability to visualize and size particles resting on the membrane in situ in the filter. In particular for oily wastewater streams, where the measured particle size distribution of feed emulsions was in the range 0.1–10 μm.

In the current investigation, the critical parameter (J_{crit}/τ_w) was not found to be linear, which was in disagreement with the findings of a number of previous research works, where the measured particle sizes were kept invariant and not influenced by the increase in the shear rate. However, there were a few researchers who encountered similar behavior of nonlinearity of the critical parameter (J_{crit}/τ_w) for flocculated feed, where they attributed that to the change of particle size as a result of the increase in the shear rate, particularly those of loosely flocculated. Thus, correlation between critical flux the product of the wall shear stress and the critical particle size was investigated on the basis of a number of most commonly used models.

REFERENCES

Bacchin, P., Aimar, P. & Field, R.W. (2006) Critical and sustainable fluxes: theory, experiments and applications. *Journal of Membrane Science*, 281, 42–69.

Baruah, G.L. & Belfort, G. (2003) A predictive aggregate transport model for microfiltration of combined macromolecular solutions and poly-disperse suspensions: model development. *Biotechnology Progress*, 19, 1524–1532.

Belfort, G., Davis, R.H. & Zydney, A.L. (1994) The behavior of suspensions and macromolecular solutions in cross-flow microfiltration. *Journal of Membrane Science*, 96 (1–2), 1–58.

Chen, V. (1998) Performance of partially permeable microfiltration membranes under low fouling conditions. *Journal of Membrane Science*, 147 (2), 265–278.

Chiu, T.Y. & James, A.E. (2005) Critical flux determination of non-circular multi-channel ceramic membranes using TiO$_2$ suspensions. *Journal of Membrane Science*, 254, 295–301.

Field, R.W., Wu, D., Howell, J.A. & Gupta, B.B. (1995) Critical flux concept for microfiltration fouling. *Journal of Membrane Science*, 100, 259–272.

Fradin, B. & Field, R.W. (1999) Cross-flow microfiltration of magnesium hydroxide suspensions: determination of critical fluxes, measurement and modelling of fouling. *Separation and Purification Technology*, 16 (1), 25–45.

Gésan-Guiziou, G., Wakeman, R.J. & Daufin, G. (2002) Stability of latex cross-flow filtration: cake properties and critical conditions of deposition. *Chemical Engineering Journal*, 85 (1), 27–34.

Howell, J.A. (1995) Sub-critical flux operation of microfiltration. *Journal of Membrane Science*, 107 (1–2), 165–171.

Huisman, I.H. (1999) Particle transport in cross-flow microfiltration. I. Effects of hydrodynamics and diffusion. *Chemical Engineering Science*, 54 (2), 271–280.

Kwon, D.Y., Vigneswaran, S., Fane, A.G. & Ben Aim, R. (2000) Experimental determination of critical flux in cross-flow microfiltration. *Separation and Purification Technology*, 19 (3), 169–181.

Li, H., Fane, A.G., Coster, H.G.L. & Vigneswaran, S. (2000) An assessment of depolarisation models of cross-flow microfiltration by direct observation through the membrane. *Journal of Membrane Science*, 172 (1–2), 135–147.

Madaeni, S.S. (1997) The effect of operating conditions on critical flux in membrane filtration of activated sludge. *Journal of Chemical Technology and Biotechnology*, 74, 539–543.

Mohhammadi, T., Pak, A., Karbbassian, M. & Golshan, M. (2004) Effect of operating conditions on microfiltration of oil-water emulsion by a kaolin membrane. *Desalination*, 168, 201–205.

Ohya, H., Kim, J.J., Chinen, A., Aihara, M., Semenova, S.I., Negishi, Y., Mori, O. & Yasuda, M. (1998) Effects of pore size on separation mechanisms of microfiltration of oily water, using porous glass tubular membrane. *Journal of Membrane Science*, 145 (1), 1–14.

Ripperger, S. & Altmann, J. (2002) Cross-flow microfiltration – state of the art. *Separation and Purification Technology*, 26 (1), 19–31.

Tien, C. (2006) *Introduction to cake filtration: analyses, experiments and applications*. 1st edition. Elsevier, Amsterdam, The Netherlands.

Wakeman, R.J. & Tarleton, E.S. (1993) Sensitivity analysis for solid liquid separation equipment selection using an expert system. *Proceedings Filtech Conference, Karlsruhe, Germany, October 1993*. The Filtration Society. pp. 43–57.

CHAPTER 9

Treatment of dye solution via low pressure nanofiltration system

Abu Zahrim Yaser & Nidal Hilal

9.1 INTRODUCTION

Many industries including palm oil mill, leather, textile, pulp and paper etc. generate large amounts of colored wastewater. It is the first pollutant that can be seen by the public and may cause reduction in photosynthetic activities, produce carcinogenic by-products in drinking water, chelate with metal ions, and are toxic to aquatic biota (Anjaneyulu et al., 2005; Neoh et al., 2012; Zahrim et al., 2009). Color is a combination of dissolved and colloidal materials (Davies and Cornwell, 2013). Color from wastewater is due to various sources. Petroleum refinery wastewater color is due to the presence of creosote, asphalt and heavy oil fractions. The agricultural based industry wastewater may be due to the presence of lignin, tannin, melanoidin, humic acid etc. For dye manufacturing, textile, leather, and pulp and paper, the colored appearance is due to the dye's compound (Anjaneyulu et al., 2005; Zahrim et al., 2009).

Nanofiltration (NF) is located between reverse osmosis (RO) and ultrafiltration (UF). Nanofiltration membranes carry quite unique properties such as pore radius size (1–5 nm) and surface charge density which influences the separation of various solutes. Over the years, nanofiltration (NF) has been studied by many researchers for the treatment of various colored wastewater for water reuse (Hilal et al., 2004a; Zahrim et al., 2011a). The nanofiltration (NF) process of pulp and paper wastewater for the purpose of reuse was investigated by Z. Beril Gönder et al. (2012). Under the optimized conditions i.e. pH 10, 25°C, 1.2 MPa, 91% chemical oxygen demand (COD), 92% total hardness and 98% sulfate removal were achieved (Gönder et al., 2012). Noghabi et al. (2012) studied nanofiltration (AFC80) of sugar industry wastewater at 1.0 MPa. At this pressure, they found that permeate flux was around 15 kg m^{-2} h^{-1} and sucrose, sodium and potassium rejection exceeded 98, 74 and 67%, respectively (Noghabi et al., 2012). Produced water is generated in the petroleum industry as a by-product of various processes in production and refining. The NF treatment of produced water has been reported by Alzahrani et al. (2013). They reported that NF membrane successfully attained 96% of overall drinking water standards, despite its inefficiency in removing boron, molybdenum and ammonia (Alzahrani et al., 2013). Recently, nanofiltration was used to treat a raw textile effluent supplied from rinsing baths of the Spanish textile industry (Aouni et al., 2012). At the operating pressure 0.5 MPa, pH 7.1; the COD, color and conductivity removal was reported to be around 95, 98, and 55%, respectively (Aouni et al., 2012). The estimated concentration of dyes at dyeing bath is around 90–8000 mg L^{-1} (Van der Bruggen et al., 2001; Zahrim et al., 2011a). Reviews on nanofiltration of real textile effluents have been written by several authors (Hilal et al., 2004a; Lau and Ismail, 2009; Zahrim et al., 2011a).

Since there are more than 8000 chemically different types of dyes for different applications (Anjaneyulu et al., 2005), the researchers also put great effort into investigating the interaction between the dye(s) molecule and membrane system. Patel and Nath (2013) investigated the effect of ultrasonic irradiation during nanofiltration of two-component dye (C.I. Reactive Black 5 and C.I. Reactive Yellow 160) and NaCl mixture. The study to determine the influence of dye's molecular mass on membranes types and cut-offs was carried out by Aouni et al. (2012). The C.I. Acid Black 201 dye is frequently used in dyeing of leather, cotton and woollen fabric (Manu and Chaudhari, 2002; Zahrim et al., 2010). Several studies have been conducted for

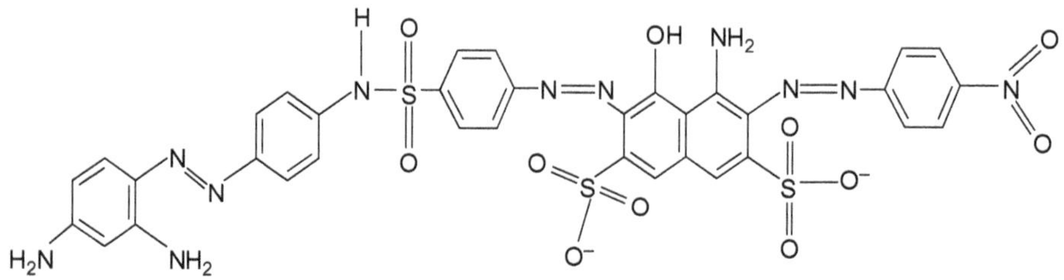

Figure 9.1.   Chemical formula for C.I. Acid Black 210 dye.

the treatment of C.I. Acid Black 210 dye such as adsorption using tannery waste (Piccin *et al.*, 2012), electrochemical oxidation (Costa *et al.*, 2009), coagulation/flocculation (Zahrim *et al.*, 2010), sand filtration (Zahrim *et al.*, 2011b), sonochemical (Li *et al.*, 2008), and biological means (Dave and Dave, 2009; Mohan *et al.*, 2007; Ozdemir *et al.*, 2008). Recently, Zahrim *et al.* (2013) investigated the effect of initial dye concentration and NF feed solution pH on low pressure nanofiltration of C.I. Acid Black 210 dye solution. In this chapter, the pressure effect on nanofiltration of C.I. Acid Black 210 dye solution with tubular membrane will be discussed. Tubular membranes are more appropriate than flat sheet membrane because of their longer lifetime and the easier membrane recovery for long term applications (Warczok *et al.*, 2004). The process was carried out in constant initial dye concentration mode, i.e., the permeate is returned back to the feedtank. The evaluation parameters are irreversible fouling factor, flux, dye concentration, total organic carbon (TOC), conductivity removal, and permeate pH.

9.2   METHODOLOGY

9.2.1   *C.I. Acid Black 210 dye*

The C.I. Acid Black 210 dye (molecular mass: 938) (commercial name: Durapel Black NT dye) (Fig. 9.1) was purchased from Town End (Leeds) Plc, (United Kingdom) and used without further purification. Sodium sulfate is generally used as a diluent (personal communication with Dr Adrian Hayes, Technical Director, Town End (Leeds) Plc). A dye mass of 20 g in powder form was dissolved in Milli-Q Plus, 18.2 MΩ cm (Millipore) water to make 5 L solution at a concentration of 100 mg L^{-1}. The color of the dye solution is visible at minimum concentration of 10 mg L^{-1}.

9.2.2   *Nanofiltration membrane – AFC80*

The NF membrane used was commercially sourced AFC80 (Paterson Candy International Ltd., UK). The AFC series are thin-film composite membranes with a polyamide active film. This film is formed from an aliphatic amide for AFC80. On the basis of the chemistry used, the membrane is expected to have free acid groups and a net negative charge, though each may have some free amine groups present. It has been reported that the mean pore diameter was around 0.68–3.51 nm (Bowen and Doneva, 2000; Warczok *et al.*, 2004). The pure water permeability measured in this study is around 28–31 L m^{-2} h^{-1} MPa^{-1}.

9.2.3   *Tubular membrane system*

The tubular membrane system was designed by CARDEV Ltd. (Hilal *et al.*, 2004b) with further modifications carried out on the system. The modified system consists of a feed tank (5 L working volume), pump, and membrane housing (Fig. 9.2). The maximum inlet pressure is 0.52–0.53 MPa and the system automatically shuts down if the pressure exceeds 0.53 MPa. The inlet pressure is

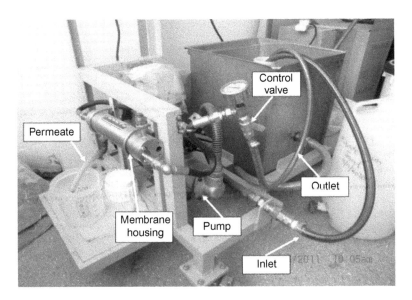

Figure 9.2. Tubular nanofiltration system.

controlled by an outlet valve. This unit is fitted with tubular nanofiltration membranes with inner diameter of 1.27 cm and effective length of 28.6 cm. Thus, the calculated area of membrane tube is around 0.0114 m². In this system, the retentate is sent back to the tank. The pump used is from Mono Pump Ltd, Manchester model CMM253/H13F and delivers a maximum transmembrane pressure around 0.43 MPa. The pump is fitted with a motor from Brook Compton Ltd., Doncaster Model KP 7575. At the maximum inlet pressure, the maximum flow rate measured is 1 m³ h⁻¹ (Akbari *et al.*, 2002; Zahrim and Hilal, 2013).

9.2.4 *Procedure to determine normalized permeation flux, product permeate quality and irreversible fouling factor*

The procedures followed for the determination of normalized permeation flux, product permeates quality and irreversible fouling factor are outlined as:

- The membrane is soaked in RO water for 24 hours before the first permeation test with RO water. The pump is switched on at maximum outlet valve opening for 2 minutes after placing 5 L RO water.
- The valve is slowly reduced until it reaches highest "unsteady" inlet pressure (inlet: 0.55 MPa, outlet: 0.40 MPa). After maximum inlet pressure is reached, reduction of valve opening causes system to automatically shut down.
- The system is then left for at least two hours to minimize the compaction effect. Permeate and retentate are returned to the feed tank.
- The outlet valve is then opened again to get the desired inlet pressure.
- The pure water flux (PWF) is determined by changing the inlet valve. The system is left until steady state is reached (50 minutes). The permeate volume is measured for 20 minutes operation. Temperature (θ) is measured every 10 minutes. A minimum of three points at various transmembrane pressures are obtained.The membrane permeability is determined by plotting pure water flux (PWF) against transmembrane pressure.

The transmembrane pressure (ΔP) is calculated as:

$$\Delta P = \frac{\text{Inlet pressure} + \text{Outlet pressure}}{2} \qquad (9.1)$$

Normalized pure water permeation flux (J_{25}) is determined using the temperature correction as stated by Akbari *et al.* (2002):

$$J_{25} = J_\theta \exp(0.0239(25 - \theta)) \tag{9.2}$$

- The pump is switched off and the RO water is then replaced with dye solution (5 L) and the system is left for 50 minutes to achieve steady state condition. The zero time is begun after 50 minutes.
- The UV/Vis absorbance, pH and conductivity of permeate taken at time 20, 40, 60, 80, 100 and 120 minutes after steady state condition is reached. The permeate is returned, after each analysis, to the feed tank to maintain initial dye concentration.
- After collecting sample at the 120th minute, about 30 mL of permeate is collected, kept at 4°C and analyzed for TOC.
- The pump is then switched off and the dye solution replaced with the tap water (approximately 10 L). The pump is set at the lowest transmembrane pressure available (maximum outlet valve opening). Both retentate and permeate are discarded.
- The pump is switched off and tap water replaced with RO water. The pump is set at the lowest transmembrane pressure (maximum outlet valve opening). Both retentate and permeate are returned to the feed tank and the system operated for one hour. The system was then left overnight.
- The membrane permeability is determined as step (5). The irreversible fouling factor defines as IFF was determined as (Mänttäri and Nyström, 2000; Warczok *et al.*, 2004):

$$\frac{\text{Initial membrane permeability} - \text{Membrane permeability after cleaning with water}}{\text{Initial membrane permeability}} \times 100\% \tag{9.3}$$

9.2.5   *Chemical cleaning*

Before dismantling the membrane, about 10 L of 0.1 mM NaOH (pH 10.30) was passed through the system without returning the retentate. The system was rinsed by about 10 L RO water. The main purpose of this procedure is to minimize the dye stain attached on the surface of the tubing/pump. Cleaning of the membrane is carried out by submerging the membrane in the solution by following the sequence: 0.01 M HNO$_3$ (pH 1.94), 0.0001 M NaOH (pH 10.30) and RO water for 60 minutes each. The water permeability is then measured again. If the pure water permeability is greater than $15 \, \text{L m}^{-2} \, \text{h}^{-1} \, \text{MPa}^{-1}$; no membrane replacement is implemented. Then the membrane is dismantled and stored in 20% glycerol (ACROS Organic) + 1% sodium metabisulfite (ACROS Organic) solution to prevent biological growth on the surface of the membrane.

9.2.6   *Physicochemical analysis*

The pH and conductivity were measured using a Jenway 3540 pH/conductivity meter. The calibration was carried out daily at 20°C. The buffer solutions (pH: 4, 7 and 0) for pH meter calibration were supplied by Fisher Scientific, UK. Residual concentration of dye (without filtering or centrifuging) was obtained with a UV/vis-spectrophotometer (UVmini-1240, Shimadzu) by measuring the absorbance (315 nm). Before the absorbance was measured, the baseline flatness and wavelength accuracy for the UV-spectrophotometer were carried out daily. The absorbance was measured using Milli-Q water as background and the concentration of dye was computed from calibration curves preliminary determined at different pH values. If the reading of absorbance was greater than 3.0, then the necessary dilution was made. After every experiment, the precision cell (10.00 mm, quartz SUPRASIL® (Hellma GmbH & Co., Germany)) was cleaned by soaking in methanol (HPLC grade, Fischer Scientific UK Ltd., UK) overnight. The values of the initial and final concentrations of the dye measured as outlined above were used to calculate the removal percentage of the dye using Equation (9.4):

$$\text{Dye removal } [\%] = 100 \times \frac{C_0 - C_f}{C_0} \tag{9.4}$$

where: C_0 is the initial dye concentration and C_f is the permeate dye concentration.

Total organic carbon (TOC) content was measured by TOC analyzer with autosampler (ASI-V) (model TOC-VCPH, Shimadzu). For estimation of the dye's isoelectric point (IEP), the pH meter was placed in the dye solution (100 mg L^{-1} while stirring (XULA, 2011). Then the 0.001 M HCl was added, the volume of HCl and the solution pH was recorded. The isoelectric point (IEP) observed was around 7.2.

9.3   RESULTS AND DISCUSSION

Fouling problems will lead to higher operation costs: higher energy demand, increased cleaning and reduced life time of the membrane elements (Hilal *et al.*, 2004a). Membrane fouling can be classified as reversible and irreversible. Reversible fouling can be cleaned easily by changing process parameters and water rinsing while irreversible fouling is difficult to clean and might require chemical cleaning like caustic soda cleaning (Tansel *et al.*, 2000; Zahrim *et al.*, 2011a). The factors that affect membrane fouling can be classified into three categories: membrane properties, dyeing wastewater and operating parameters (Cheryan, 1998; Zahrim *et al.*, 2011a). In this study, the irreversible fouling factor for pressure of 0.36, 0.39 and 0.44 MPa is 8.6, 10 and 12%, respectively (Fig. 9.3). It can be shown that the increase in pressure caused an increase in irreversible fouling (Fig. 9.3). From the previous study, it is believed that at pH 10, the dye is in aggregate (Zahrim *et al.*, 2010) and hydrophobic form (Koyuncu, 2003). The higher pressure supplied induces a greater shear stress that may lead to the dye aggregates breakage (Crozes *et al.*, 1997). Then the diffusion of smaller dye aggregates through the membrane is higher than convection (interception) transport and hence caused more fouling (Aydiner *et al.*, 2010). The dye-membrane surface interaction might include multiple effect interaction such as covalent bond, van der Waals, hydrogen bond, and hydrophobic interactions (Zahrim *et al.*, 2011a). Generally, cake layer formation, adsorption and pore blocking (i.e., complete, standard and intermediate blocking) are the expected mechanisms for irreversible fouling (Bowen *et al.*, 1995; Tansel *et al.*, 2000; Zahrim *et al.*, 2011a). At higher pressure, the contribution of cake formation, complete blocking and standard blocking mechanism may be little since the dye aggregates become smaller (Schafer *et al.*, 2005; Zahrim *et al.*, 2011a). Furthermore, during nanofiltration process, only a monolayer of dye aggregates is speculated. Thus, the adsorption (Gomes *et al.*, 2005; Zahrim and Hilal, 2013) and standard blocking (Bowen *et al.*, 1995) could be the dominant mechanism. Standard blocking has possibility since the dye aggregates/molecules could be less than AFC80 membrane pore (Bowen and Doneva, 2000; Warczok *et al.*, 2004). As an example, the dye molecules at acid dye bath is estimated about 0.8 nm (Van der Bruggen *et al.*, 2005). The adsorption of dye aggregates could occur on the membrane surface, in the pores and the back of the membrane (e.g., support layer) (Schafer *et al.*, 2005). De *et al.* (2009) carried out an investigation into leather dyeing wastewater using nanofiltration membrane. They found that the deposition thickness increased from 13.3 to 14.6 µm as the transmembrane pressure increased from 0.83 to 1.24 MPa (De *et al.*, 2009). Fouling at constant pressure of 0.39 MPa characterized by monolayer adsorption shows weak interaction between dye aggregates (Zahrim *et al.*, 2013).

Figure 9.4 shows that the increase in pressure also increases the normalized flux. At transmembrane pressures of 0.36, 0.39 and 0.44 MPa, the corresponding normalized flux is 0.78, 0.89 and 1.34, respectively. A similar pattern was observed by Wu *et al.* (1998) during AFC80 membrane nanofiltration of simulated textile wastewater (total dye concentration around 2400 mg L^{-1}). They stated that at the pressure of 1.0, 2.6 and 2.8 MPa, the flux is 7.0, 8.5 and 12.0 L m^{-2} h^{-1},

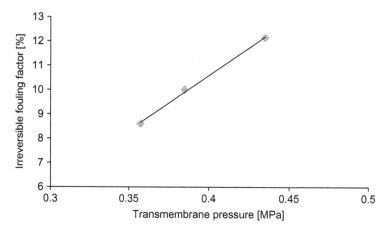

Figure 9.3.   Effect of transmembrane pressure on the irreversible fouling factor (experimental conditions: [dye]: 100 mg L^{-1}, solution pH: 10.0).

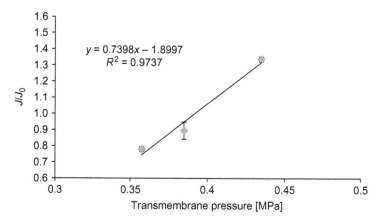

Figure 9.4.   Effect of transmembrane pressure on the flux (experimental conditions: [dye]: 100 mg L^{-1}, solution pH: 10.0).

respectively (Wu et al., 1998). It can be noted that the flux is linearly related to the pressure suggesting that the pressure applied in this study is within a "pressure controlled-region". This fact is strengthened by the figures that show a linear relationship for dye and TOC removal at different transmembrane pressure (Figs. 9.5 and 9.6). In any case, a "mass transfer controlled region" should be avoided in real operation because the increase in pressure only results in build-up of multiple dye aggregate layers (He et al., 2008).

Color is an important water reuse parameter for certain dyeing processes (Zahrim et al., 2011a). In this study, the dye removal slightly increased from 98.9% at low transmembrane pressure (0.36 MPa) to 99.2% at high transmembrane pressure (0.44 MPa) dye removal (Fig. 9.5). A similar pattern also was observed by Koyuncu (2003) who studied nanofiltration of industrial reactive dye effluent (~440 mg L^{-1} dye) at the range of 0.8 to 2.4 MPa. However, using a higher concentration of dye (i.e., 2400 mg L^{-1}), Wu et al. (1998) found that the dye removal is slightly decreased as the pressure increases (from 1.0 to 2.8 MPa). It is believed that the high pressure applied could induce the solubilization of retained dye.

The TOC removal increased with increasing pressure with the highest removal, about 98% at 0.44 MPa (Fig. 9.5). Lower TOC removal compared to dye removal (dye molecular mass: 938)

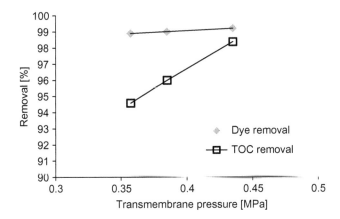

Figure 9.5.   Effect of transmembrane pressure on the dye and TOC removal (experimental conditions: [dye]: 100 mg L^{-1}, solution pH: 10.0).

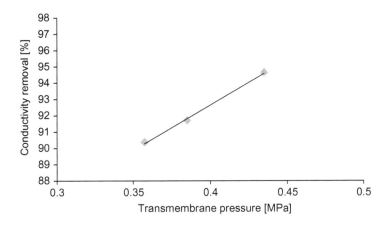

Figure 9.6.   Effect of transmembrane pressure on the conductivity removal (experimental conditions: [dye]: 100 mg L^{-1}, solution pH: 10.0, initial conductivity: 272 µS cm^{-1}).

might be due to the presence of low molecular mass impurities such as various amine compounds and acetic acid (molecular mass: 60) in the permeate (Blus, 1999; Qin *et al.*, 2007). This result suggests that low molecular mass impurities might have more adsorption tendency at high pressure (Weng *et al.*, 2009). In another study, Weng *et al.* (2009) reported that only 60, 75 and 85% of acetic acid was retained during nanofiltration of acetic acid + xylose at pressure 0.5, 0.98 and 1.47 MPa, respectively. Sabate *et al.* (2008) reported 20–40% retention of amine compounds at an operating pressure 0.8 MPa.

   In this study, the conductivity in the permeate is predominantly affected by the presence of sodium sulfate. At transmembrane pressures of 0.36, 0.39 and 0.44 MPa, the corresponding conductivity removal is 90.4, 91.7 and 94.6%, respectively, (Fig. 9.6) might be due to higher adsorption tendency for dye-sodium sulfate aggregates. In another study, Zhang *et al.* (2005) reported 18–95% rejection of sodium sulfate alone at pressure 0.25 MPa. Inlet pressure was found to have minimal effect on permeate pH (Fig. 9.7).

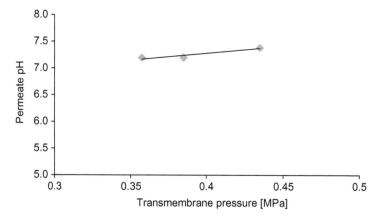

Figure 9.7.   Effect of transmembrane pressure on the permeate pH (experimental conditions: [dye]: 100 mg L^{-1}, solution pH: 10.0).

9.4   CONCLUSION

Results from this work show that the increase in transmembrane pressure caused increase in irreversible fouling. The adsorption and standard blocking could be the dominant mechanism. The increase in pressure also increases the normalized flux; without any buildup of dye layer. Thus, the transmembrane pressure selected in this study is suitable for low dye concentration solution treatment. The effect of low transmembrane pressure (0.35–0.45 MPa) for highly concentrated dye solution is worth being investigated. Although the pressure range in this study only exerts little effect in dye removal, higher transmembrane pressure plays an important role in the removal of impurities and conductivity (sodium sulfate). Permeate pH is not significantly affected by the operating pressure. Based on these findings, medium transmembrane pressure (~0.40 MPa) is recommended for the dye solution treatment.

REFERENCES

Akbari, A., Remigy, J.C. & Aptel, P. (2002) Treatment of textile dye effluent using a polyamide-based nanofiltration membrane. *Chemical Engineering and Processing*, 41 (7), 601–609.

Alzahrani, S., Mohammad, A.W., Hilal, N., Abdullah, P. & Jaafar, O. (2013) Comparative study of NF and RO membranes in the treatment of produced water. Part I: Assessing water quality. *Desalination*, 315 (0), 18–26.

Anjaneyulu, Y., Chary, N.S. & Raj, D.S.S. (2005) Decolourization of industrial effluents – available methods and emerging technologies – a review. *Reviews in Environmental Science and Bio/Technology*, 4 (4), 245–273.

Aouni, A., Fersi, C., Cuartas-Uribe, B., Bes-Pia, A., Alcaina-Miranda, M.I. & Dhahbi, M. (2012) Reactive dyes rejection and textile effluent treatment study using ultrafiltration and nanofiltration processes. *Desalination*, 297 (0), 87–96.

Aydiner, C., Kaya, Y., Gonder, Z.B. & Vergili, I. (2010) Evaluation of membrane fouling and flux decline related with mass transport in nanofiltration of tartrazine solution. *Journal of Chemical Technology and Biotechnology*, 85 (9), 1229–1240.

Blus, K. (1999) Synthesis and properties of acid dyes derived from 7-amino-1-hydroxynaphthalene-3-sulphonic acid. *Dyes and Pigments*, 41 (1–2), 149–157.

Bowen, W.R. & Doneva, T.A. (2000) Atomic force microscopy studies of nanofiltration membranes: surface morphology, pore size distribution and adhesion. *Desalination*, 129 (2), 163–172.

Bowen, W.R., Calvo, J.I. & Hernandez, A. (1995) Steps of membrane blocking in flux decline during protein microfiltration. *Journal of Membrane Science*, 101 (1–2), 153–165.

Cheryan, M. (1998) *Ultrafiltration and microfiltration handbook.* Technomic Publishing Company, Inc., Lancaster, UK, Basel, Switzerland.

Costa, C.R., Montilla, F., Morallón, E. & Olivia, P. (2009) Electrochemical oxidation of Acid Black 210 dye on the boron-doped diamond electrode in the presence of phosphate ions: effect of current density, pH, and chloride ions. *Electrochimica Acta,* 54 (27), 7048–7055.

Crozes, G.F., Jacangelo, J.G., Anselme, C. & Laine, J.M. (1997) Impact of ultrafiltration operating conditions on membrane irreversible fouling. *Journal of Membrane Science,* 124 (1), 63–76.

Dave, S.R. & Dave, R.H. (2009) Isolation and characterization of *Bacillus thuringiensis* for Acid Red 119 dye decolourisation. *Bioresource Technology,* 100 (1), 249–253.

Davies, M.L. & Cornwell, D.A. (2013) *Introduction to environmental engineering.* 5th edition. McGraw-Hill International Edition. New York, NY.

De, S., Das, C. & DasGupta, S. (2009) *Treatment of tannery effluents by membrane separation technology.* Nova Science Publishers, Inc., New York, NY.

Gönder, Z.B., Arayici, S. & Barlas, H. (2012) Advanced treatment of pulp and paper mill wastewater by nanofiltration process: effects of operating conditions on membrane fouling. *Separation and Purification Technology,* 76 (3), 292–302.

Gomes, A.C., Goncalves, I.C. & de Pinho, M.N. (2005) The role of adsorption on nanofiltration of azo dyes. *Journal of Membrane Science,* 255 (1–2), 157–165.

He, Y., Li, G.M., Wang, H., Zhao, J.F., Su, H.X. & Huang, Q.Y. (2008) Effect of operating conditions on separation performance of reactive dye solution with membrane process. *Journal of Membrane Science,* 321 (2), 183–189.

Hilal, N., Al-Zoubi, H., Darwish, N.A., Mohammad, A.W. & Abu Arabi, M. (2004a) A comprehensive review of nanofiltration membranes: treatment, pretreatment, modelling, and atomic force microscopy. *Desalination,* 170 (3), 281–308.

Hilal, N., Busca, G., Talens-Alesson, F. & Atkin, B.P. (2004b) Treatment of waste coolants by coagulation and membrane filtration. *Chemical Engineering and Processing,* 43 (7), 811–821.

Koyuncu, I. (2003) Direct filtration of Procion dye bath wastewaters by nanofiltration membranes: flux and removal characteristics. *Journal of Chemical Technology and Biotechnology,* 78 (12), 1219–1224.

Lau, W.J. & Ismail, A.F. (2009) Polymeric nanofiltration membranes for textile dye wastewater treatment: preparation, performance evaluation, transport modelling, and fouling control – a review. *Desalination,* 245 (1–3), 321–348.

Li, M., Li, J.T. & Sun, H.W. (2008) Sonochemical decolorization of Acid Black 210 in the presence of exfoliated graphite. *Ultrasonics Sonochemistry,* 15, 37–42.

Mänttäri, M. & Nyström, M. (2000) Critical flux in NF of high molar mass polysaccharides and effluents from the paper industry. *Journal of Membrane Science,* 170 (2), 257–273.

Manu, B. & Chaudhari, S. (2002) Anaerobic decolorisation of simulated textile wastewater containing azo dyes. *Bioresource Technology,* 82 (3), 225–231.

Mohan, S.W., Rao, N.C. & Sarma, P.N. (2007) Simulated acid azo dye (Acid Black 210) wastewater treatment by periodic discontinuous batch mode operation under anoxic-aerobic-anoxic microenvironment conditions. *Ecological Engineering,* 31 (4), 242–250.

Neoh, C., Yahya, A., Adnan, R., Abdul Majid, Z. & Ibrahim, Z. (2012) Optimization of decolorization of palm oil mill effluent (POME) by growing cultures of *Aspergillus fumigatus* using response surface methodology. *Environmental Science and Pollution Research,* 20 (5), 1–12.

Noghabi, M.S., Razavi, S.M.A., Mousavi, S.M., Elahi, M. & Niazmand, R. (2012). Effect of operating parameters on performance of nanofiltration of sugar beet press water. *Procedia Food Science,* 1 (0), 160–164.

Ozdemir, G., Pazarbasi, B., Kocyigit, A., Omeroglu, E.E., Yasa, I. & Karaboz, I. (2008) Decolorization of Acid Black 210 by *Vibrio harveyi* TEMS1, a newly isolated bioluminescent bacterium from Izmir Bay, Turkey. *World Journal of Microbiology & Biotechnology,* 24 (8), 1375–1381.

Patel, T.M. & Nath, K. (2013) Alleviation of flux decline in cross flow nanofiltration of two-component dye and salt mixture by low frequency ultrasonic irradiation. *Desalination,* 317 (0), 132–141.

Piccin, J.S., Gomes, C.S., Feris, L.A. & Gutterres, M. (2012) Kinetics and isotherms of leather dye adsorption by tannery solid waste. *Chemical Engineering Journal,* 183 (0), 30–38.

Qin, J.J., Oo, M.H. & Kekre, K.A. (2007) Nanofiltration for recovering wastewater from a specific dyeing facility. *Separation and Purification Technology,* 56 (2), 199–203.

Sabate, J., Pujola, M., Labanda, J. & Llorens, J. (2008) Influence of pH and operation variables on biogenic amines nanofiltration. *Separation and Purification Technology,* 58 (3), 424–428.

Schafer, A.I., Andritsos, N., Karabelas, A.J., Hoek, E.M.V., Scneider, R. & Nystrom, M. (eds.) (2005) Fouling in nanofiltration. Elsevier, Oxford, UK, Amsterdam, The Netherlands.

Tansel, B., Bao, W.Y. & Tansel, I.N. (2000) Characterization of fouling kinetics in ultrafiltration systems by resistances in series model. *Desalination*, 129 (1), 7–14.

Van der Bruggen, B., De Vreese, I. & Vandecasteele, C. (2001) Water reclamation in the textile industry: nanofiltration of dye baths for wool dyeing. *Industrial & Engineering Chemistry Research*, 40 (18), 3973–3978.

Van der Bruggen, B., Cornelis, G., Vandecasteele, C. & Devreese, I. (2005) Fouling of nanofiltration and ultrafiltration membranes applied for wastewater regeneration in the textile industry. *Desalination*, 175 (1), 111–119.

Warczok, J., Ferrando, M., Lopez, F. & Guell, C. (2004) Concentration of apple and pear juices by nanofiltration at low pressures. *Journal of Food Engineering*, 63 (1), 63–70.

Weng, Y.-H., Wei, H.-J., Tsai, T.-Y., Chen, W.-H., Wei, T.-Y., Hwang, W.-S., Wang, C.-P. & Huang, C.-P. (2009) Separation of acetic acid from xylose by nanofiltration. *Separation and Purification Technology*, 67 (1), 95–102.

Wu, J.N., Eiteman, M.A. & Law, S.E. (1998) Evaluation of membrane filtration and ozonation processes for treatment of reactive-dye wastewater. *Journal of Environmental Engineering-ASCE*, 124 (3), 272–277.

XULA (Xavier University of Lousiana) 2011. Titration curve of an amino acid [Online]. [Accessed 26 January 2011].

Zahrim A.Y. & Hilal, N. (2013) Treatment of highly concentrated dye solution by coagulation/flocculation-sand filtration and nanofiltration. *Water Resources and Industry*, 3, 23–34.

Zahrim, A.Y., Rachel, F.M., Menaka, S., Su, S.Y., Melvin, F. & Chan, E.S. (2009). Decolourisation of anaerobic palm oil mill effluent via activated sludge-granular activated carbon. *World Applied Sciences Journal*, 5, 126–129.

Zahrim, A.Y., Hilal, N. & Tizaoui, C. (2010) Evaluation of several commercial synthetic polymers as flocculant aids for removal of highly concentrated C.I. Acid Black 210 dye. *Journal of Hazardous Materials*, 182 (1–3), 624–630.

Zahrim, A.Y., Tizaoui, C. & Hilal, N. (2011a) Coagulation with polymers for nanofiltration pre-treatment of highly concentrated dyes: a review. *Desalination*, 266 (1–3), 1–16.

Zahrim, A.Y., Tizaoui, C. & Hilal, N. (2011b) Removal of highly concentrated industrial grade leather dye: study on several flocculation and sand filtration parameters. *Separation Science and Technology*, 46 (6), 883–892.

Zahrim A.Y., Hilal, N. & Tizaoui, C. (2013) Tubular nanofiltration of highly concentrated C.I. Acid Black 210 dye. *Water Science and Technology*, 67 (4), 901–906.

Zhang, S., Jian, X. & Dai, Y. (2005) Preparation of sulfonated poly(phthalazinone ether sulfone ketone) composite nanofiltration membrane. *Journal of Membrane Science*, 246 (2), 121–126.

CHAPTER 10

Fractionating the value-added products from skimmed coconut milk using membrane separation technique

Ching Yin Ng, Abdul Wahab Mohammad & Law Yong Ng

10.1   INTRODUCTION

The abundance of coconut by-products after oil extraction in the coconut industry has created the environmental issue in some areas near to coconut processing plants. Coconut is well known for its great versatility as seen in nutraceutical, pharmaceutical, medicine, food and beverage industries (Yong *et al.*, 2009). Precious and valuable virgin coconut oil (VCO) can be obtained from coconut, and the remaining residues after VCO production contain some very nutritional components which are highly beneficial to the consumers. These valuable components can be separated by suitable separation techniques or other appropriate processes. The components that can be extracted from the coconut include coconut protein, plant hormones, vitamins, minerals, amino acids etc. The skimmed coconut milk consists of total proteins (approximately 70%) and numerous carbohydrates, sugars, vitamins and minerals (Seow and Gwee, 1997). Therefore, these components can be separated and/or concentrated and used as functional food, food supplement and food formulation through fractionation and concentration processes by membrane separation technique. The great versatility of these valuable compounds from the coconut has attracted the attention of many researchers and industrialists to produce more profitable and value-added coconut products.

Fractionation is a common separation process used to obtain multiple components with higher purity from a complex mixture due to their sizes and physical-chemical properties. Membrane technology is one of the promising methods which manages to fractionate the mixture due to the molecular size of components in the solution. For more than a decade, the membrane separation technique has been applied in different areas such as food industry, dairy manufacturing, medical, pharmaceutical field and other industries. The outputs of the membrane process are considered as better in some aspects when compared to other fractionating processes. The purity and quality of the products can be assured whilst the aroma of the aromatic compounds in the food and dairy products can be retained by the membrane process (Daufin *et al.*, 2001; Engel *et al.*, 2002; Jiao, *et al.*, 2004). This is the most significant advantage in employing the membrane separation method for the food and dairy industry. A combination of ultrafiltration (UF) and nanofiltration (NF) has been widely employed in the food industry to concentrate and extract valuable components from the natural solutions. Many researchers have reported similar topics in their studies (Hagenmaier *et al.*, 2006; Mohammad *et al.*, 2012; Rao *et al.*, 1994; Yin *et al.*, 2013). In addition, the employment of NF membrane can also contribute to the selective separation of finer charged molecules like protein hydrolysates due to the size-exclusion and Donnan-exclusion mechanisms (Bourseau *et al.*, 2009).

In this study, high value-added plant hormones (kinetin and zeatin) and coconut protein are the targeted products after the filtration process using UF and NF membranes. The feed solution of the UF process is skimmed coconut milk, while the permeate of the UF process will be used as the feed for the NF process in order to retain the plant hormones compounds. The concentrated protein solution after the UF process can be a great functional food in food-related industries. Figure 10.1 shows the molecular structures of kinetin and zeatin compounds. These hormones

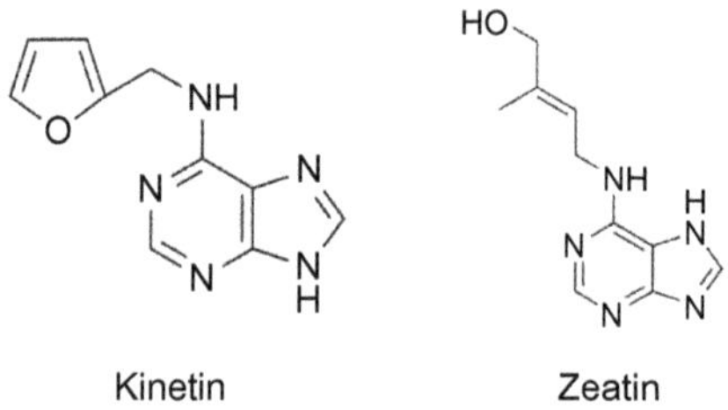

Figure 10.1.   Molecular structures of plant hormones, cytokinins group components (kinetin and zeatin).

belong to the plant hormone family and are named as cytokinins group. Zeatin and kinetin have been proved to possess the anti-aging property. These plant hormones can be used to accelerate the cell division and they possess the anti-viral and anti-bacteria characteristics. The kinetin and zeatin compounds can be potentially applied in pharmaceutical, medicinal and nutraceutical field. The efficiency of membrane separation (using UF and NF) in fractionating and concentrating the skimmed coconut milk will be investigated in terms of their permeate flux decline and membrane rejections towards coconut protein, kinetin and zeatin.

10.2   EXPERIMENTAL

10.2.1   *Materials*

The standard bovine serum albumin (BSA), imunoglobulin (IgG) and protein assay dye reagent concentrate were supplied by Bio-Rad, USA. The reagents and materials used in high performance liquid chromatography (HPLC) analysis were listed as below. Prior to the sample analysis using HPLC, the standard curves of kinetin and zeatin were plotted. The plant hormones/cytokinins standards (kinetin and zeatin) were purchased from Sigma-Aldrich (Steinheim, Germany). All of the kinetin and zeatin standards were dissolved in HPLC-grade methanol with the concentration ranging from 10 to 500 μM. These kinetin and zeatin standards were stored between 0 to 4°C before the HPLC analysis. The chemicals used in HPLC analysis include HPLC-grade methanol (Tedia, USA), formic acid (Tedia, USA) and triethylamine (TEA) (Merck, Germany). TEA has been used to adjust the pH value of the buffer solution (0.1% formic acid) to pH 3.2. The concentration of kinetin and zeatin were determined using a high performance liquid chromatography (HPLC) system (Agilent Technologies 1200 Series, Santa Clara, USA). HPLC analysis is a suitable technique to determine concentration of kinetin and zeatin. UV detector in HPLC was used to detect and separate the kinetin and zeatin in samples. The data were processed by the accompanying system software (ChemStation for LC 3D Systems). Prior to HPLC analysis, the samples were filtered by a 0.45 μm Whatman glass microfiber filter. 10 μL filtered samples were injected into a C18 reverse phase column (Zorbax SB-C18 100Å, 150 mm in length, 2.1 mm in diameter, Agilent Technologies, Santa Clara, USA). The initial HPLC running condition was a methanol (0.1%)-formic acid buffer (10:90, v/v). The column thermostat was set at 25°C. The flow rate was 0.3 mL min^{-1} throughout the whole separation process. To achieve high accuracy detection of cytokinins (kinetin and zeatin) compounds, the wavelength was fixed at 265 nm. Table 10.1 shows the cytokinins (kinetin and zeatin) standard response characteristics using HPLC analysis.

10.2.2   *Preparation of skimmed coconut milk*

The skimmed coconut milk used in this work belongs to the *Malayan Tall* variety. The fresh coconut milk was extracted from solid grated coconut using a coconut milk extractor. The extracted white coconut endosperm liquid was filtered through micron-sized sieve cloth before any processes. The purpose of this step is to remove any large solid particles presented in the solution. Next, the

Table 10.1.   Cytokinins standard response characteristics using HPLC.

Details	Cytokinins standard	
	Kinetin	Zeatin
Mean retention time		
– Minute [min][a]	26.8	18.4
– Relative standard deviation (*RSD*) [%]	0.19	0.002
Peak area[a]		
– Mean	7327.34	612.17
– RSD [%]	1.7	0.5
Equation of calibration curves	$y = 160.4x - 749.2$	$y = 122.4x - 53.31$
Regression (*R*²)[b]	0.977	0.999
Linear range [mg L⁻¹]	2–100	2–100
LOQ [μM][c]	1.7	3.47

[a] Data were measured with repeated injection ($n = 5$) of plant hormone standard at a concentration of 20 mg L⁻¹ each.
[b] In the calibration equation, x represents concentration of the analyte [mg L⁻¹] and y the peak area (mili absorption unit in second [mAu s]).
[c] Limit of quantification (*LOQ*) was estimated based on signal-to-noise ration (S/N) = 10.

filtered coconut milk was pasteurized to prolong the shelf-life of coconut milk as it is a perishable liquid. In this step, the coconut milk was heated at 60°C for 15 minutes and this can reduce the microbial load to 10% (Hagenmaier, 1980). The high temperature of pasteurization should be avoided as protein molecules start to denature at 70°C. The warm coconut milk was processed with coconut cream separation process using 125 L capacity cream separator (Elecrem, France). The warm skimmed coconut milk makes the separation process easier and more effective. There are two outputs from the cream separator: concentrated coconut cream and skimmed coconut milk. The skimmed coconut milk was used as the feed solution throughout this work. A membrane filtration process was applied to further fractionate and concentrate the components inside the skimmed coconut milk to obtain high value-added coconut products.

10.2.3   *Ultrafiltration and nanofiltration*

The UF membrane employed in this study is made of polysulfone (PSF) material with a molecular weight-cut-off (MWCO) of 10,000 Da from Koch, USA. The abbreviation used for the UF membrane is PSF10. The employed NF membrane (NF1) is produced by Amfor Inc., China. All of the membranes used were soaked in ultrapure water overnight to remove any preservative layer and dirt particles prior to the experimental runs. The membrane compaction was conducted at a higher pressure than the operating pressure. This was aimed to enhance the permeate flux and permeability of the membranes (Persson *et al.*, 1995). The pure water was used to conduct the permeability test of the membrane. The purpose of this step is to confirm that the membrane is in good condition before filtration process using skimmed coconut milk.

The UF process was performed by a cross-flow filtration unit, which was equipped with a 4 L stainless steel jacketed feed tank with an embedded mixer (Ika, Germany) and a variable feed pump (Hydra-cell Pump, MN, USA). The temperature of the feed was controlled by circulating the water through a jacketed feed tank. The cross-flow velocity (*CFV*) and trans-membrane pressure (*TMP*) were controlled using a feed flowmeter (F-400, Blue-White, USA) and a permeate needle valve (Swagelok, UK). A digital balance (GF 6100, A&D, Japan) was connected to a computer to record the permeate flux in the continuous manner. The pressure at 0.2 MPa and temperature at 60°C were set during ultrafiltration of skimmed coconut milk using cross-flow system. The value of *CFV* was 2.728 L min⁻¹ while the stirring rate was 400 rpm throughout the UF process

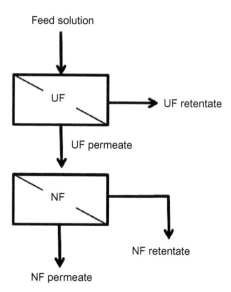

Figure 10.2.   Schematic diagram of fractionating the skimmed coconut milk using membrane separation.

using PSF10 membrane. The active membrane surface area of cross-flow system is 32.06 cm². The UF process was conducted under a batch concentration (the retentate was returned to the feed tank) mode. The volume reduction factor (*VRF*) of 2 was applied in each batch of the UF cross-flow filtration process. The masses of collected permeates were recorded every 60 seconds throughout the filtration process with the help of electronic balance. The UF permeate was further fractionated through NF process.

The NF process was performed using a dead-end filtration module (Sterlitech HP4750 stirred cell, WA., USA) with an active membrane surface area of 15.20 cm². The constant pressure value was set at 0.5 MPa throughout the NF process. Other operating conditions were similar to UF. A compressed nitrogen cylinder was connected to the dead-end cell to supply the desired pressure. Figure 10.2 is a schematic diagram showing the application of UF and NF processes in fractionating the skimmed coconut milk to produce the value-added products. The masses of NF permeates were also recorded for every 60 seconds, for a duration of 60 minutes throughout the filtration process. The membrane performances have been studied in terms of their flux decline and membrane retention capability. The mathematical equation used for the solution flux (*J*) in unit [L m⁻² h⁻¹] is shown by:

$$J = V/At \tag{10.1}$$

where *V* is determined by the volume of collected permeate per unit liter [L]; *A* is the active surface area of membrane [m²] and *t* is the time interval of filtration [hour].

The measurement of membrane performances were evaluated in terms of volume reduction factor (*VRF*) and retention factor at the end of each run (*RF*f), for proteins (BSA, IgG) and cytokinins (kinetin, zeatin). The calculation methods were defined as:

$$VRF = V_0/(V_0 - V_p - V_p') \tag{10.2}$$

$$RF_f = 1 - (C_{pf}/CR) \tag{10.3}$$

$$CF = C_{i,\text{retentate}}/C_{i,\text{feed}} \tag{10.4}$$

where V_0, V_p and V_p' represent initial feed volume, the average permeate volume over the filtration and the final permeate volume; *CR* and C_{pf} are the retentate concentration and final permeate concentration for proteins (BSA and IgG) or cytokinins (kinetin and zeatin). Concentration factor (*CF*) is used to express degree of concentration of target compounds. It is defined as the ratio of the concentration of a component i in the retentate ($C_{i,\text{retentate}}$) to the concentration of the same component in the feed ($C_{i,\text{feed}}$). All of these performance indicators are important towards understanding the overall methodology and the results are illustrated in the Table 10.2.

10.2.4   *Analysis*

The targeted components in this study are: bovine serum albumin (BSA), immunoglobulin (IgG), kinetin and zeatin. BSA and IgG belong to dominant types of protein in skimmed coconut milk, whilst kinetin and zeatin are plant hormones compounds. The protein (BSA and IgG) compositions have been analyzed according to the Biuret method using Genesys 10 scanning UV-VIS spectrophotometer (Boston, MA). Different concentrations of standard bovine serum albumin (BSA) and immunoglobulin G (IgG) had been prepared and analyzed using a spectrophotometer to obtain a standard plot prior to the analysis of real samples. The molecular weight of proteins in solution was analyzed by sodium dodecyl sulfate-polyacrylamide gel electrophoresis (SDS-PAGE).

Samples of skimmed coconut milk such as feed, permeate and retentate, obtained from UF and NF, were kept for further analysis. The basic nutrient contents of samples were determined by proximate analysis according to Association of Official Analytical Chemists (AOAC) method (AOAC, 1990). The contents of fat, protein, carbohydrate, ash and moisture in samples can be known through this proximate analysis. The retention factor of targeted components (BSA, IgG, kinetin and zeatin) were calculated using Equation (10.3) to verify the performance of membrane separation in this study. The behaviors of solution flux during the filtration processes can be analyzed by plotting the curve of flux versus time.

10.3   RESULTS AND DISCUSSION

10.3.1   *Concentrate and fractionate the skimmed coconut milk using ultrafiltration and nanofiltration*

As discussed, skimmed coconut milk is a complex mixture which consists of various molecular sizes and properties compounds. In order to separate the desired compound within the skimmed coconut milk, fractionation processes were carried out. Figure 10.2 shows the application of UF and NF in the fractionation of skimmed coconut milk. UF membrane with bigger pore size was used to retain or concentrate the proteins in skimmed coconut milk. The protein content in skimmed coconut milk obtained was around 70% of total coconut protein composition (Hagenmaier, 1980; Kwon *et al.*, 1996a). The enrichment of protein in coconut milk has been attracting the attention of experts from different fields to further investigate and produce new products from the coconut, especially the protein products. The coconut protein products can be potentially used as an alternative in the formulation of milk.

During the UF process, batch concentration mode was carried out using the cross-flow system. The skimmed coconut milk was concentrated until a *CF* of 2 was achieved. Concentrating the coconut protein using UF process would foul the applied membrane and reduce the solution permeate flux. Figure 10.3 displays the normalized flux decline for UF and NF processes. The UF normalized flux showed lower flux compared to NF. The deposition and accumulation of foulants onto the membrane surfaces and inside the pores contributed to the permeate flux obtained. Some components which have sizes close to the membrane pore size possibly seal the exposed membrane surface pores. Besides, the smaller sized components in skimmed coconut milk could possibly deposit inside the membrane pores and greatly reduce the amount of other components to pass through the membrane. However, the flux decline rate obtained by NF membrane was greater than that of UF membrane. This shows that the fine molecules in the UF permeate fouled the NF membrane within a very short time of filtration. Thus, the normalized flux of NF dropped drastically.

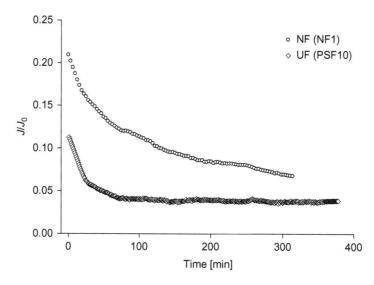

Figure 10.3.   Permeate normalized flux along the entire filtration time for UF and NF.

10.3.2   *Performance indicator of membrane separation*

The Bradford protein assay method was conducted to determine the protein concentration in UF feed, retentate and permeate. The proteins considered in this study were BSA and IgG as they are the predominant protein fractions in the coconut milk (Kwon *et al*., 1996a; Samson *et al*., 1971). Retention of proteins by UF membrane was calculated using Equation (10.3).

Table 10.2 shows the volume reduction factor (*VRF*) of feed, concentration factor (*CF*) of solutes and retention factor (*RF*$_f$) of solutes in UF and NF processes. In the UF of skimmed coconut milk, the VRF of 2 was set. The soluble protein of BSA and IgG were concentrated to about 1.12 times in the retentate of UF process. The retention factors obtained for the proteins are 0.9836 (for BSA) and 0.9981 (for IgG). The amount of protein in the UF permeate was too little and negligible compared to the UF feed. NF process was conducted using the UF permeate (which was conducted previously) as its feed solution. The *VRF* of 10 was achieved by NF after approximately 5 hours of filtration time. Concentration factors (*CF*) of kinetin and zeatin obtained using NF membrane (NF1) were around 2.25 and 3.0, respectively, while the retention factor (*RF*$_f$) obtained for kinetin and zeatin was 0.9238 and 0.9511, respectively. The retention of zeatin in this study was higher than that of kinetin using the membrane NF1. This can be explained by the molecular structure of zeatin (Fig. 10.1). The reactive hydroxyl bond (–OH) in the zeatin molecule can be readily combined with any reactive compounds within the coconut solution. Hence, the sizes of compounds formed can be larger than the kinetin molecules itself. As a result, more zeatin molecules can be retained by the NF membrane when compared to the kinetin molecules. The concentration and retention factor results verified that the application of UF and NF processes is a feasible technique to fractionate complex coconut solution to obtain various value-added products.

10.3.3   *Proximate analyses of samples produced*

Table 10.3 displays the results of proximate analyses of the UF and NF samples from feed, retentate and permeate streams. Through these analyses, the amount of protein, fat, carbohydrate, ash and moisture content in every sample was reported in unit g/100 g or percentage. The moisture or water content constituted the highest percentage in all analyzed samples. From the Table 10.3, it can be clearly seen that the fat content only presents in the UF feed stream or the so-called skimmed

Table 10.2. The volume reduction factors (*VRF*), concentration factors (*CF*) and retention factors (*RF*$_f$) of solutes in UF and NF processes, which were performed by PSF10 and NF1 membranes, respectively.

Process	VRF	Concentration factor (*CF*)				Retention factor (*RF*$_f$)			
		BSA	IgG	Kinetin	Zeatin	BSA	IgG	Kinetin	Zeatin
UF (PSF10)	2	1.116	1.118	–	–	0.9836	0.9981	0.5400	0.0618
NF (NF1)	10	–	–	2.25	3.00	–	–	0.9238	0.9511

Table 10.3. Basic nutrient contents in UF and NF samples by proximate analysis.

Parameter, unit	UF feed	UF retentate	UF permeate/NF feed	NF retentate	NF permeate
Protein [g/100 g]	2.1	6.5	0.2	0.3	0.2
Fat [g/100 g]	0.8	0	0	0	0
Carbohydrate [g/100 g]	4.5	3.5	2.9	2.7	1.3
Ash [g/100 g]	1.0	0.8	0.8	0.8	0.6
Moisture [g/100 g]	91.6	89.2	96.1	96.2	97.9
Energy [kJ/100 g]	142	79	50	50	25

coconut milk, with a concentration of about around 0.8%. From Table 10.3, the composition of protein in the UF feed was about 2.1%. After the skimmed coconut milk is concentrated through UF process, the concentration of protein was increased to 6.5% in the UF retentate stream. According to the result obtained, there was only 0.2% of protein in the UF permeate. Water content is the major constituent in all of the NF streams (retentate and permeate). Other constituents such as protein, fat, carbohydrate and ash in NF samples were much lower in percentage than UF samples except water content.

10.3.4 *SDS-PAGE analysis*

Gel-electrophoresis method using SDS-PAGE to determine the molecular weight of protein is presented in Figure 10.4. The left lane indicates molecular marker (10–260 kDa), the middle lane indicates the feed of UF and the right lane indicates the UF retentate. From the SDS-PAGE profile, it is easy to observe that the protein bands presented in the UF feed were the same as the UF retentate. But, the bands presented in the UF retentate showed higher intensity than UF feed. This means that the concentration of protein fractions in the UF retentate were higher than the UF feed. Thus, the concentration of coconut proteins through the employment of UF process was successful. Four major bands ranging from 16 kDa to 150 kDa dominated the total protein composition in the unreduced form. Kwon et al. (1996a) reported that the prominent bands belonged to albumin and globulin proteins. The albumin and globulin fractions were the predominant protein fractions, accounting for 60 and 30% of the total protein, respectively (Angelia et al., 2010; Samson et al., 1971). In the previous study, the total coconut protein was found to have a similar pattern as shown in Figure 10.4 (Kwon et al., 1996b). The coconut proteins are separated into four bands which indicate that the protein molecular weights are greater than 150, 55, 34 and 17 kDa.

10.4 CONCLUSION

From the obtained results, it can be concluded that the utilization of ultrafiltration membrane to concentrate the protein content in skimmed coconut milk can be successfully achieved with

kDa

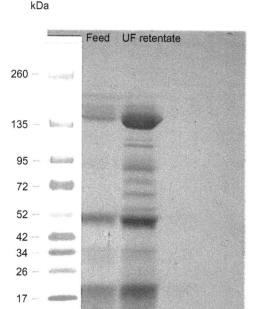

Figure 10.4.   Different molecular weights of protein bands in the UF feed and UF retentate through SDS-PAGE analysis.

high efficiency. The retention of major coconut proteins (BSA and IgG) can reach about 98–99% using ultrafiltration membrane (PSF10). The stable permeate flux was obtained under batch concentration with a volume reduction factor of 2 in UF process. The retention of zeatin (0.98) was higher than that of kinetin (0.68) compound using NF1 membrane. SDS-PAGE result shows 4 major bands of protein fractions are found in a molecular weight range from 16 to 150 kDa, in both of the UF feed and retentate. The intensity of protein fractions in UF retentate was higher than that of the UF feed. The proposed fractionation technique appeared to be a feasible process to obtain the desired compounds. In brief, the concentrated coconut protein and high-value added biological products (kinetin and zeatin) can be successfully produced by the fractionation of skimmed coconut milk using membrane technology.

ACKNOWLEDGEMENT

The authors would like to acknowledge the financial grant funded by Universiti Kebangsaan Malaysia via grant 02-01-02-SF1021.

REFERENCES

Angelia, M.R.N., Garcia, R.N., Caldo, K.M.P., Prak, K., Utsumi, S. & Tecson-Mendoza, E.M. (2010) Physicochemical and functional characterization of cocosin, the coconut 11S globulin. *Food Science and Technology Research*, 16, 225–232.
AOAC (1990) Official Method of Analysis. The Association of Official Agricultural Chemists, Arlington, VA.
Bourseau, P., Vandanjon, L., Jaouen, P., Chaplain-Derouiniot, M., Massé, A., Guérard, F., Chabeaud, A., Fouchereau-Péron, M., Le Gal, Y., Ravallec-Plé, R., Bergé, J.P., Picot, L., Piot, J.M., Batista,

I., Thorkelsson, G., Delannoy, C., Jakobsen, G. & Johansson, I. (2009) Fraction of fish protein hydrolysates by ultrafiltration and nanofiltration: impact on peptidic populations. *Desalination*, 244, 303–320.

Daufin, G., Escudier, J.P., Carrere, H., Berot, S., Fillaudeau, L. & Decloux, M. (2001) Recent and emerging applications of membrane processes in the food and dairy Industry. *Food and Bioproducts Processing*, 79, 89–102.

Engel, E., Lombardot, J.-B., Garem, A., Leconte, N., Septier, C., Le Quéré, J.-L. & Salles, C. (2002) Fractionation of the water-soluble extract of a cheese made from goats' milk by filtration methods: behaviour of fat and volatile compounds. *International Dairy Journal*, 12, 609–619.

Hagenmaier, R. (ed.) (1980) *Coconut aqueous processing*. 2nd edition. Publications, University of San Carlos, Cebu City, Philippines.

Hagenmaier, R.D., Cater, C.M. & Mattil, K.F. (2006) Coconut skim milk as an intermediate moisture product. *Journal of Food Science*, 40, 717–720.

Jiao, B., Cassano, A. & Drioli, E. (2004) Recent advances on membrane processes for the concentration of fruit juices: a review. *Journal of Food Engineering*, 63, 303–324.

Kwon, K., Park, K.H. & Rhee, K.C. (1996a) Fractionation and characterization of proteins from coconut (*Cocos nucifera* L.). *Journal of Agricultural and Food Chemistry*, 44, 1741–1745.

Kwon, K.S., Bae, D., Park, K.H. & Rhee, K.C. (1996b) Aqueous extraction and membrane techniques improve coconut protein concentrate functionality. *Journal of Food Science*, 61, 753–756.

Mohammad, A., Ng, C., Lim, Y. & Ng, G. (2012) Ultrafiltration in food processing industry: review on application, membrane fouling, and fouling control. *Food and Bioprocess Technology*, 5, 1143.

Persson, K.M., Gekas, V. & Trägårdh, G. (1995) Study of membrane compaction and its influence on ultrafiltration water permeability. *Journal of Membrane Science*, 100, 155–160.

Rao, H.G.R., Grandison, A.S. & Lewis, M.J. (1994) Flux pattern and fouling of membranes during ultrafiltration of some dairy products. *Journal of the Science of Food and Agriculture*, 66, 563–571.

Samson, A.S., Khaund, R.N., Cater, C.M. & Mattil, K.F. (1971) Extractability of coconut proteins. *Journal of Food Science*, 36, 725–728.

Seow, C.C. & Gwee, C.N. (1997) Coconut milk: chemistry and technology. *International Journal of Food Science & Technology*, 32, 189–201.

Yin, N.C., Mohammad, A.W. & Yong, N.L. (2013) Membrane performance and potential separation of cytokinins during ultrafiltration of skimmed coconut milk. *Advanced Science Letters*, 19, 3620–3624.

Yong, J.W.H., Ge, L., Ng, Y.F. & Tan, S.N. (2009) The chemical composition and biological properties of coconut (*Cocos nucifera* L.) water. *Journal of Molecules*, 14, 5144–5164.

CHAPTER 11

Recovery of rubber from skim latex using membrane technology

Khairul Muis Mohamed Yusof, Jaya Kumar Veellu, Ahmad Jaril Asis, Zainan Abdullah, Mohamed Kheireddine Aroua & Nik Meriam Nik Sulaiman

11.1 INTRODUCTION

Natural rubber field latex (NRFL) obtained from the *Hevea brasiliensis* tree remains one of the most important agricultural products in Malaysia. There are two main types of natural rubber processing factories in Malaysia, namely for the production of latex concentrate (NR latex) and the Standard Malaysian Rubber (SMR). NR latex as an industrial raw material is supplied as concentrated latex, where the dry rubber content (DRC) is about 60%. Currently there are three main methods to concentrate the latex, namely centrifugation, creaming and evaporation. Centrifugation is the most preferable method to concentrate NRFL into latex concentrate and accounts for some 98% of the total concentrate produced in Malaysia (Cheng, 1988).

Natural rubber skim latex (NRSL) which contains about 3–5% DRC is discharged as waste during centrifugation of NRFL into latex concentrate. It is well known in the industry that rubber can be recovered from NRSL by coagulation using sulfuric acid. The recovered rubber is then dried in the open before the rubber is sold as skim crepe rubber. However, the skim rubber obtained from this coagulation method is of an inferior quality as it contains a high proportion of entrained non-rubber constituents and acid content. The skim rubber also possesses undesirable physical properties including the generation of obnoxious odor, leading to low economic value in skim rubber. Furthermore, the method also produces highly acidic effluent which may cause pollution to the environment if the effluent is not treated in a proper manner before being discharged to the environment (Devaraj et al., 2003).

Studies by Devaraj (2004) and Devaraj et al. (2005) reported that membrane technology could be an alternative method of concentrating NRFL into latex concentrate. In those studies, NRFL with a DRC of 28% was concentrated to 50% using tubular cross flow ultrafiltration (UF) where polymeric PVDF membranes were used. This method yields a clear serum which can be further used for extraction of many valuable bio-chemicals such as proteins and carbohydrate sugar (L-Quebrachitol) (Rhodes and Wiltshire, 1932). In another study by Devaraj and Zairossani (2007) it was reported that using UF polymeric membrane (PVDF), the DRC of NRSL can be increased from 5 to 30% but it takes more than 10 hours concentration process.

Polymeric membranes are known for their rapid fouling in presence of high solids content of the feed and at high temperatures polymeric membranes are not able to operate ($>50°C$). In comparison to polymeric membranes, ceramic membranes are slower to foul and can be cleaned using more extreme membrane cleaning methods (Panglisch, 2009). Moreover, the operating life span of ceramic membrane is usually five times more than polymeric membrane thus enables higher productivity (Panglisch, 2009). This study is focused on evaluating the use of ceramic membrane (tubular and hollow fiber type) in concentrating NRSL and to study the effect of transmembrane pressure (TMP) on the concentration process and degree of concentration. A suitable sequence of cleaning in place (CIP) procedure was established to regenerate the fouled ceramic membrane. To the best of our knowledge, there is no study in the open literature on the evaluation of ceramic membranes in concentrating NRSL.

11.2   EXPERIMENTAL

11.2.1   *Materials*

The source of feed was NRSL and the samples were obtained from centrifugation process of NRFL into latex concentrate at Sime Darby Plantation Latex Factory, Batu Anam, Johor, Malaysia. To ensure the NRSL stability is maintained during UF concentration run, the latex was preserved with ammonia and ammonium laureate.

11.2.2   *Experimental design*

Ceramic type ultrafiltration membranes which were commercially available were selected for this evaluation with a range of suitable pore sizes. The schematic diagram of the system is shown in Figure 11.1. Two types of ceramic membrane configuration were evaluated which were ceramic multi-tubular (MT) and ceramic hollow fiber (HF) membranes. For both MT and HF membranes, two pore sizes were evaluated. The membranes were thus coded as PMT1 (smaller pore size) and PMT2 (larger pore size) for MT membranes, and PHF1 (smaller pore size) and PHF2 (larger pore size) for HF membranes, respectively. Both membranes were mounted onto stainless steel housing and connected to a diaphragm pump for circulation. Optimum *TMP* was identified by comparing permeate flux at different preset *TMP*s for MT membrane and HF membrane. A different set of CIP procedures and chemicals were evaluated to give maximum membrane recovery after the concentration run.

11.2.3   *NR skim latex concentration run*

Membrane recovery was first determined before the start of any experiments. This was done by comparing water flux of unused membrane with water flux measured after using the membrane. Membrane recovery [%] is calculated by dividing the water flux of used membrane by that of unused membrane. If the recovery is below 80% the membrane is considered fouled and has to be

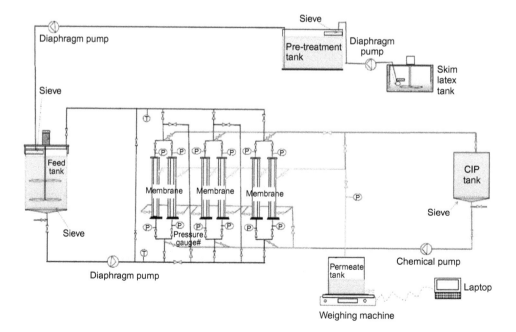

Figure 11.1.   Schematic diagram of an ultrafiltration system.

cleaned before run with NR skim latex. After water flux test was completed, 60 kg skim latex was transferred into feed tank. The skim latex was pretreated with ammonia and ammonium laureate to increase the stability. Samples were taken from the feed and tested for *DRC*, total solids content (*TSC*), ammonia content, volatile fatty acid (VFA) number, and pH. The pretreated NR skim latex was then fed into the membranes via diaphragm pump and the feed was allowed to be circulated for 5 minutes before setting to pre-selected *TMP*. The system has to be stabilized so that the membrane is fully soaked with the feed and compacted (Devaraj *et al.*, 2003). The retentate was recycled back into the feed tank. The permeate was diverted into a permeate tank with a digital balance connected to a laptop. The permeate flow rate was recorded every 10 minutes.

After the retentate reached the desired concentration, the feed pump was stopped. A retentate sample was taken for *DRC*, *TSC*, ammonia content, VFA, and pH analysis. The remaining latex in the system was then drained off. The system was flushed with a sufficient amount of water until there were no traces of latex in the system. The fouled membrane has to be cleaned with cleaning chemical after each trial. The membrane was flushed with sodium hydroxide (NaOH) to remove all organic foulants on the membrane surface. After flushing ends, water flux test was carried out to determine the membrane recovery. If the recovery is below 80%, the membrane should be soaked overnight in a cleaning solution (CS1) in a separate container.

11.3   RESULTS AND DISCUSSION

11.3.1   *Optimum TMP for concentration process*

The effect of *TMP* on permeate flux for both ceramic multi tubular and hollow fiber membrane is shown in Figure 11.2 and Figure 11.3. The optimum *TMP* recorded for multi tubular membrane and hollow fiber membrane is at *TMP* of 0.3 MPa (3 bar) and 0.15 MPa (1.5 bar), respectively. Concentration process at lower pressure caused a decline in permeate flux. This is because the driving force became less at the lower pressure. The cake layer formation is higher at low pressure and feed flow rate (Devaraj *et al.*, 2003; Devaraj and Zairossani, 2007). It was observed that permeate flux drop when backpressure i.e. partially closed outlet valve is used to increase the *TMP* of the system. This can be explained by concentration polarization effect to permeate flux. At higher *TMP* due to backpressure, feed flow rate is reduced causing less shearing effect on the membrane surface, which increases concentration polarization and subsequently reduces permeate flux (Devaraj and Zairossani, 2007).

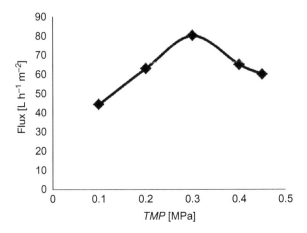

Figure 11.2.   Effect of trans-membrane pressure (*TMP*) on permeate flux – multi tubular membrane.

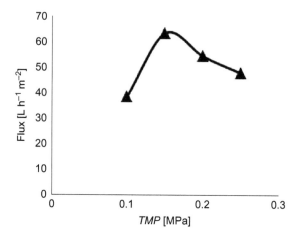

Figure 11.3.   Effect of trans-membrane pressure (*TMP*) on permeate flux – hollow fiber membrane.

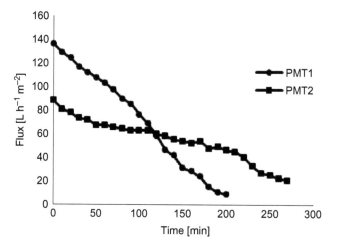

Figure 11.4.   Variation in permeate flux during natural rubber (NR) skim latex concentration process – multi tubular membrane.

11.3.2   *NR skim latex permeate flux profile*

Comparison of flux versus time between two types of pore sizes PMT1 and PMT2 for MT membrane is shown in Figure 11.4. A typical decline in permeate flux was observed for both concentration run. An optimum *TMP* of 0.3 MPa (3 bar) was set throughout the process. MT with pore size of PMT1 gives better flux with initial average flux of 130 L h^{-1} m^{-2} during the first 30 minutes concentration run as compared with PMT2 (83 L h^{-1} m^{-2}). More obvious permeate flux decline was recorded after 50% permeate removal. Average flux during last 30 minutes for PMT1 and PMT2 were 12 L h^{-1} m^{-2} and 23 L h^{-1} m^{-2}, respectively. Average flux for the concentration runs and the *DRC* are shown in Table 11.1. *DRC* for PMT1 increased from 3.9% for feed to 23.0% for retentate and for PMT2 from 4.1% for feed to 24.7% for retentate after the concentration run.

The concentration trial for ceramic HF membrane was done at a constant *TMP* of 0.15 MPa (1.5 bar). Ceramic HF membrane gave different permeate flux decline profile compared with MT membrane as shown in Figure 11.5. The flux for PHF1 was maintained constant even after 7 hours of concentration runs. However, the average flux recorded was 59.23 L h^{-1} m^{-2} and took longer

Table 11.1. Membrane performance after natural rubber (NR) skim latex run.

Membrane and pore size	Average flux [$L\,h^{-1}\,m^{-2}$]	Feed DRC [%]	Retentate DRC [%]
PMT1	88.40	3.9	23.0
PMT2	62.20	4.1	24.7
PHF1	59.23	3.5	12.1
PHF2	64.38	3.7	13.8

DRC: dry rubber content.

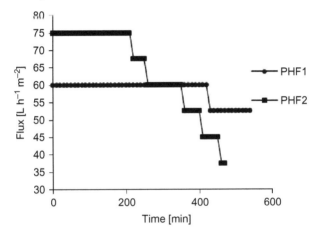

Figure 11.5. Variation in permeate flux during natural rubber (NR) skim latex concentration process – hollow fiber membrane.

concentration time as compared with MT membrane. The DRC of the retentate was low. DRC for PHF1 increased from 3.5% to 12.1%. Slight improvement was shown by PHF2 concentration run. The flux remained constant during the initial 3.7 hours of concentration run and the flux started to decline gradually towards the end of the run. The average flux recorded was $64.38\,L\,h^{-1}\,m^{-2}$ and the DRC increased from 3.7 to 13.8%.

11.3.3 *Membrane cleaning and flux recovery*

NRSL has a narrow particle size distribution ranging from 0.2 to 0.6 μm with the mean value of 0.5 μm (Devaraj *et al.*, 2005). The smaller rubber particles of skim latex could contribute to membrane fouling together with NR proteins which have a molecular weight ranging from 5 to 50 kD and could be the other membrane foulant (Devaraj *et al.*, 2005).

From the study, flushing method is suitable to clean the fouled ceramic membrane compared with backwashing method. The effects of cleaning chemical composition were investigated in this chapter. Flushing with NaOH for a specified time followed by overnight soaking in a cleaning solution (CS1) gave better membrane recovery with 88.9% improvement as compared with other methods and chemicals Table 11.2.

11.4 CONCLUSION

A pretreated NRSL with ammonia and ammonium laureate was concentrated using ceramic type membranes to a certain DRC with average flux of more than $50\,L\,h^{-1}\,m^{-2}$. MT membrane with

Table 11.2.   Membrane cleaning procedures.

Recovery after concentration run [%]	Cleaning procedure	Recovery after cleaning [%]	Improvement [%]
7.3	Backwash with NaOH for X min	23.6	16.3
10.0	Backwash with cleaning solution (CS1) for X min	11.1	1.2
1.2	Flushing with cleaning solution (CS2) for Y min	3.6	2.4
3.4	Flushing with NaOH for Y min	44.7	41.3
4.2	Flushing with NaOH for X min	62.1	57.9
3.2	Flushing with NaOH for X min and overnight soaking with cleaning solution (CS1)	92.1	88.9

Note: X, Y, CS1 and CS2 are proprietary information and cannot be disclosed.

pore size of PMT1 recorded higher average flux ($88.4\,\mathrm{L\,h^{-1}\,m^{-2}}$) against HF membranes. The optimum *TMP* for a maximum permeate flux for MT and HF membrane was recorded at 0.3 MPa (3 bar) and 0.15 MPa (1.5 bar), respectively. A suitable cleaning protocol was established where the fouled membrane was flushed with NaOH followed by overnight soaking with a cleaning solution CS1 to achieve membrane recovery of more than 80%. More effective CIP procedures need to be further studied to achieve shorter cleaning time thus increasing productivity. Future studies need to focus on finding a suitable use of concentrated NRSL as value added raw material.

ACKNOWLEDGEMENT

The authors would like to thank the Senior Vice President I, Rubber Operation and Mechanisation Department of Sime Darby Plantation Sdn. Bhd., for granting permission to carry out this study at Sime Darby Latex Sdn. Bhd. and to Head, R&D Centre for allowing the publication of this study.

REFERENCES

Cheng, S.F. (1988) Types, composition, properties, storage and handling of natural rubber latex concentrates. Notes on NR examination glove manufacture. Rubber Research Institute of Malaysia, Kuala Lumpur, Malaysia. pp. 1–12.

Devaraj, V. (2004) *Concentration of natural rubber field latex by ultrafiltration*. M. Eng. Sc. Thesis, University of Malaya, Kuala Lumpur, Malaysia.

Devaraj, V. & Zairossani, M.N. (2005) The use of membrane separation technology to achieve environment friendliness in natural rubber processing. *Proceeding of International Rubber Conference, 24–28th October 2005, Yokohama, Japan*. John Wiley & Sons. pp. 107–121.

Devaraj, V. & Zairossani, M.N. (2007) Alternative and environment friendly skim latex processing for value-added products recovery. *Proceedings Malaysian Rubber Board Latex Seminar, 18 May 2006 Kuala Lumpur, Malaysia*.

Devaraj, V., Meriam, N.S., Nambiar, J. & Yusof, A. (2003) Environmentally friendly natural rubber latex concentration via membrane separation technology. *Proceedings of the 5th International Membrane Science and Technology Conference, 10–14 November 2003, University of New South Wales, Sydney, NSW, Australia*.

Panglisch, S. (2009) Advances in ceramic membrane filtration. *Proceedings of the AWWA Membrane Technology Conference, 17th March 2009, Memphis, TN*.

Rhodes, E. & Wiltshire, J.L. (1932) Quebrachitol – a possible by-product from latex. *Journal Rubber Research Institute of Malaya*, 3 (3), 160–171.

CHAPTER 12

Application of membrane separation technology for biodiesel processing

Mohammad Mahdi A. Shirazi, Ali Kargari, Ahmad Fauzi Ismail & Takeshi Matsuura

12.1 INTRODUCTION

Energy use has become a crucial concern in recent years due to a rapid increase in energy consumption and worldwide demand in both developed and developing countries. In addition to that, environmental issues such as climate change and global warming caused by utilization of the conventional energy resources (i.e., petroleum, natural gas, coal etc.) have also highlighted the necessity to search for alternative energies.

Currently, conventional energy sources constitute almost 80% of global energy consumption (Hasheminejad *et al.*, 2011). Consequently, renewable and clean energy resources such as wind, solar, biomass, hydropower and geothermal as promising CO_2-free alternatives are of growing interest and will play an important role in the world's future (Kargari and Takht Ravanchi, 2012; Sanaeepur *et al.*, 2014). Renewable energy resources, i.e. solar energy, wind energy and geothermal sources, and bioenergy obtained by chemical conversion of biomass, can be used to generate energy, either in thermal or electrical form, again and again. However, one of the promising sources of renewable energies is bioenergy, or more specifically, biofuels (Noureddin *et al.*, 2014), including biogas, biodiesel, bioethanol, and biomass gasifier. It is to be noted that among various biofuels, the biodiesel has gained more worldwide attention (Noureddin *et al.*, 2014).

Biodiesel, referred to as monoalkyl esters are alcohols of lower molecular weights (e.g. ethanol and methanol) derived from vegetable oils and animal fats in the presence of catalysts (either basic or acidic agents, or either homogeneous or heterogeneous catalysts). The biodiesel, as a good alternative fuel for diesel engines has been gaining great importance worldwide for its proper quality exhaust, sustainability and biodegrability (Sharma *et al.*, 2008). Transesterification reaction is the most used method for commercial biodiesel production (Helwani *et al.*, 2009; Leung *et al.*, 2010). The use of homogeneous alkaline catalysts, i.e. mostly NaOH and KOH, provide a higher reaction rate and conversion for the transesterification of triglycerides to biodiesel. The problems of reversibility encountered in stepwise reactions to biodiesel production were overcome using higher alcohol molar ratios to shift the reaction to completion (Noureddin *et al.*, 2014). However, it should be noted that this higher molar ratio and further amount of basic catalyst will lead to higher soap generation which consequently leads to more complicated downstream processing (Shirazi M.M.A. *et al.*, 2013a).

Biodiesel generation was intended to mainly address the issue of fuel supply security; however, recently, attention has been centered on the use of biodiesel for the transportation sector in order to minimize the overall production of CO_2 from petro-based fuels combustion (Salvi *et al.*, 2013; Rajasekar and Selvi, 2014). Moreover, biodiesel does not increase green-house gases (GHGs) in the atmosphere due to its closed cycle. In other words, it is said to be carbon-neutral, as biodiesel yielding plants take away more CO_2 than that contributed to the atmosphere when used as source of transportation energy. Table 12.1 summarizes the liquid biofuels consumption, i.e. biodiesel and bioethanol, in the transportation sector, 2006–2012, for typical countries.

Table 12.1. Biofuel, e.g. biodiesel and bioethanol, consumption in typical countries [TJ] (PBL Report, 2013).

Country	2006	2008	2010	2012
USA	473793	819755	1012973	1070660
Canada	5789	29306	51560	68520
Japan	400	500	1800	1800
Germany	144818	107561	123947	120873
UK	8029	33072	47202	37051
Brazil	270201	502514	588900	517495
China	42200	49188	50696	63217
India	5038	6191	7611	11736

Table 12.2. Physicochemical properties of biodiesel obtained from different types of oil (Atadashi *et al.*, 2011a).

Oil source	Kinematic viscosity $[mm^2 s^{-1}]$	Cetane No.	*LHV* $[MJ L^{-1}]$	Cloud point [°C]	Flash point [°C]	Density $[kg L^{-1}]$	Sulfur [wt%]
Peanut	4.9	54	33.6	5	176	0.883	–
Soybean	4.5	45	33.5	1	178	0.885	–
Babassu	3.6	63	31.8	4	127	0.879	–
Palm	5.7	62	33.5	13	164	0.880	–
Sunflower	4.6	49	33.5	1	183	0.860	–
Rapeseed	4.2	51–59.7	32.8	–	–	0.882	–

It is indicated in the literature that the higher heating values (HHVs) of biodiesels are relatively high, and ranged from 39 to 41 MJ kg^{-1}. These values are slightly lower when compared with those of petro-based gasoline (i.e., 46 MJ kg^{-1}), petro-diesel (i.e., 43 MJ kg^{-1}) or petroleum (42 MJ kg^{-1}), but greater than coal (32–37 MJ kg^{-1}) (Demirbas, 2009).

Biodiesel termed as clean fuel does not contain carcinogenic substances and its sulfur content level is also lower than its content in petro-based diesel. The highly biodegradable potential of biodiesel and its superb lubricating property as well, makes it to be an excellent fuel. Furthermore, its similarities and renewability in physicochemical properties to mineral diesel, revealed its potentials and practical usability as fuel for the replacement of petro-based diesel in the close future. It is to be noted that a few other physicochemical properties of biodiesel are of great concern and require to be enhanced to make it fit for use in clean form, i.e. 100% biodiesel. These properties include among others; engine power, reduced emission of NO$_x$, increased calorific value, and low temperature properties improvement. Biodiesel decreases long term engine wear in compression ignition engines, and the lubricant property of the biodiesel is about 66% better than mineral-diesel (PBL Report, 2013). The oxidation stability of the biodiesel is also important to prevent it from degradation when stored over time. Recently, biodiesel was found to be compatible in blended form with mineral-diesel in the ratio 20 (biodiesel): 80 (mineral-diesel) (Chauhan *et al.*, 2013). Table 12.2 presents an overview of physicochemical properties of biodiesel obtained from various types of oil.

It is to be noted that having a good and complete transesterification is not enough, and the produced biodiesel should be purified for it contains impurities such as glycerol (i.e. the main co-product in transesterification reaction), water, soap, excess catalyst and unreacted oil (Atadashi *et al.*, 2011a). Otherwise, it can cause serious damage to engines (Table 12.3).

Table 12.3.   Major drawbacks of contaminants of biodiesel for engine (Atadashi *et al.*, 2011b).

Impurity	Drawbacks
Glycerol	Decantation, fuel tank bottom deposits, injector fouling, storage problem, settling problems, and severity of engine durability problems
Water	Reduces the heat of combustion, corrosion of system, failure of fuel pump, formation of ice crystals, bacteriological growth, and pitting in the piston
Methanol	Deterioration of natural rubber seals, lower flash points, lower viscosity, and corrosion
Soap/catalyst	Pose corrosion problem, damage injectors, plugging of filters, and weakening of engine
Free fatty acids	Less oxidation stability, corrosion of vital engine components
Glycerides	Crystallization, turbidity, higher viscosity, and deposition at piston, valves and nozzles

A number of unit operations have been used to separate/purify crude biodiesel including water washing, dry washing, and more recently membrane separation technology. In this chapter, conventional biodiesel separation methods, as well as their advantages and disadvantages, are reviewed. The chapter will be continued with emphasis on the most suitable practical and environmental-friendly method for effective biodiesel production, i.e. membrane separation technology.

12.2 CONVENTIONAL BIODIESEL REFINERY METHODS

After the completion of the transesterification reaction, biodiesel should be processed through separation and then purification steps. The separation step may be carried out via gravitational settling or centrifugation. The crude biodiesel is then refined (i.e., purified) and dried to meet the stringent international standard specifications, which are shown in Table 12.4. Otherwise, the impurities can reduce the biodiesel fuel quality and affect the engine's performance. Due to the soap formation in presence of NaOH and KOH when making biodiesel, the biodiesel refining is complicated; however, a higher conversion yield could be achieved. As well, the dissolved soap and methanol can act as co-solvent between the biodiesel and glycerol and consequently dissolve a higher amount of glycerol in biodiesel, leading to decrease in biodiesel yield and its quality.

The first step usually employed to recover biodiesel after the transesterification reaction is separation of crude biodiesel from the co-product, i.e. glycerol. Due to the density difference, the phase separation of biodiesel-glycerol is fast. This density difference (i.e., ~0.88 g cm^{-3} and ~1.05 g cm^{-3} for biodiesel and glycerol, respectively) is sufficient to employ a simple gravity separation step to separate the non-polar phase, i.e. biodiesel, from the polar phase, i.e. glycerol. In other words, prior to any unit operations of biodiesel, the implementation of a pre-treatment step in order to thoroughly separate biodiesel and glycerol is inevitable. Hence, the gravitational settling pre-treatment could be applied as the first and simplest step immediately taken after the completion of the transesterification reaction. Therefore, achieving faster decantation could significantly contribute to reducing the overall biodiesel production costs.

Shirazi M.M.A. *et al.* (2013a) studied a simple but effective strategy to enhance the gravitational settling of biodiesel-glycerol mixture. This idea is based on acceleration of glycerol decantation through salt (i.e. NaCl) addition to the mixture of biodiesel-glycerol. It is to be noted that this was the first attempt to enhance glycerol decantation though NaCl addition. In this work, after the completion of the reaction, different NaCl quantities (i.e., 0, 0.5, 1, 3, 5 and 10 g) were added to the mixture and the experiments were continued until equilibrium was reached inside the decanting vessel, i.e. a measured cylinder. The equilibrium was defined as the condition in which the interface between biodiesel (i.e., upper phase) and the glycerol (i.e., lower phase) did not change. In this work, gas chromatography (GC), density measurement, interfacial tension

Table 12.4.   International biodiesel standard specifications (Atadashi *et al.*, 2011b).

Properties	Unit	EN14214[a]	ASTM[b]
Ester content	% [m m^{-1}]	96.5	–
Cetane	–	$\geq$51	>47
Flash point	°C	>101	>130
Water and sediment	vol%	0.05	<0.05
Kinematic viscosity [at 40°C]	mm^2 s^{-1}	3.5–5	1.9–6.0
Sulfated ash	% [m m^{-1}]	0.02	0.02
Sulfur	mg kg^{-1}	$\leq$10	–
Carbon residue	% [m m^{-1}]	–	0.05
Acid number	mg KOH g^{-1}	0.50	0.50
Triglyceride	% [m m^{-1}]	0.2	0.2
Free glycerin	% [m m^{-1}]	0.02	0.02
Total glycerin	% [m m^{-1}]	0.25	0.24
Phosphate content	mass%	0.001	0.001
Methanol content	% [m m^{-1}]	0.2	–
Sodium/potassium	mg L^{-1}	5	5

[a]EN 14214: European Standard for requirements and test methods for FAME, a common biodiesel type.
[b]ASTM: American Society for Testing and Materials.

(IFT) and ion chromatography (IC) were used to analyze the biodiesel and glycerol phases and decantation behavior (Shirazi M.M.A. *et al.*, 2013a).

   The authors discussed the phase separation mechanism in the biodiesel-glycerol system, which is simultaneous coalescing and settling (counter-current motion) of glycerol droplets in the continuous biodiesel phase. Figure 12.1 shows the effect of NaCl addition on the decantation time. As can be observed, NaCl addition led to faster glycerol settling and consequently decreased residence time in the pre-treatment step, which eventually contributed to decreased operating time and production costs. However, they concluded that higher quantities of salt, although leading to lower decantation time, cannot lead to the best efficiency. In better words, a higher amount of NaCl can decrease the decantation time; however, it may decrease the methyl ester yield. Therefore, the authors concluded that the 1 g is the most efficient quantity of NaCl to be added which accelerated the decantation process by 100%. (Shirazi M.M.A *et al.*, 2013a).

   As mentioned earlier, the transesterification is the most common method which is carried out in the presence of a basic catalyst (such as KOH) and an alcohol (such as methanol). To complete a transesterification reaction stoichiometrically, a 3:1 M ratio of methanol to triglycerides is needed. Nevertheless, due to the reversible nature of the reaction, significantly excessive methanol is usually considered to shift the reaction toward the product side. (Atadashi *et al.*, 2013; Talebian-Kiakalaieh *et al.*, 2013). Therefore, the presence of excess methanol, besides existing residual catalyst and thermal energy, can affect the phase separation of biodiesel-glycerol mixture. Noureddin *et al.* (2014) studied the interactive effects of prominent parameters including temperature, NaCl addition and methanol concentration on the decantation behavior of biodiesel-glycerol mixture using Box-Behnken design matrix and response surface methodology (RSM). The obtained results revealed that in low temperature ranges, the major parameter influencing decantation speed was density difference, while at higher range of temperatures, viscosity variation played the main role. The authors concluded that at optimum conditions, i.e. temperature of 45°C, 1 g NaCl addition and 20% excess methanol, the decantation time decreased by 200%, and the methyl ester yield at such conditions was measured at >90% (Sharma *et al.*, 2008).

   The next step in biodiesel refining, i.e. after gravitation settling, is the purification step. There are two major conventional methods for this step, including wet-washing and dry-washing

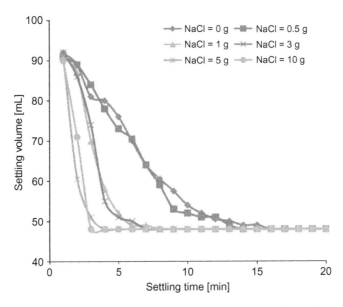

Figure 12.1. The effect of NaCl addition on the decantation time of glycerol in continues biodiesel phase (Shirazi M.M.A. *et al.*, 2013a).

methods. It is to be noted that the main objective of the purification step, either wet-washing or dry-washing, is to remove glycerol, soap, excess alcohol, residual catalyst, and other impurities.

Water-washing, which is a type of wet-washing, is generally carried out to remove the above-mentioned impurities. In this process, distilled water, deionized water, or even tap water can be used to remove contaminants from biodiesel (Atadashi *et al.*, 2011a). In the water-washing process, water at 60–80°C will be well mixed with biodiesel and after some time of agitation will be allowed to settle and then heated to evaporate residual water. This is the most conventional method that has been used for biodiesel purification. Besides its advantages, there are two major disadvantages for the water-washing of biodiesel. The first one is a large amount of water consumption, i.e. 3–10 L fresh water per 1 L of biodiesel (Atadashi *et al.*, 2011a; 2011b; 2013; Talebian-Kiakalaieh *et al.*, 2013), which is a drawback due to the limited availability of fresh water resources (Shirazi *et al.*, 2012). Furthermore, higher water consumption can produce large amount of highly polluted saponified wastewater with high level of COD, BOD, TDS and TSS values (Atadashi *et al.*, 2011b). For example, in our laboratory (MPRL), we measured the 35,600 mg O_2 L^{-1} for COD, 30500 mg O_2 L^{-1} for BOD, 840 mg L^{-1} for TDS and 440 mg L^{-1} for TSS, respectively, for a biodiesel water-washing wastewater. Therefore, one of the today's most challenging and also neglected aspects of conventional biodiesel production with water-washing downstream processing is the high volume of this highly polluting saponified effluent. If this is not addressed properly and given the increasing volume of biodiesel produced globally, such issues could severely jeopardize the supposedly clean and environmental-friendly nature of biodiesel. In this regards, we, in MPRL, are working on possibility of membrane-based treatment of water-washing biodiesel.

Further to water, the biodiesel could be washed by acid. Acid is added to biodiesel and the co-product, i.e. glycerol, to neutralize the basic agents, e.g. catalyst and decompose the soap formed. This process is followed with water-washing to purify biodiesel from contaminants. Various acids could be used in this procedure. For example, Sharma and Singh (2009) studied washing biodiesel with 10% phosphoric acid (H_3PO_4) through bubble wash technique after decantation step. The process is followed by further water-washing using hot distilled water. The authors indicated that using the mentioned procedure, pure biodiesel with meeting international standard was obtained (Sharma and Singh, 2009). Cayli and Kusefoglu (2008) studied biodiesel wet-washing using water and acid solution, i.e. hot water at 70°C and 5% H_3PO_4 at 50°C. The authors dried the biodiesel layer in a vacuum and checked with ceric ammonium nitrate reagent for glycerol removal. It is to be noted that water content has to be reduced to a limit of 0.05% (v/v) to meet the ASTM D6751 standard specification. Organic solvents such as petroleum-ether also have been used to refine biodiesel. This process is usually followed with the use of large amount of hot water to remove residual soap and catalyst (Cayli and Kusefoglu, 2008). Therefore, the weak point of wastewater generation still exists.

The dry-washing method commonly used to refine crude biodiesel is usually achieved through the use of silicates (e.g. Magnesol or Trisyl), ion-exchange resins (e.g. Amberlite or Purolite), cellulosic, and activated carbon and fibers. These adsorbents consist of acidic and basic sites and have strong affinity for polar compounds such as methanol, glycerides, soap and glycerol. This method, i.e. dry-washing, is followed with the use of a filtration step to remove adsorbents. Dry-washing is usually carried out at temperature of 60°C and is mostly completed within 30 min (Atadashi et al., 2011a; Atadashi et al., 2011b). This process has the ability to lower the amount of glycerides and total glycerol to a reasonable level. Comparing to wet-washing, this method has the advantage of being waterless, has strong affinity to polar compounds and no liquid wastewater stream. However, using a large amount of adsorbent, an extra regeneration step for adsorbents, and limited adsorbing capacity are some of the weak points of the dry-washing method.

12.3   MEMBRANE-BASED BIODIESEL REFINERY

12.3.1   *What is a membrane?*

Fundamentally, membranes are semi-permeable barriers that separate different species of solution by allowing restricted passage of some components of a stream in a selective manner (Takht Ravanchi et al., 2009). A membrane can be homogenous or heterogeneous, symmetric or asymmetric in structure (Hosseinkhani et al., 2014; Mirtalebi et al., 2014; Shirazi et al., 2015), solid or liquid (Kazemi et al., 2013; Madaeni et al., 2008), and carry a positive or negative charge or be neutral or bipolar (Mohammadi et al., 2015). Transport through a membrane could be affected by either convection or diffusion of individual molecules, induced by an electric field or concentration, pressure or temperature gradient (Lau and Ismail, 2009). Takht-Ravanchi et al. (2009) tabulated various membrane processes and their various aspects, comprehensively. It is worth quoting that most of the commercially available membranes are made of polymers and inorganic materials (Khulbe et al., 2010), such as ceramics (Kim and Van der Bruggen, 2010), carbon and zeolites (Feng et al., 2015; Salleh and Ismail, 2015). Table 12.5 presents a comparative overview of polymeric and inorganic membranes.

Membrane processes are well-established technologies in water/wastewater treatment (Shirazi M.J.A. et al., 2013; Shirazi M.M.A. et al., 2013b), desalination (Lee et al., 2011; Shirazi M.M.A. et al., 2014), biorefinery (Shuit et al., 2012), gas separation (Rezakazemi et al., 2014), and etc. However, commercial applications of membranes and membrane processes are mostly limited to liquid filtration and relatively inert gas separation. Hence, the application of membrane processes to refine new feed streams, i.e. mostly non-aqueous fluids, is an emerging scope in technology of membranes. One of such emerging fields is biodiesel refining.

12.3.2   *Membrane-based biodiesel processing*

Application of membranes and membrane processes for biodiesel processing is usually designed in two categories, e.g. the membrane reactor to transesterify fats and lipids to biodiesel, and the separative membrane to purify the crude biodiesel from its impurities (Table 12.3).

A membrane reactor (MR) is also known as a membrane-based reactive separator. In other words, a MR is defined as a device that combines reaction and separation in a single unit

Table 12.5. A comparative overview of polymeric and inorganic membranes.

Membrane	Advantages	Disadvantages	Applications	Current status
Polymeric	– Cheap – Good quality control – Flexibility in fabrication – High structural integrity	– Weak structure – Sensitive to pH, temperature and harsh environment – Prone to denature – Short life	– Water/wastewater treatment – Gas separation – Biorefinery	Wide applications, specifically mature in liquid filtration
Inorganic	– High thermal stability – Non-sensitive to pH and harsh environment – Long term durability	– Brittle – Expensive	– Water/wastewater treatment – Biorefinery	Small scale applications

(Shuit *et al.*, 2012). Development of MRs and their successful application for producing biodiesel have renewed the strong interest to develop alternative fuel to replace petro-diesel fuel. MRs can serve different purposes such as intensify the contact between reactants and catalysts, selectively remove the products from the reaction mixture, and control the addition of reactants to the mixture (Andric *et al.*, 2010). It is to be noted that MRs can be employed to avoid the equilibrium conversion limits of conventional reactors. Moreover, the MRs can efficiently improve the maximum achievable conversion of reversible reactions (Sanaeepour *et al.*, 2012). Besides, MRs provide the potential of higher selectivities and/or yields in many different processes as well as being safe and more environmentally friendly (Rios *et al.*, 2004; Uemiya, 2004). Various aspects of biodiesel production using MRs by examining the fundamental concepts of the membrane reactor technology and relevant operating conditions are reviewed by Shuit *et al.* (2012). However, in this chapter the application of selective membranes and related processes to refine biodiesel is discussed and investigated.

12.3.3   *Membrane experiments for biodiesel refining*

Wang *et al.* (2009) studied the application of ceramic membrane (Pall Membrane Co., USA) with various pore sizes, i.e. 0.6, 0.2 and 0.1 µm to remove the residual soap and free glycerol from crude biodiesel. In this work, refined palm oil-methanol (MeOH) (6:1 mol/mol) and potassium hydroxide (KOH, 1.2 wt%) were used to produce biodiesel through the transesterification reaction. In this work, a ceramic membrane tube (outer diameter (OD): 26 mm, inner diameter (ID): 25 mm) with 19 channels of diameter of 3 mm giving a filtration area of 0.045 m^2 was used for the experiments. In order to select a suitable membrane, first, the particle size distribution of crude biodiesel was analyzed by zeta-potential analyzer (at 25°C), and the mean diameter of particles was recorded as a reference for selection of suitable membrane. A known amount of crude biodiesel sample was cross-filtrated by all membranes at the pressure of 0.15 MPa and temperature of 60°C. The initial flux permeate, defined as the flux of permeate (at first 3 min) was recorded and the permeate samples were analyzed for the content of potassium, sodium, calcium, magnesium and free glycerol. Table 12.6 presents the impurities content of permeate samples for each applied membrane. Then, the most suitable membrane, i.e. microfilter membrane with 0.1 µm pore size (at various feed pressures between 0.05 and 0.20 MPa and temperatures of 30 and 70°C) was used for the next experiments (Wang *et al.*, 2009).

In this work, the authors concluded that due to the immiscibility of free glycerol and biodiesel and also the surface activity of soap, the soap existed in the form of reversed micelle which was very similar to the form of phospholipids in the hexane miscella, the size of which was larger than a single molecule. Moreover, most of the impurities such as residual catalysts are soluble in the glycerol, and the glycerol droplets and hydrophilic ends of soaps can form larger

Table 12.6.   Contents of impurities in the permeate samples and the control sample (Wang *et al.*, 2009).

Samples	Metals [mg kg^{-1}]				Free glycerol [wt%]	Flux [L m^{-2} h^{-1}]
	K$^+$	Na$^+$	Ca^{2+}	Mg^{2+}		
Feed	160	8.98	1.45	0.33	0.261	–
P^a of 0.6 μm	4.25	0.68	0.70	0.25	0.0276	675
P of 0.2 μm	2.20	0.88	0.55	0.26	0.0257	480
P of 0.1 μm	1.7	1.36	0.95	0.15	0.0152	360
Control^b	2.46	1.41	0.64	0.18	0.017	–

[a]P: Permeate.
[b]Water-washed biodiesel sample.

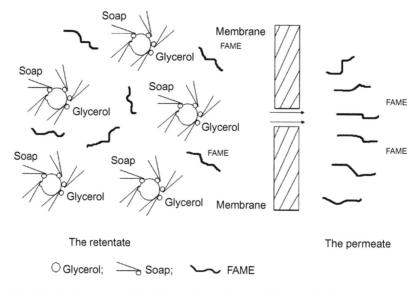

Figure 12.2.   Mechanism of separation of contaminants from crude biodiesel using microfiltration (Wang *et al.*, 2009).

micelles with mean size of 2.21 μm, based on zeta-potential analyzer. Therefore, such large molecules are easier to be removed by the microfilters. Figure 12.2 presents the hypothesis of the above-mentioned mechanism for contaminant removal ;using membrane separation technology from crude biodiesel. In this work, the authors concluded that using membrane to refine crude biodiesel showed the advantage of no wastewater generation compared to the conventional water-washing one, and methanol, due to its good solubility in both polar and non-polar phases was the right agent for membrane cleaning (Wang *et al.*, 2009).

In another work, Gomes *et al.* (2010) studied on the performance of ceramic membranes which they used for glycerol removal from crude biodiesel. In this work, three different tubular-type ceramic membranes made of α-Al$_2$O$_3$/TiO$_2$ with 250 mm length, 7 mm diameter and 0.005 m^2 filtration area and with pore sizes of 0.2, 0.4 and 0.8 μm were used for experiments under various operating pressures, i.e. 0.1 to 0.3 MPa. A commercial MF-UF experimental filtration set-up (Fig. 12.3) was used in this work. The authors used a synthetic sample of mixture of biodiesel, glycerol and methanol. The total MF process time, approximately 2 h and under constant temperature (60°C), was determined to enable stabilization flux in all experiments. Figure 12.4a shows the permeate flux versus filtration time for the 0.2-μm membrane and the three corresponding

M1 – Membrane module
P1 – Circulating pump
T1 – Feed tank
F1 – Flow meter
V1 – Pressure valve
V2, V3, V4, V5, V6 – Drain valves
T2 – Thermometer
B1 – Analytical balance
PG1, PG2 – Pressure gauges
FS1 – Flow controller
PS – Pressure controller
C1 – Circulating bath temperature
       control

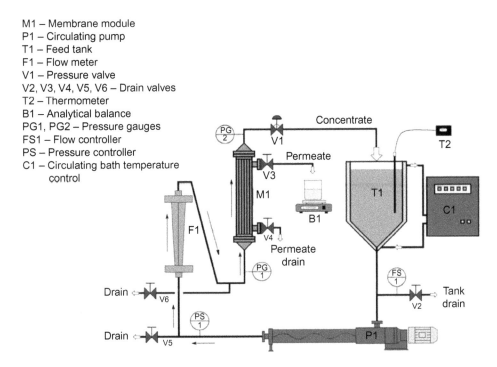

Figure 12.3.   General scheme of the experimental set-up in Gomes *et al.* (2010) work.

pressures. As could be observed, at 0.1 MPa this membrane gave the lowest stabilized permeate flux, i.e. ~12 kg h^{-1} m^{-2}, which is six-fold smaller than the highest value obtained during experiments. However, this condition resulted in the largest glycerol rejection, i.e. 99.6%. The 0.4-μm membrane gave the highest permeate fluxes for all the operating pressures (Fig. 12.4b) when compared to the other membranes. The highest stabilized permeate flux of this membrane was ~83 kg h^{-1} m^{-2} at 0.2 MPa. In this condition, 99.3% glycerol rejection was achieved. Despite the initial high permeate fluxes for 0.8-μm membrane, as could be observed in Figure 12.4c, it presented the largest permeate flux reduction. The authors concluded that this was due to higher fouling effect when the membrane pore size is large (Gomes *et al.*, 2010).

As mentioned earlier, during transesterification reaction an alcohol should be used. As the applied alcohol can act as the co-solvent between the polar phase (glycerol) and the non-polar phase (biodiesel), the concentration of excess alcohol in the final mixture can affect the glycerol separation. In this work, authors also studied the effect of ethanol concentration, i.e. 5, 10 and 20% on the permeate flux and glycerol rejection. Results, as shown in Figure 12.4d, indicated that the highest ethanol concentration in the initial mixture afforded high initial fluxes; however, its flux decrease rate was the greatest. The authors indicated that the ethanol concentration affected the glycerol droplet size distribution, since the same operating pressure, the glycerol content in the permeate for a higher ethanol concentration was over four-fold that for a lower concentration. The best performance in this work was that the membrane with pore size of 0.2 μm at 0.2 MPa and with 99.4% glycerol rejection. Moreover, the concentration of excess alcohol can significantly affect the microfiltration performance (Gomes *et al.*, 2010).

Recently, Alves *et al.* (2013) studied comparatively the performance of water-washing and micro/ultrafiltration processes for biodiesel purification. In this work, the authors used polymeric membranes made of mixed cellulose esters (MCE, with 0.22 and 0.30 μm pore sizes) and polyethersulfone (PES, with 10 and 30 kDa nominal MWCO) for experiments through a dead-end apparatus. As a novel strategy, the authors added water to the crude biodiesel sample at

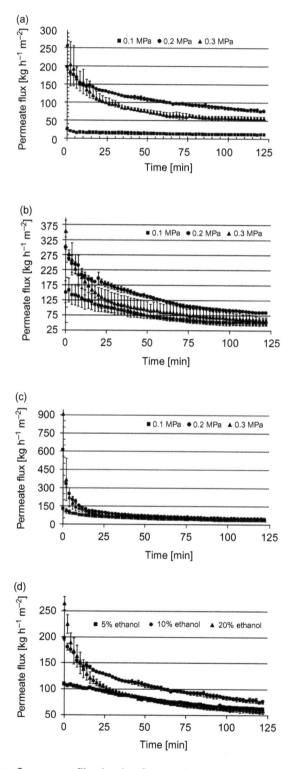

Figure 12.4.   Permeate fluxes versus filtration time for ceramic membranes in Gomes *et al.* (2010) work.

Table 12.7. Characterization of crude and water-washed biodiesel in Alves *et al.* (2013) work.

		Biodiesel sample	
Parameter	Unit	Crude	Washed
Density	$kg\,m^{-3}$	880	880
Kinematic viscosity	$mm^2\,s^{-1}$	5	5
Water content	$mg\,L^{-1}$	525.4	1026.2
Acid value	$mg_{KOH}\,g^{-1}$	0.08	0.14
Saponification value	$mg_{KOH}\,g^{-1}$	191.5	179.2
Amount of soap	$g_{soap}\,g_{sample}^{-1}$	2.6×10^{-3}	–
Free glycerol	$\%\,(wt\,wt^{-1})$	0.029	0.007

concentrations of 0.1 and 0.2 wt%. Deionized water was mixed with the crude biodiesel using a magnet stirrer for 1 h at ambient temperature prior to the filtration experiments. At first step, i.e. after transesterification reaction and the settling process to separate biodiesel-glycerol mixture, the crude biodiesel and the water-washed sample were characterized, as shown in Table 12.7. As could be observed, water-washing did not change density and viscosity values of crude biodiesel. According to international standards, the limit for water content in biodiesel is $500\,mg\,L^{-1}$; however, obviously water-washing increased the water content and water-washed biodiesel has more water than the limit allowed by the legislation. This behavior can be associated with the interaction between mono and diglycerides molecules and water, since water solubility in the ester phase is very small. It is to be noted that the mono and diglycerides, left from an incomplete reaction, can act as an emulsifier, allowing the water to be mixed with the biodiesel. For the first step, the authors concluded that except for the water content, all the other analyzed parameters of the water-washed biodiesel are in accordance the limits imposed by international standards. However, as mentioned earlier, the generation of highly polluted wastewater should be noted (Alves *et al.*, 2013).

Figure 12.5 shows the obtained permeate flux of biodiesel through the MF membranes at 0.1 and 0.2 MPa. As could be observed, a flux decline is in the first 2 min of operation for the 0.22-μm membrane (Fig. 12.5a). However, for the 0.33-μm membrane steady declines are observed (Fig. 12.5b) for 1 and 5 min at 0.1 and 0.2 MPa of operating pressure, respectively. The stabilized flux is greater at 0.2 MPa than at 0.1 MPa, showing that greater operating pressure enables greater fluxes within the analyzed pressure range. Ultrafiltration tests were carried out under constant operating pressure of 0.4 MPa, using 10 and 30 kDa membranes. Besides the higher pressure, the flux with UF membranes was smaller than with the MF ones and a less pronounced flux decline was observed in this case (Alves *et al.*, 2013).

The physico-chemical characteristics of permeate samples showed that the filtration process, either MF or UF, did not change the density and the viscosity of biodiesel. Moreover, the results of this work show that the applied membranes were not able to remove the excess catalyst and free acid content of crude biodiesel. Only the UF membrane of 30 kDa (at 0.4 MPa) was not able to reduce the amount of soap and free glycerol content detected in the crude biodiesel sample. Hence, an additional test was carried out with the same membrane (i.e. 30 kDa) at 0.3 MPa operating pressure. The authors indicated that this reduction in the applied pressure increased the membrane performance for biodiesel refining. However, the 30 kDa membrane did not show promising values for biodiesel separation. It could be probably due to the more open pore size in comparison with 10 kDa membrane. The operating pressure in MF experiments did not change the biodiesel quality. Regarding the free glycerol content, only the 10-kDa membrane was able to reduce the glycerol content according to the international legislation limit for biodiesel, i.e. less than 0.02 wt% (Table 12.8). The authors confirmed these results by gas chromatography (GC) analysis. Results indicated that UF process with the 10 kDa membrane is able to reduce the glycerol content to the desired level, as mentioned above (Alves *et al.*, 2013).

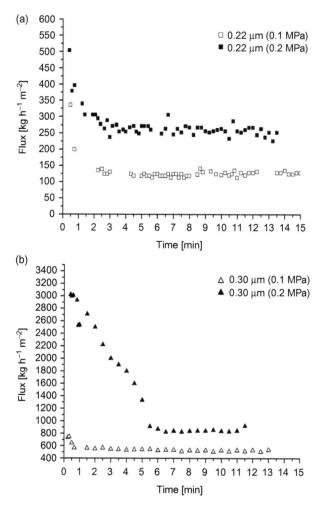

Figure 12.5.   Permeate flux of biodiesel at 0.1 and 0.2 MPa of operating pressures via MF membranes of 0.22 μm (a) and 0.30 μm (b) pore size, in Alves *et al.* (2013) work.

In this work, the authors carried out further filtrations using the 10-kDa membrane and adding water to the crude biodiesel sample prior to the experiment. According to the obtained results, water addition reduced the stabilized flux; however, a higher rejection of free glycerol was achieved. In other words, with water addition decrease in flux was associated with higher rejections. This higher glycerol rejection can be explained by the fact that glycerol and water, which are completely soluble, formed an immiscible phase with the biodiesel phase. The molecules of water joined to glycerol and these larger molecules were unable to pass through the membrane pores. However, the authors reported that with increase the filtration time, the glycerol content of the permeate phase increased. This was due to the dead-end configuration of experimental apparatus. Therefore, it could be concluded that such module configuration, i.e. dead-end, is not suitable for biodiesel refining (Alves *et al.*, 2013).

In the Saleh *et al.* (2010) work, a polymeric UF membrane (100 kDa) made of polyacrylonitrile (PAN) was used for glycerol removal from biodiesel samples, which was prepared by transesterification of canola oil and methanol. In this work, the effect of different materials present in the reaction, e.g. soap, water and methanol, on the final glycerol content was studied. Operating conditions were set at 25°C temperature and 0.552 MPa pressure for all tests conducted.

Table 12.8. Characterization of permeate samples after filtering biodiesel in Alves *et al.* (2013) work.

	Parameter	
Experiment	Soap $[10^{-3} \, g_{soap} \, g_{sample}^{-1}]$	Free glycerol $[\%wt \, wt^{-1}]$
0.22 μm at 0.1 MPa	1.3	0.022
0.22 μm at 0.2 MPa	1.3	0.025
0.3 μm at 0.1 MPa	1.6	0.026
0.3 μm at 0.2 MPa	1.6	0.026
10 kDa at 0.4 MPa	1.0	0.020
30 kDa at 0.4 MPa	2.7	0.031
30 kDa at 0.3 MPa	2.4	0.029

Table 12.9. Free glycerol in the permeate and retentate samples after UF process in Saleh *et al.* (2010) work.

		Free glycerol [mass%] at 25°C and 0.552 MPa							
	Time [min]	FAME only	FAME+ 0.06% water	FAME + 0.1% water	FAME + 0.2% water	FAME + 1% soap	FAME + 1% methanol	FAME + 1% soap + 1% methanol + 0.06% water	FAME + 1% soap + 0.06% water
	Feed	0.037	0.036	0.032	0.040	0.029	0.039	0.047	0.030
Permeate	15	0.027	0.030	0.020	0.013	0.025	0.031	0.041	0.021
	30	0.027	0.029	0.018	0.012	0.029	0.035	0.038	0.020
	60	0.032	0.028	0.017	0.013	0.028	0.033	0.038	0.020
	120	0.032	0.028	0.017	0.014	0.029	0.031	0.041	0.020
	180	0.033	0.027	0.017	0.013	0.030	0.031	0.040	0.018
Retentate	15	0.037	0.039	0.033	0.045	0.029	0.037	0.043	0.025
	30	0.037	0.040	0.034	0.037	0.032	0.037	0.042	0.025
	60	0.036	0.040	0.034	0.036	0.032	0.036	0.040	0.026
	120	0.034	0.041	0.032	0.034	0.029	0.034	0.043	0.028
	180	0.034	0.039	0.035	0.035	0.032	0.034	0.041	0.031

The authors characterized the samples of feed, permeate and retentate for the glycerol content using CG method based on the ASTM D6584 standard (Table 12.9). Results of this work showed low concentrations of water had considerable effect on glycerol rejection from crude biodiesel, even at approximately 0.08 mass%. It is to be noted that this is 4 orders of magnitude less than the amount of water required in a conventional water-washing process. The authors suggested that the separation mechanism for free glycerol was due to the removal of an ultrafine dispersed glycerol-rich phase present in the crude biodiesel. This hypothesis was confirmed by the presence of particulates in the crude biodiesel. The size of the particles and the free glycerol separation both increased with increase the water content (mass%) of the crude biodiesel phase. In this work, the results indicated that the trends of separation and particle size versus water content in biodiesel phase were very similar and exhibited a sudden increase to 0.08 wt% water in the crude biodiesel. This hypothesis supports the conclusion in this work that water increased the size of the distributed glycerol droplets in the crude biodiesel leading to its separation by UF membrane. The authors finally concluded that the technology for biodiesel refining, i.e. UF process, was found to use 2.0 g of water per 1 L of treated biodiesel versus the current 10 L of water per 1 L of treated biodiesel in water-washing (Saleh *et al.*, 2010).

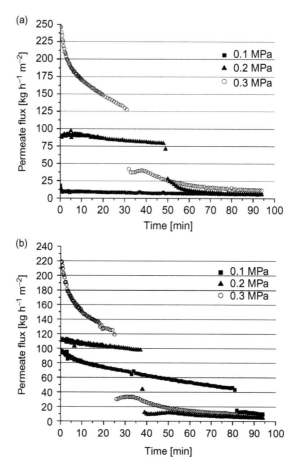

Figure 12.6.   Permeate flux vs. filtration time at various operating pressures and 50°C feed temperature for 20% acidified water content, (a) for 0.1-μm and (b) 0.05-μm membranes, in Gomes *et al.* (2013) work.

As mentioned earlier, during the transesterification, a basic catalyst (usually KOH) is used for reaction, therefore, the final mixtures, e.g. the polar and non-polar phases, are basic. Therefore, using acidified water can be effective in membrane-based refining of biodiesel. Gomes *et al.* (2013) studied the effect of acidified water addition to the crude biodiesel (produced by ethyl transesterification of degummed soybean oil) on the glycerol removal efficiency. Experiments were carried out using MF (0.2, 0.1 μm and 0.05 μm) and UF (20 kDa) membranes made of α-Al_2O_3/TiO_2. Experimental conditions in this work were set on 0.1, 0.2 and 0.3 MPa trans-membrane pressures, 50°C feed temperature, and acidified-water content of 10, 20 and 30%. Figure 12.6 presents the behavior of permeation flux vs. filtration time for 0.1-μm and 0.05-μm membranes at various trans-membrane pressures for 20% acidified water content in the feed stream. The main characteristic observed in this curve and for these membranes is the sharp drop in the permeate flux with the MF membranes under 0.2 and 0.3 MPa pressures. The authors indicated that with the UF membrane (i.e. 20 kDa), this behavior was observed for the three applied pressures. The highest initial flux for the 0.2-μm membrane was achieved at 0.2 MPa. As this membrane had larger pore size, the highest pressure promotes a greater ease of pore clogging by glycerol agglomerates that are retained, so that under 0.3 MPa pressure the initial permeate flux was lower than under 0.2 MPa. Under 0.1 MPa, no sharp reduction was verified and the flux, although low, remained almost constant throughout the operation. The authors concluded that for

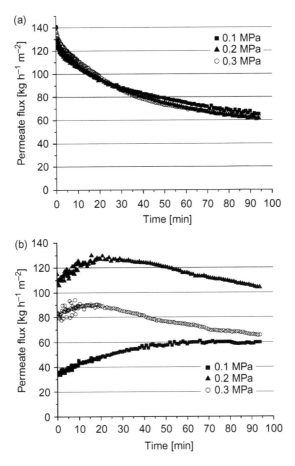

Figure 12.7.   Permeate flux vs. filtration time at various operating pressures and 50°C feed temperature for 10% acidified water content, (a) for 0.1-μm and (b) 20 kDa membranes, in Gomes *et al.* (2013) work.

membrane-based biodiesel refining using MF membranes, microfiltration depends more on the aqueous phase retained on the membrane surface accumulates, and provides an extra resistance responsible for the significant flux reduction (Gomes *et al.*, 2013).

With the membranes of 0.1 and 0.05 μm (Fig. 12.6) a similar flux was observed under 0.2 and 0.3 MPa pressures. In the highest pressure value, the initial flux was quite high, presenting a great reduction of flux before the sharp drop. Under 0.2 MPa, although the flux has remained constant for about half the time of operation, the drop also occurred. Results of this work indicated that besides the dependence on the pressure, the compression also depends on the concentration of the aqueous phase in the feed stream, since when the initial flux is very high, the reduction in the flux has occurred at an earlier filtration time (Gomes *et al.*, 2013). With the 0.05-μm membrane and pressure of 0.1 MPa (Fig. 12.6b) the flux remained constant at a relatively high value during most of the filtration. The authors concluded that as the difference between the retained agglomerates and the membrane pore size is greater, the pore clogging is more difficult and the fouling is lower. Also, the filtration behavior of 0.1-μm and 20 kDa membranes in the presence of 10% acidified water in the feed stream and under various operating pressures, i.e. 0.1, 0.2 and 0.3 MPa, and at 50°C feed temperature is presented in Figure 12.7. Finally, the authors concluded that the UF membranes, i.e. 0.05-μm and 20 kDa, under 0.1 MPa pressure promoted permeate flux above

Table 12.10.   SEM images of electrospun nano-fibrous polystyrene filters before and after thermal treating, in Shirazi M.J.A. *et al.* (2014) work.

$60\,kg\,h^{-1}\,m^{-2}$ and glycerol content in the permeate lower than 0.02%, being thus selected as the most appropriate for biodiesel refining (Gomes *et al.*, 2013).

12.3.4   *Other membrane applications for biodiesel processing*

Further to above mentioned applications, i.e. biodiesel refinery, membrane processes also can be used for other subjects related to biodiesel processing. Shirazi M.J.A. *et al.* (2014) studied the possibility of a new and low cost source for biodiesel production, i.e. edible oil mill effluent. The authors investigated the application of membrane-based coalescing filtration for separation of edible oil and water emulsion. In this work, electrospun nano-fibrous filters were used for filtration process. The novelty of this work is focused on a simple and versatile surface treatment method, i.e. thermal treatment, to improve the filtration efficiency of electrospun polystyrene filters (Table 12.10). Results indicated that thermal treatment of electrospun nano-fibrous filters significantly had affected the edible oil separation through coalescing phenomenon (Shirazi M.J.A. *et al.*, 2014).

Table 12.11.   Electrospinning conditions investigated in the Shirazi M.M.A. *et al.* (2013b) work.

Polymer	Polystyrene, 190000 Mw
Solvent	N,N-dimethylformamide (DMF), 99.9%
Polymer/solvent concentration	20 wt%
High voltage	18 kV
Needle	Stainless steel, 18 gauge
Tip-collector distance	17 cm
Injection flow rate	0.1 mL min^{-1}
Time	30 min
Humidity	25–30%

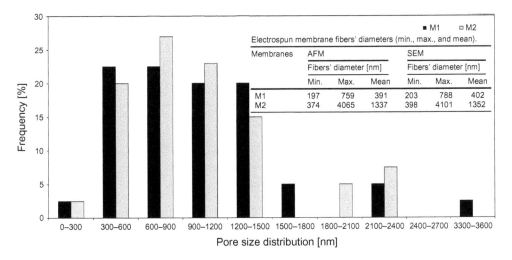

Electrospun membrane fibers' diameters (min., max., and mean).

Membranes	AFM			SEM		
	Fibers' diameter [nm]			Fibers' diameter [nm]		
	Min.	Max.	Mean	Min.	Max.	Mean
M1	197	759	391	203	788	402
M2	374	4065	1337	398	4101	1352

Figure 12.8.   Pore size distribution and fibers' diameter of electrospun MF membranes in Shirazi M.M.A. *et al.* (2013b) work.

As mentioned earlier, water-washing has been proved to meet the international standard specifications laid out for biodiesel and is conventionally used in most biodiesel plants around the world. However, it gives rise to higher water consumption and consequently a high amount of wastewater with high values of BOD and COD. Hence, purification of this environmentally-hazardous wastewater should be taken into serious consideration (Shirazi M.J.A. *et al.*, 2014). In another work, Shirazi M.M.A. *et al.* (2013b) studied the microfiltration of water-washing effluent generated during biodiesel purification through wet-washing by fresh water. In this work, two electrospun nano-fibrous membranes made of polystyrene were used for MF experiments through a dead-end module. Table 12.11 shows the electrospinning conditions in this work. Also, thermal treatment through contact heating was used for surface modification of electrospun MF membranes. At the first step, the authors characterized the as-spun microporous membranes using scanning electron (SEM) and atomic force (AFM) microscopes. The results of characterization for pore size distribution as well as fibers' diameter are presented in Figure 12.8. As could be observed, thermal treatment led to an increase in the fiber diameters, through merging smaller fibers, and a decrease in the mean pore size, respectively. Moreover, AFM analysis showed that smoother surface was achieved after contact heating of electrospun membrane. In the next step, these membranes were used for treatment of wastewater generated through biodiesel water-washing. Results of the MF process are shown in Table 12.12. As could be observed, the novel MF membranes could effectively be used for reduction of polluting capacity of water-washing effluent. In the case of BOD and COD, the values decreased from 170 and 445 mg O_2 L^{-1} to 125 and 184 mg O_2 L^{-1}, and 77 and 112 mg O_2 L^{-1} for M1 and M2 membranes, respectively.

Table 12.12.   Characteristics of biodiesel wash water effluent, (A) raw wastewater sample, (B) filtrated
using M1 membrane (before thermal treatment), and (C) filtrated using M2 membrane (after
thermal treatment), in Shirazi M.M.A. *et al.* (2013b) work.

Sample	COD [mg O_2 L^{-1}]	BOD [mg O_2 L^{-1}]	TS [mg L^{-1}]	TDS [mg L^{-1}]	TSS [mg L^{-1}]
A	445	170	740	700	40
B	184	125	60	40	20
C	112	77	56	28	28

COD: Chemical oxygen demand.
BOD: Biological oxygen demand.
TS: Total solids.
TDS: Total dissolved solid.
TSS: Total suspended solid.

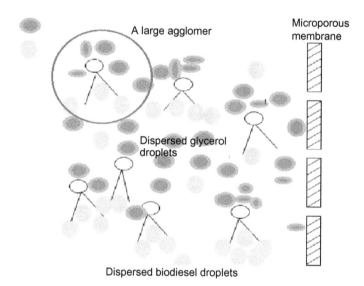

Figure 12.9.   General scheme of the proposed mechanism involved in the treatment of saponified wastewater
in biodiesel water-washing process, investigated in Shirazi M.M.A. *et al.* (2013b) work.

These results, i.e. proposed mechanism for treatment of biodiesel wastewater using electrospun
membranes, could be explained as follows (Shirazi M.M.A. *et al.*, 2013b).
    Biodiesel wastewater contains various impurities such as soap, glycerol, catalyst residue,
alcohol and unreacted oil. Moreover, biodiesel droplets could be found in this highly polluted
wastewater. It is to be noted that all these impurities, except biodiesel, are soluble in water.
Moreover, the soap (i.e. saponified coagulum) has two components, hydrophilic and hydrophobic ends. Therefore, both biodiesel and glycerol droplets can be attached to the related ends of
soap molecules, which is ultimately the final mixture that contains water (aqueous phase) and
large agglomers (Fig. 12.9). As in contrast to the conventional commercial membranes which
mostly have a 2D structure, the electrospun membranes have a 3D structure (Shirazi M.M.A.
et al., 2013b). Therefore, the separation mechanism of this kind of membranes is triplicate, i.e.
screening-depth filtration-adsorption. As the size of formed agglomers in the wastewater are
much larger than the membranes' pore size, and due to both depth filtration mechanism and
adsorption of impurities on the nano-fibers' surface, it was found out clearly that electrospun
membranes could be used efficiently for treating the biodiesel's water-washing effluent. In better

words, the multi-objective separation mechanism of electrospun membranes can be the cause of their superior separation performance.

12.4 FUTURE PERSPECTIVE

High demands for clean fuels and the problems associated with fossil fuels make the use of new clean fuels such as biodiesel, biogas, hydrogen, etc. necessary. Among these fuels, biodiesel has found more attention because of very similar properties to the diesel fuels. The major problems with this biofuel are the source of feedstock (i.e. waste cooking oils and non-edible vegetable oils) and the large amount of wastewater during the production and purification steps. Fortunately, new genetically modified types of plants for this purpose have been introduced which can be irrigated by wastewaters. The perspective of a biodiesel production plant could be a fully integrated system containing several membrane units for purification of raw materials, i.e. oil degumming, membrane reactors for heterogeneous transesterification reactions, oil dewatering and water de-oiling steps for purification of raw biodiesel and separation of byproducts such as glycerol and water, separation of biodiesel washing wastewaters, etc. The recovered waters from different wastewater treatment steps could be integrated for reuse in washing steps. Then, the development of membrane knowledge including fabrication of new generation of polymeric and inorganic membranes for these applications is critical. In this regard, low-cost and low-fouling electrospun membranes can play an important role especially for the fabrication of super hydrophobic membranes for biodiesel and its wastewater processing.

REFERENCES

Alves, M.J., Nascimento, S.M., Pereira, I.G. & Martins, M.I. (2013) Biodiesel purification using micro and ultrafiltration membranes. *Renewable Energy*, 58, 15–20.

Andric, P., Meye, A.S., Jensen, P.A. & Dam-Johnsen, K. (2010) Reactor design for minimizing product inhibition during enzymatic lignocellulose hydrolysis. II. Quantification of inhibition and suitability of membrane reactors. *Biotechnology Advances*, 28, 407–425.

Atadashi, I.M., Aroua, M.K. & Abdul Aziz, A. (2010) High quality biodiesel and its diesel engine application: a review. *Renewable and Sustainable Energy Reviews*, 14, 1999–2008.

Atadashi, I.M., Aroua, M.K. & Abdul Aziz, A. (2011a) Biodiesel separation and purification: a review. *Renewable Energy*, 36, 437–443.

Atadashi, I.M., Aroua, M.K., Abdul Aziz, A.R. & Sulaiman, N.M.N. (2011b) Refining technologies for the purification of crude biodiesel. *Applied Energy*, 88, 4239–4251.

Atadashi, I.M., Aroua, M.K., Abdul Aziz, A.R. & Sulaiman, N.M.N. (2013) The effects of catalysts in biodiesel production: a review. *Journal of Industrial and Engineering Chemistry*, 19, 14–26.

Cayli, G. & Kusefoglu, S. (2008) Increased yields in biodiesel production from used cooking oils by a two steps process: comparison with one step process by using TGA. *Fuel Processing Technology*, 89, 118–122.

Chauhan, B.S., Kumar, N., Cho, H.M. & Lim, H.C. (2013) A study on the performance and emission of a diesel engine fueled with karanja biodiesel and its blends. *Energy*, 56, 1–7.

Demirbas, A. (2009) Progress and recent trends in biodiesel fuels. *Energy Conversion and Management*, 50, 14–34.

Feng, C., Khulbe, K.C., Matsuura, T., Farnood, R. & Ismail, A.F. (2015) Recent progress in zeolite/zeotype membranes. *Journal of Membrane Science and Research*, 1, 49–72.

Gomes, M.C.S., Pereira, N.C. & Davantel de Barros, S.T. (2010) Separation of biodiesel and glycerol using ceramic membranes. *Journal of Membrane Science*, 352, 271–276.

Gomes, M.C.S., Arroyo, P.A. & Pereira, N.C. (2013) Influence of acidified water addition on the biodiesel and glycerol separation through membrane technology. *Journal of Membrane Science*, 431, 28–36.

Hasheminejad, M., Tabatabaei, M., Mansourpanah, Y. & Javani, A. (2011) Upstream and downstream strategies to economize biodiesel production. *Bioresource Technology*, 102, 461–468.

Helwani, Z., Othman, M.R., Aziz, N., Fernando, W.J.N. & Kim, J. (2009) Technologies for production of biodiesel focusing on green catalytic techniques: a review. *Fuel Processing Technology*, 90, 1502–1514.

Hosseinkhani, O., Kargari, A. & Sanaeepour, H. (2014) Facilitated transport of CO_2 through Co (III) S-EPDM ionomer membrane. *Journal of Membrane Science*, 469, 151–161.

Kargari, A. & Takht Ravanchi, N. (2012) Carbon dioxide: capturing and utilization. In: Liu, G. (ed.) *Greenhouse gases-capturing, utilization and reduction*. InTech, Rijeka, Croatia.

Kazemi, P., Peydayesh, M., Bandegi, A., Mohammadi, T. & Bakhtiari, O. (2013) Pertraction of methylene blue using a mixture of D2EHPA/M2EHPA and sesame oil as a liquid membrane. *Chemical Papers*, 67, 722–729.

Khulbe, K.C., Feng, C. & Matsuura, T. (2010) The art of surface modification of synthetic polymeric membranes. *Journal of Applied Polymer Science*, 115, 855–895.

Kim, J. & Van der Bruggen, B. (2010) The use of nanoparticles in polymeric and ceramic membrane structures: review of manufacturing procedures and performance improvement for water treatment. *Environmental Pollution*, 158, 2335–2349.

Lau, W.J. & Ismail, A.F. (2009) Polymeric nanofiltration membranes for textile dye wastewater treatment: preparation, performance evaluation, transport modelling, and fouling control – a review. *Desalination*, 245, 321–348.

Lee, K.P., Arnot, T.C. & Mattia, D. (2011) A review of reverse osmosis membrane materials for desalination-development to date and future potential. *Journal of Membrane Science*, 370, 1–22.

Leung, D.Y.C., Wu, X. & Leung, M.K.H. (2010) A review on biodiesel production using catalyzed transestrification. *Applied Energy*, 87, 1083–1095.

Madaeni, S.S., Aalami-Aleagha, M.E. & Daraei, P. (2008) Preparation and characterization of metallic membrane using wire arc spraying. *Journal of Membrane Science*, 320, 541–548.

Mirtalebi, E., Shirazi, M.M.A., Kargari, A., Tabatabei, M. & Ramakrishna, S. (2014) Assessment of atomic force and scanning electron microscopes for characterization of commercial and electrospun nylon membranes for coke removal from wastewater. *Desalination and Water Treatment*, 52, 6611–6619.

Mohammadi, S., Kargari, A., Sanaeepour, H., Abbassian, K., Najafi, A. & Mofarrah, E. (2015) Phenol removal from industrial wastewaters: a short review. *Desalination and Water Treatment*, 53 (8), 2215–2234.

Noureddin, A., Shirazi, M.M.A., Tofeily, J., Kazemi, P., Motaee, E., Kargari, A., Mostafaei, M., Akia, M., Karout, A., Jaber, R., Hamieh, T. & Tabatabaei, M. (2014) Accelerated decantation of biodiesel-glycerol mixtures: optimization of a critical stage in biodiesel biorefinery. *Separation and Purification Technology*, 132, 272–280.

PBL Report (2013) Trends in global CO₂ emissions: 2013 Report. PBL Netherlands Environmental Assessment Agency, The Netherlands.

Rajasekar, E. & Selvi, S. (2014) Review of combustion characteristics of CI engines fueled with biodiesel. *Renewable and Sustainable Energy Reviews*, 35, 390–399.

Rezakazemi, M., Ebadi Amooghin, A., Montazer-Rahmati, M.M., Ismail, A.F. & Matsuura, T. (2014) State-of-the-art membrane based CO₂ separation using mixed matrix membranes (MMMs): an overview on current status and future directions. *Progress in Polymer Science*, 39, 817–861.

Rios, G.M., Belleville, M.P., Paolucci, D. & Sanchez, J. (2004) Progress in enzymatic membrane reactors – a review. *Journal of Membrane Science*, 242, 189–196.

Saleh, J., Trembly, A.Y. & Dube, M.A. (2010) Glycerol removal from biodiesel using membrane separation technology. *Fuel*, 89, 2260–2266.

Salleh, W.N.W. & Ismail, A.F. (2015) Carbon membranes for gas separation processes: recent progress and future perspective. *Journal of Membrane Science and Research*, 1, 2–15.

Salvi, B.L., Subramanian, K.A. & Panwar, N.L. (2013) Alternative fuels for transportation vehicles: a technical review. *Renewable and Sustainable Energy Reviews*, 25, 404–419.

Sanaeepour, H., Hosseinkhani, O., Kargari, A., Ebadi Amooghin, A. & Raisi, A. (2012) Mathematical modeling of a time-dependent extractive membrane bioreactor for denitrification of drinking water. *Desalination*, 289, 58–65.

Sanaeepur, S., Sanaeepur, H., Kargari, A. & Habibi, M.H. (2014) Renewable energies: climate-change mitigation and international climate policy. *International Journal of Sustainable Energy*, 33, 203–2012.

Sharma, Y.C. & Singh, B. (2009) Development of biodiesel: current scenario. *Renewable and Sustainable Energy Reviews*, 13, 1646–1651.

Sharma, Y.C., Singh, B. & Upadhyay, S.N. (2008) Advancements in development and characterization of biodiesel: a review. *Fuel*, 87, 2355–2373.

Shirazi, M.J.A., Bazgir, S. & Shirazi, M.M.A. (2014) Edible oil mill effluent; a low-cost source for economizing biodiesel production: electrospun nanofibrous coalescing filtration approach. *Biofuel Research Journal*, 1, 39–42.

Shirazi, M.J.A., Bazgir, S., Shirazi, M.M.A. & Ramakrishna, S. (2013) Coalescing filtration of oily wastewaters: characterization and application of thermal treated electrospun polystyrene filters. *Desalination and Water Treatment*, 51, 5974–5986.

Shirazi, M.M.A., Kargari, A. & Shirazi, M.J.A. (2012) Direct contact membrane distillation for seawater desalination. *Desalination and Water Treatment*, 49, 368–375.

Shirazi, M.M.A., Kargari, A., Tabatabaei, M., Akia, M., Barkhi, M. & Shirazi, M.J.A. (2013a) Acceleration of biodiesel-glycerol decantation through NaCl-assisted gravitational settling: a strategy to economize biodiesel production. *Bioresource Technology*, 134, 401–406.

Shirazi, M.M.A., Kargari, A., Bazgir, Tabatabaei, M., Shirazi, M.J.A., Abdullah, M.S., Matsuura, T. & Ismail, A.F. (2013b) Characterization of electrospun polystyrene membrane for treatment of biodiesel's water-washing effluent using atomic force microscopy. *Desalination*, 329, 1–8.

Shirazi, M.M.A., Kargari, A., Bastani, D. & Fatehi, L. (2014) Production of drinking water from seawater using membrane distillation (MD) alternative: direct contact MD and sweeping gas MD approaches. *Desalination and Water Treatment*, 52, 2372–2381.

Shirazi, M.M.A., Kargari, A. & Tabatabaei, M. (2015) Sweeping gas membrane distillation (SGMD) as an alternative for integration of bioethanol processing: study on a commercial membrane and operating parameters. *Chemical Engineering Communications*, 202, 457–466.

Shuit, S.H., Ong, Y.T., Lee, K.T., Subhash, B. & Tan, S.H. (2012) Membrane technology as a promising alternative in biodiesel production: a review. *Biotechnology Advances*, 30, 1364–1380.

Takht Ravanchi, M., Kaghazchi, T. & Kargari, A. (2009) Application of membrane separation processes in petrochemical industry: a review. *Desalination*, 235, 199–244.

Talebian-Kiakalaieh, A., Amin, N.A.S. & Mazaheri, H. (2013) A review on novel processes of biodiesel production from waste cooking oil. *Applied Energy*, 104, 683–710.

Uemiya, S. (2004) Brief review of steam reforming using a metal membrane reactor. *Topics in Catalysis*, 29, 79–84.

Wang, Y., Wang, X., Liu, Y., Ou, S., Tan, Y. & Tang, S. (2009). Refining of biodiesel by ceramic membrane separation. *Fuel Processing Technology*, 90, 422–427.

CHAPTER 13

Mathematical modeling of nanofiltration-based deionization processes in aqueous media

Mohammad M. Zerafat, Mojtaba Shariaty-Niassar, S. Jalaledin Hashemi, Samad Sabbaghi & Azadeh Ghaee

13.1 INTRODUCTION

Nanofiltration (NF) membrane properties stand between those of reverse osmosis (RO) and ultra-filtration (UF) giving them special advantages in separation processes and proposing them as a suitable alternative for RO in many applications due to lower energy requirement and extended flux rates (Szymczyk *et al.*, 2006). Ionic rejection by NF is mainly governed by: (i) transport through concentration polarization (CP) layer, (ii) partitioning phenomena at membrane interfaces and, (iii) solute transport through membrane pores (Deon *et al.*, 2007).

Several attempts have been made to model NF separation processes. The TMS (Torell, Meyer and Sievers) model is one of the first approaches in this regard (Garcia-Aleman and Dickson, 2004). Space charge (SC) modeling was proposed by Gross & Osterle (1969) as a modification to TMS. Although, successful NF predictions by SC is reported (Wang *et al.*, 1995), the application is limited due to complex calculation requirements especially in mixed electrolyte solutions.

Today, the most prevalent NF models are derived from SC by assuming radial homogeneity of ionic concentration and potential across the pores, which is valid in the case of small surface charge densities and sufficiently narrow pores, maintained under most NF conditions. The most widely used pore transport models have also been built upon the extended Nernst-Planck (ENP) equation. Tsuru *et al.* (1991) were the first to propose an ENP-based model for NF. Bowen and Mukhtar (1996) suggested a hybrid model based on ENP with a Donnan condition at membrane-electrolyte interfaces and the hindered nature of transport through the membranes. Bowen *et al.* (1997) modified the hybrid model proposing "Donnan-steric-pore model" (DSPM). Several modifications have been suggested for DSPM since, taking into account the effects of pore size distribution on the rejection of uncharged solutes and NaCl for hypothetical NF membranes (Bowen and Welfoot, 2002) incorporating dielectric constant variations between bulk and pore to calculate ionic distribution between bulk and pore solutions. Different approaches are proposed to include dielectric exclusion (DE) in the modeling, namely DSPM&DE (Bandini and Vezzani, 2003) and SEDE (steric-electric-dielectric exclusion) (Szymczyk and Fievet, 2005). Finite difference linearization of pore concentration gradient is also proposed to simplify the solution of a 3-parameter model (ε, X_d, and ε_p) for electrolyte rejection by removing the requirement of numerical integration for ENP leading to the linearized transport model (LTM) (Bowen *et al.*, 2002).

Further development of existing models entails the inclusion of more complex or just neglected phenomena governing the separation behavior in order to improve the physical relevance of process description. As an example, SEDE is modified considering the variations of volumetric charge density by including a charge isotherm in the model (SEDE-VCh) (Silva *et al.*, 2011).

In the present chapter, the modeling of the NF separation process is considered by a modified version of ENP (MENP) based on the inclusion of activity coefficient variations for solute transport through the membrane active layer (Zerafat *et al.*, 2013) and advection-diffusion (AD) for CP which has been previously considered by linearized assumptions (Bhattacharjee *et al.*, 1999). Activity coefficient variations can have a significant influence on model predictions especially in

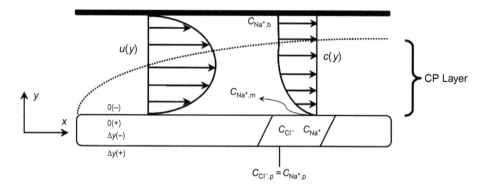

Figure 13.1.   The schematic of a cross-flow membrane filtration channel.

concentrated electrolyte solutions. Partitioning phenomena (steric, Donnan and dielectric exclusion) are also considered at membrane external interfaces. In the case of pure salt solutions (NaCl and Na_2SO_4), the competition of ionic species on the rejection of chloride ions will be investigated based on simulation results.

13.2   THE MATHEMATICAL MODEL

Prediction of permeate flux and rejection of ionic solutes from a multi-component mixture involves a simultaneous solution of 3 transport models: (i) mass transport in CP, (ii) equilibrium partitioning at membrane interfaces and, (iii) pore ionic transport. The coupled ionic transport processes in a cross-flow membrane filtration channel are depicted in Figure 13.1. The CP model is applied to the filtration channel, describing local variations of permeate flux and surface concentration. The pore transport model provides a detailed analysis of ionic transport across the membrane. Interaction among ions in a multi-component mixture is considered by examining activity coefficient variations due to the reduction in ionic strength, negligible in dilute solutions. The pore transport model is then coupled to CP through membrane surface boundary conditions. Partitioning phenomena (steric, Donnan and dielectric exclusion) are also considered at membrane external interfaces.

13.2.1   *CP in multi-ionic systems*

CP as a type of reversible fouling mechanism is a serious challenge restricting both the possible membrane application fields and life-time at the same time. Efforts have been performed for the inclusion of CP through a coupled model applied for multi-component electrolyte solutions in cross-flow NF (Bhattacharjee et al., 1999; 2001). This model predicts local variations of ionic concentrations, flux and ionic rejections along a rectangular cross-flow filtration channel by a coupled solution of the convective-diffusion equation for transport of ions in the boundary layer adjacent to the membrane surface and ENP for ionic migration through membrane pores. This phenomenon is a common field between all pressure-driven membrane separation processes (MF, UF, RO and NF), which can be modeled by simple film theory (FT):

$$J_v = \frac{D_i}{\delta} \ln\left(\frac{C_{i,m} - C_{i,p}}{C_{i,b} - C_{i,p}}\right) \tag{13.1}$$

where δ is the boundary layer thickness, D_i is the diffusion coefficient, $C_{i,m}$ is the concentration of the ith ion at the membrane surface, $C_{i,b}$ is the concentration of the ith ion in the bulk solution, $C_{i,p}$ the permeate concentration and J_v the permeate flux. More rigorous descriptions of CP are also applied among which convective-diffusion (CD) formulation has been given special consideration (Kim and Hoek, 2005). For a multi-component ionic system in a cross-flow unit, the general form of CD for each component in the polarization layer is:

$$u(y)\frac{\partial C_i}{\partial x} + J_v \frac{\partial C_i}{\partial y} = D_i \frac{\partial^2 C_i}{\partial y^2} \qquad (13.2)$$

where u is the axial velocity, C_i is the concentration of the ith ion and D_i the diffusion coefficient (Fig. 13.1). This equation can be solved through an implicit finite difference discretization approach given as:

$$\frac{u_{i,n}}{\Delta x} C_{i,n} = \left(\frac{u_{i,n}}{\Delta x} + \frac{2D_i}{\Delta y^2} \right) C_{i,n+1} + \left(\frac{J_{v(i,n)}}{2\Delta y} - \frac{D_i}{\Delta y^2} \right) C_{i+1,n+1}$$

$$- \left(\frac{J_{v(i,n)}}{2\Delta y} + \frac{D_i}{\Delta y^2} \right) C_{i-1,n+1} \qquad (13.3)$$

Linear or free flow axial velocity profiles may be assumed for NF of dilute electrolyte solutions in case that boundary layer is quite thin compared with channel height (Bhattacharjee *et al.*, 2001), while more generalized equations can be introduced based on simplified versions of Navier-Stokes equation in narrow channels:

$$J_v \frac{\partial u}{\partial y} = \frac{\partial P}{\partial x} + \frac{\mu}{\rho} \frac{\partial^2 u}{\partial y^2} \qquad (13.4)$$

The boundary conditions at the channel inlet and bulk solution are, respectively:

$$C_i|_{x=0} = C_{i,b} \qquad (13.5a)$$

$$C_i|_{y=\delta} = C_{i,b} \qquad (13.5b)$$

where, $C_{i,b}$ denotes the bulk ionic concentration. A flux continuity is also maintained at the membrane surface:

$$\frac{\partial C_i}{\partial y}\bigg|_{y=0} = -\frac{J_v}{D_i}(C_{i,m} - C_{i,p}) \qquad (13.6)$$

13.2.2   *Equilibrium partitioning*

Non-steric mechanisms involved in the exclusion at membrane interfaces are less understood and thus described through the assumption of equilibrium partitioning at the pore entrance or exit by using partition coefficients that relate intra-pore to interfacial bulk concentrations. Partitioning phenomena consist of steric, Donnan and dielectric exclusion mechanisms:

$$\frac{C_i(0^+)}{C_i(0^-)} = \frac{\gamma_i(0^-)}{\gamma_i(0^+)} \varphi_i \exp\left(-\frac{\Delta W_i}{k_B T} \right) \exp\left(-\frac{z_i F}{RT} \Delta \psi_D \right) \qquad (13.7)$$

Activity coefficients are considered in some studies (Geraldes and Alves, 2008; Szymczyk and Fievet, 2005) while others have neglected their effect based on the assumption of dilute solutions. The energy barrier due to dielectric constant reduction (ΔW_i) impedes ionic transport, due to the higher solvation energy in a medium with reduced dielectric constant (Parsegian, 1969). Ionic solvation energy is given by the Born model (Born, 1920):

$$\Delta W_i = \frac{z_i^2 e^2}{8\pi\varepsilon_0 a_i} \left(\frac{1}{\varepsilon_p} - \frac{1}{\varepsilon_b} \right) \qquad (13.8)$$

where a_i is the Stokes radius of the solute, ε_0 is the permittivity of free space, ε_p and ε_b are dielectric constants in the pore and bulk solution, respectively. Inside narrow membrane pores, the induced energy due to the membrane dielectric constant (ε_m) is said to be influential; thus, Born model is used in the modified form in some studies (Hagmeyer and Gimbel, 1998):

$$\Delta W_i = \frac{z_i^2 e_0^2}{8\pi\varepsilon_0 a_i}\left(\frac{1}{\varepsilon_p} + \frac{0.393}{\frac{r_p}{r_i}\varepsilon_m}\left(1 - \frac{\varepsilon_m}{\varepsilon_p}\right)^2 - \frac{1}{\varepsilon_b}\right) \tag{13.9}$$

Pore dielectric constant (ε_p) is also evaluated by assuming a single layer of solvent on pore surfaces with a different dielectric constant and an internal cylinder with bulk dielectric properties, which results in the following relation upon averaging on the pore radius (Bowen *et al.*, 2002):

$$\varepsilon_p = 80 - 2(80 - \varepsilon^*)\left(\frac{d}{r_p}\right) + (80 - \varepsilon^*)\left(\frac{d}{r_p}\right)^2 \tag{13.10}$$

where d is the diameter of water molecule (0.28 nm) and ε^* is the dielectric constant of the single solvent layer.

For a single salt solution, pore concentrations can be calculated using the following relations (Kumar *et al.*, 2013):

$$c_1(0^+) = \frac{-X_d + \sqrt{X_d^2 + 4\varphi_1'\varphi_2' c_1^{(0^-)} c_2^{(0^-)}}}{2} \tag{13.11a}$$

$$c_2^{(0^+)} = \frac{+X_d + \sqrt{X_d^2 + 4\varphi_1'\varphi_2' c_1^{(0^-)} c_2^{(0^-)}}}{2} \tag{13.11b}$$

where, X_d denotes the membrane charge density and φ the combined portioning coefficient.

Figure 13.2 shows the variations of average pore dielectric constant and the resulting solvation energy barrier for three different ions (Na^+, Cl^- and SO_4^{2-}). The results show that, at larger pore diameters the single solvent layer loses its significance and as a result the solvation energy barrier falls dramatically especially at $d > 1$ nm.

13.2.3   Ionic transport across membrane pores

The extended Nernst-Planck (ENP) is the governing equation for steady-state flux J_i of charged species through membrane pores combining individual contributions of diffusion, convection and electrical mobility and electrolyte activity variations as follows:

$$J_i = K_{i,C} C_i J_v - [D_{i,p}]\left[\mathrm{grad}(C_i) + \frac{z_i C_i F}{RT}\mathrm{grad}(\psi) + C_i\,\mathrm{grad}\,(\ln\gamma_i) + \frac{C_i v_i}{RT}\mathrm{grad}\,(P)\right] \tag{13.12}$$

where, $K_{i,C}$ is the convection hindrance factor.

Generally, pressure differential is ignored but the activity coefficient can just be ignored in dilute solutions. The activity coefficient is considered as a function of ionic strength in electrolyte solutions and several relations have been suggested to describe their behavior among which Debye-Hückel theory provides the simplest form as follows:

$$\ln\gamma_i = -A|z_+ z_-|\sqrt{I} \tag{13.13}$$

where, A is a constant equal to $\sim 0.5(M^{-1/2})$.

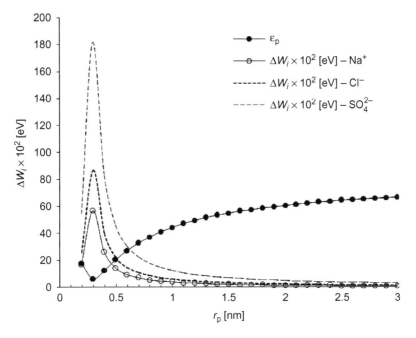

Figure 13.2.   Pore dielectric constant (ε_p) variations and the resulting solvation energy barrier (ΔW_i) by varying the pore radius in the NF and RO range.

For the special case of a single salt solution (e.g. NaCl), the electrical potential gradient along the pore length can be derived as:

$$\frac{\partial \psi_m}{\partial y} = \frac{\alpha(I)\dfrac{J_v}{D_{1,p}}(K_{1,c}C_1 - C_{1,P}) + \beta(I)\dfrac{J_v}{D_{2,p}}(K_{2,c}C_2 - C_{2,P})}{\gamma(I)\dfrac{F}{RT}I} \qquad (13.14)$$

where, the derivation procedure and α, β and γ are given in a previous publication (Zerafat *et al.*, 2013). The following boundary conditions can be used to solve Equation (13.14):

$$C_i|_{y=0} = C_{i,m} \qquad (13.15a)$$

$$C_i|_{y=-\Delta y} = C_{i,p} \qquad (13.15b)$$

where Δy is the membrane pore length (also membrane thickness), and $C_{i,0}$ and $C_{i,\Delta y}$ are the ionic pore concentrations at the pore entrance and exit, respectively.

13.3   RESULTS AND DISCUSSION

The equations mentioned in the mathematical model section are employed in this study to model the transport behavior of NaCl and Na$_2$SO$_4$ single salt electrolyte solutions constituting a two-component ionic mixture, which involve a 1-D model for ionic transport within the membrane.

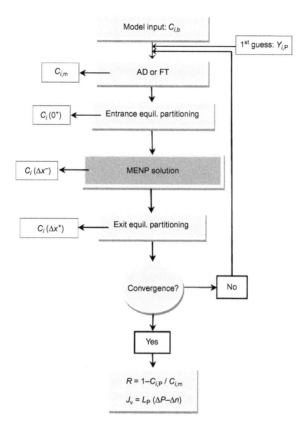

Figure 13.3.   The full procedure for implementing the iterative procedure of the coupled model.

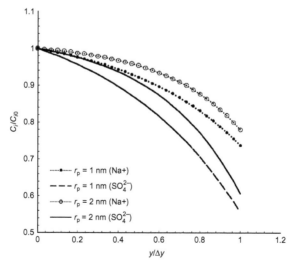

Figure 13.4.   Comparison of normalized ionic concentration regarding Na_2SO_4 for two different pore sizes (1 and 2 nm). Ionic radius for Na^+ and SO_4^{2-} are assumed as 0.184 and 0.231 nm, respectively. Ionic diffusivity for Na^+ and SO_4^{2-} is $1.33 \times 10^{-9}\ m^2\ s^{-1}$ and $1.06 \times 10^{-9}\ m^2\ s^{-1}$, respectively. Pressure differential is 1.0 MPa.

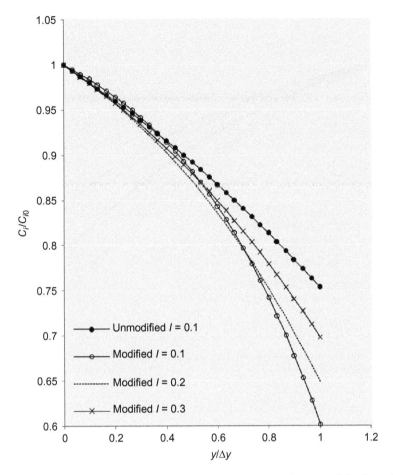

Figure 13.5. Normalized ionic concentration as a function of distance from pore inlet for various ionic strengths (Na_2SO_4).

This is performed by solving the ODE (13.12) for both components to obtain their concentration in the pores, while maintaining the electro-neutrality condition. ODE (13.12) is solved numerically using Adams-Bashforth-Molton (ABM) technique, which entails iterating on the rejection appearing on both sides. The ionic fluxes and permeate concentrations corresponding to a given feed side surface concentration and permeate flux are predicted by solving the MENP equation. The full procedure for implementing the iterative procedure of the coupled model is given in Figure 13.3.

Figure 13.4 illustrates the normalized ionic concentration inside the pores as a function of distance from the pore inlet predicted by ENP before any modification, performed for a pure Na_2SO_4 solution. It can be seen that the rejection is enhanced by reducing the membrane pore size from 2 to 1 nm due to the fact that at larger pore sizes the effect of pure screening phenomenon is decreased.

Figures 13.5 and 13.6 are a comparison of the ENP equation solved before modification and the modified results for different ionic strengths (Na_2SO_4 and NaCl, respectively). The results show that at higher ionic strengths, reduced rejections is experienced which is the case for NF for concentrated electrolyte solutions.

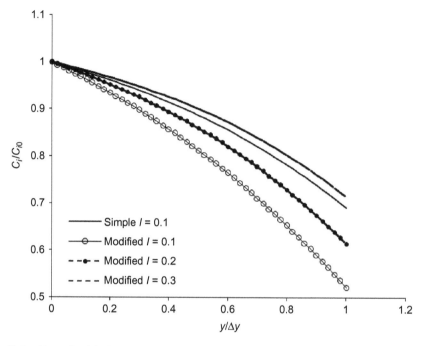

Figure 13.6.   Normalized ionic concentration as a function of distance from pore inlet for various ionic strengths (NaCl).

13.3.1   *Effect of X_d on rejection*

No physical model is yet proposed for evaluating the effective membrane charge density (X_d); thus this is still a fitting parameter to describe the observed rejection by varying the solution concentration through mathematical models to estimate X_d by using the rejection experimental data. However, the values predicted by present models are very large and physically unrealistic (Schaep and Vandecasteele, 2001). This can be due to other neglected separation mechanisms for the sake of simplification during the modeling procedure.

Figure 13.7 shows the variations of membrane rejection during transport from pores for various charge density values based on the model before modifications were performed. The results show the increasing trend observed by increasing the charge density.

Figure 13.8 shows the variations of membrane rejection during transport inside the pores for various charge density values based on the model after modifications performed (MENP). The results show the increasing trend observed by increasing the charge density and also higher rejection values at similar X_ds. The variations of rejection for various charge density values are compared for the simple and modified models in Figure 13.9. The results show a linear increasing trend by increasing membrane charge density and the modified model predicts lower X_d values at the same rejection which can be considered as an improvement when fitting the rejection experimental data to evaluate the membrane charge density.

Membrane charge density can also affect the membrane observed rejection through partitioning phenomena. Figure 13.10 shows that at each feed concentration, the corresponding values for the partitioned intra-pore concentrations have opposing behaviors. In the case that the membrane charge density owns a negative value, the negative charge ion (here, Cl^-) is decreased while the positive charge ion (Na^+) is enhanced. Figure 13.11 illustrates the effect of X_d on partitioning at various pore diameters within the NF range.

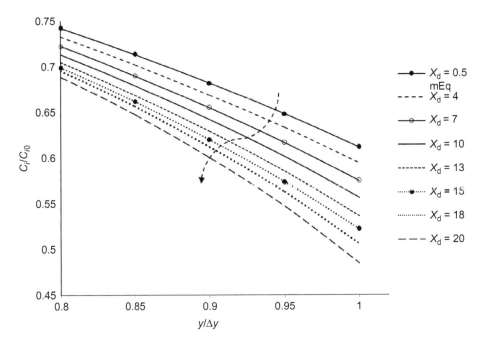

Figure 13.7.   Enhancement of membrane rejection by increasing the effective charge density values (simple model).

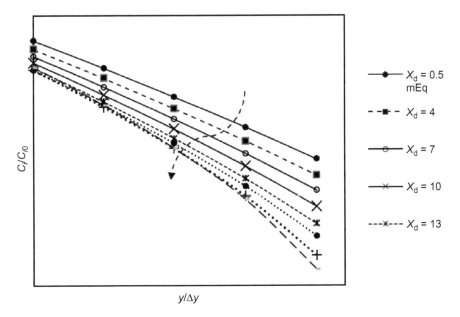

Figure 13.8.   Enhancement of membrane rejection by increasing the effective charge density values (MENP).

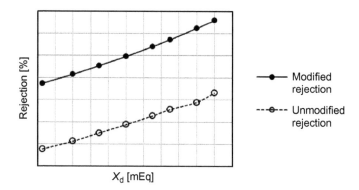

Figure 13.9.   The linearly increasing trend of membrane rejection by increasing the effective charge density values being compared for the simple and the modified models.

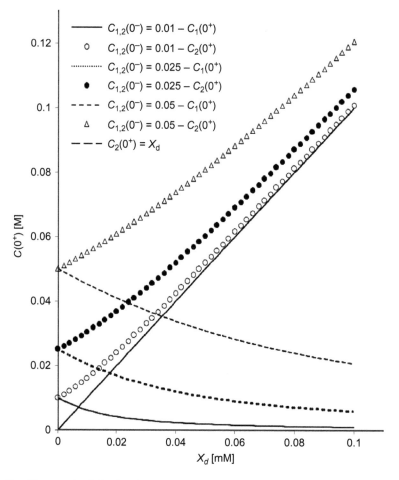

Figure 13.10.   The opposing behavior of negative and positive charge ions by increasing the effective charge density at various feed concentrations ($r_p = 2$ nm, $C_{1,2}(0^-)$ is the ions concentration at the pore entrance).

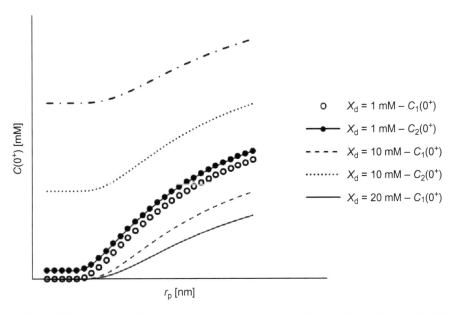

Figure 13.11. Effect of on partitioning at various pore diameters within the NF & RO range ($C_1(0^-) = 25$ mM).

13.4 CONCLUSION

The effect of activity coefficient variations is investigated on the retention behavior of NF membranes. The results show that the modified version of ENP shows reduced values as compared to the basic ENP equation. This results in lower X_d values predicted when being fitted to experimental rejection data.

REFERENCES

Bandini, S. & Vezzani, D. (2003) Nanofiltration modeling: the role of dielectric exclusion in membrane characterization. *Chemical Engineering Science*, 58, 3303–3326.
Bhattacharjee, S., Kim, A.S. & Elimelech, M. (1999) Concentration polarization of interacting solute particles in cross-flow membrane filtration. *Journal of Colloid and Interface Science*, 212, 81–99.
Bhattacharjee, S., Chen, J.C. & Elimelech, M. (2001) Coupled model of concentration polarization and pore transport in crossflow nanofiltration. *AIChE Journal*, 47 (12), 2733–2745.
Born, M. (1920) Volumen und Hydratationswärme der Ionen. *Zeitschrift für Physikalische Chemie*, 1 (1), 45–48.
Bowen, W.R. & Mukhtar, H. (1996) Characterization and prediction of separation performance of nanofiltration membranes. *Journal of Membrane Science*, 112, 263–274.
Bowen, W.R. & Welfoot, J.S. (2002) Modelling the performance of membrane nanofiltration-critical assessment and model development. *Chemical Engineering Science*, 57, 1121–1137.
Bowen, W.R., Mohammad, A.V. & Hilal, N. (1997) Characterization of nanofiltration membranes for predictive purposes-use of salts, uncharged solutes and atomic force microscopy. *Journal of Membrane Science*, 126 (1), 91–105.
Bowen, W.R., Welfoot, J.S. & Williams, P.M. (2002) Linearized transport model for nanofiltration: development and assessment. *AIChE Journal*, 48, 760–773.
Deon, S., Dutournie, P. & Bourseau, P. (2007) Transfer of monovalent salts through nanofiltration membranes: a model combining transport through pores and the polarization layer. *Industrial & Engineering Chemistry Research*, 46, 6752–6761.

Garcia-Aleman, J. & Dickson, J.M. (2004) Mathematical modeling of nanofiltration membranes with mixed electrolyte solutions. *Journal of Membrane Science*, 235, 1–13.

Geraldes, V. & Alves, A.M.B. (2008) Computer program for simulation of mass transport in nanofiltration membranes. *Journal of Membrane Science*, 321, 172–182.

Gross, R.J. & Osterle, J.F. (1968) Membrane transport characteristics of ultraflne capillaries. *Journal Chemical Physics*, 49, 228–234.

Hagmeyer, G. & Gimbel, R. (1998) Modelling the salt rejection of nanofiltration membranes for ternary ion mixtures and for single salts at different pH values. *Desalination*, 117, 247–256.

Kim, S. & Hoek, E.M.V. (2005) Modeling concentration polarization in reverse osmosis processes. *Desalination*, 186 (1), 111–128.

Kumar, V.S., Hariharan, K.S., Mayya, K.S. & Han, S. (2013) Volume averaged reduced order Donnan steric pore model for nanofiltration membranes. *Desalination*, 322, 21–28.

Parsegian, A. (1969) Energy of an ion crossing a low dielectric membrane: solutions to four relevant electrostatic problems. *Nature*, 22, 844–846.

Schaep, J. & Vandecasteele, C. (2001) Evaluating the charge of nanofiltration membranes. *Journal of Membrane Science*, 188, 129–36.

Silva, V., Geraldes, V., Brites Alves, A.M., Palacio, L., Prádanos, P. & Hernández, A. (2011) Multi-ionic nanofiltration of highly concentrated salt mixtures in the seawater range. *Desalination*, 277, 29–39.

Szymczyk, A. & Fievet, P. (2005) Investigating transport properties of nanofiltration membranes by means of a steric, electric and dielectric exclusion model. *Journal of Membrane Science*, 252, 77–88.

Szymczyk, A., Sbaï, M. & Fievet, P. (2006) Transport properties and electrokinetic characterization of an amphoteric nanofilter. *Langmuir*, 22, 3910–3919.

Tsuru, T., Nakao, S. & Kimura, S. (1991) Calculation of ion rejection by extended Nernst-Planck equation with charged reverse osmosis membranes for single and mixed electrolyte solutions. *Journal of Chemical Engineering of Japan*, 24, 511–517.

Wang, X., Tsuru, T., Nakao, S. & Kimura, S. (1995) Electrolyte transport through nanofiltration membranes by the space-charge model and the comparison with Teorell-Meyer-Sievers model. *Journal of Membrane Science*, 103 (1–2), 117–133.

Zerafat, M.M., Shariati-Niassar, M., Hashemi, S.J., Sabbaghi, S., Ismail, A.F. & Matsuura, T. (2013) Mathematical modeling of nanofiltration for concentrated electrolyte solutions. *Desalination*, 320, 17–23.

CHAPTER 14

Simulation of heat and mass transfer in membrane distillation (MD) processes: the effect of membrane pore space description

Abdussalam O. Imdakm, Elmahdi M. Abousetta & Khalid Mahroug

14.1 INTRODUCTION

Membrane distillation (MD) processes, in general, are thermal separation processes in which water vapor is generated by the evaporation of aqueous solution in direct contact with one side of the employed hydrophobic porous membrane (feed side), to be transported through membrane pores, and finally it will condense at the other side of the membrane (permeate side). In membrane distillation literature, various models of MD processes and physical setups have been developed in order to improve MD processes output, depending on the convenience of the MD separation process of interest, e.g., direct contact membrane distillation process (DCMD), vacuum membrane distillation (VMD), air gap membrane distillation (AGMD), and sweeping gas membrane distillation (SGMD), each of these has its own advantage and disadvantage, and its useful applications in many areas of industrial interest.

Among all MD processes published in the literature, direct contact membrane distillation (DCMD) and vacuum membrane distillation (VMD) processes have shown the most frequent interest. This is because of their simplicity in terms of construction and operating conditions, lower operating temperature and pressure, lower energy consumption, and their relatively high permeation rate. These MD configurations are commonly applied for processes in which water vapor is the main flux component, as in distillation, for removal of volatile components (VOC'S), and for waste water treatment (Calabro et al., 1991; 1994; Wu et al., Xu, 1991). In these processes, heated feed and cold permeate streams are in direct contact with the applied membrane surfaces (feed and permeate), and the separation process is carried out at atmospheric pressure as in DCMD or under moderate vacuum as in VMD processes, utilizing hydrophobic porous membrane. They are characterized by a simultaneous heat and mass transfer across the membrane pores and solid phase, resulting from the difference in temperature (vapor pressure) and/or composition in the fluid layers adjacent to membrane surfaces (feed and permeate).

When the aqueous feed solution is very low in solute concentration or pure water, the solution vapor pressure at liquid-vapor interface is generally governed by fluid temperatures adjacent to the membrane surface feed and permeate, $t_{s,f}$ and $t_{s,p}$, respectively as in DCMD, or $T_{s,f}$ and imposed vacuum condition at the permeate side as in VMD. These bulk-membrane interfacial surface temperatures differ significantly from their corresponding measured bulk temperature, $t_{b,f}$ and $t_{b,p}$, their values cannot be measured directly and they are influenced significantly by MD process operating conditions (process dynamics), membrane pore space topology (inter-connectivity), and employed hydrophobic membrane physical properties. This phenomenon is known in the literature as the temperature polarization, and the temperature polarization coefficient, τ, is defined in the literature (Lawson and Lloyd, 1996; 1997; Phattaranawik and Jiraratananon, 2001; Phattaranawik et al., 2003a; Schofield et al., 1987; 1990a; 1990b) as: $\tau = (t_{s,f} - t_{s,p})/(t_{b,f} - t_{b,p})$ for DCMD process, where $t_{b,f}$ is the feed side bulk temperature, $t_{s,f}$, is the feed side membrane surface temperature, $t_{b,p}$ and $t_{s,p}$ are the equivalent permeate side temperatures, and for the VMD process it is defined, in many cases as: $\tau = (T_{s,f} - T_v)/(T_{b,f} - T_v)$, where T_v is the temperature set on the vacuum side. The concept of the temperature polarization factor is used, in many cases, as a tool for evaluating the effects of input parameters on maximizing the mass flux. For example, when the value of τ approaches zero for VMD process, this means $T_{s,f} \to T_v$, and for DCMD process this means $t_{s,f} \to t_{s,p}$, and thus the MD processes in both cases are limited by mass transfer through the liquid phase boundary layer and the resistance in the membrane is negligible. However, when $\tau \to 1$, $T_{s,f} \to T_{b,f}$ in VMD and $t_{s,f} \to t_{b,f}$ and $t_{s,p} \to t_{b,p}$ in DCMD, this implies that in these MD processes, the resistance in the membrane is dominant and the liquid phase boundary layer resistance is negligible.

Modeling of MD processes to predict process vapor flux and related transport properties of interest have received a great deal of attention from many investigators; e.g., Schofield *et al.* (1987; 1990a; 1990b), Phattaranawik *et al.* (2001; 2003a; 2003b), and Lawson and Lloyd (1996, 1997). These models, are based on the dusty-gas model (Mason, and Malinauskas 1983; Mason *et al.* 1967), which is a general model developed to describe mass transport in porous media, assuming a geometrical structure of the porous medium in the form of a collection of an assembly of spherical molecules. The vapor flux expressions applied in this model were described by one or a combination of the following gas transport mechanisms, founded on the classical kinetic theory of gases for single cylindrical pore (Lawson and Lloyd 1997; Mason, and Malinauskas 1983; Mason *et al.* 1967): Knudsen diffusion, viscous flow (Poiseuille), and surface and molecular diffusion, depending on applied pore level operating conditions (pressure and temperature), average mean free path of transporting molecules, λ, and pore size, r_{pore}. In the simplest general form the vapor flux, J, can be expressed as: $J = C(P_{s,f} - P_{s,p})$, where C is the vapor flux coefficient of the applied membrane. Note molecular diffusion through membrane pores, and molecular transport as a result of surface diffusion transport mechanism, are rarely taken into consideration in MD processes.

Recently, several models were developed to predict MD vapor flux and related transport properties, assuming a geometrical structure of the membrane pore space (Lagana *et al.*, 2000; Phattaranawik *et al.*, 2003b; Present, 1958). In these models, vapor flux is described by the same vapor flux transport mechanisms applied in the dusty-gas model mentioned above, and the membrane pore space is either described by a single or a bundle of capillary tubes, known as the capillary tube model (CT). In these models, the membrane pores are assumed to be infinitely long cylindrical pores of experimentally reported effective pore radii. In many cases, the membrane pore size distribution was assumed to be uniform. This assumption is founded on the simplicity of vapor flux calculation, arguing that in most MD processes of practical interest, the membrane pore size distribution is very narrow and is equivalent to a uniform pore size distribution. Only a few attempts have been made to relate membrane pore space description (pore size distribution, and inter-connectivity), to the precision of MD's vapor flux calculation.

In previous studies (Imdakm and Matsuura, 2004; 2005; Imdakm *et al.*, 2007), we have developed a Monte Carlo simulation model to describe DCMD and VMD processes for predicting MD vapor flux. In these MC simulation models, the membrane pore space is described by a three-dimensional network (NW) model of inter-connected bonds (pores) and sites (nodes). To illustrate the influence of membrane pore space description and how it may affect the resultant vapor flux and related transport properties, a comparison is made in this study between the NW and CT models considering both DCMD and VMD processes. In both models, NW and CT, the membrane pores are simulated by cylindrical tubes of distributed effective sizes, based on experimentally reported membrane pore size distribution, or they may be assumed to be of uniform radii. Vapor flux through membrane pores is assumed to be governed by gas transport mechanisms founded on the kinetic theory of gases for a single cylindrical pore (Present, 1958).

14.2   THEORY

14.2.1   *Pore level vapor flux transport mechanism(s)*

As mentioned above, the calculation method applied in this study is a Monte Carlo simulation method (Imdakm and Matsuura, 2004; 2005; Imdakm and Sahimi, 1991; Imdakm *et al.*, 2007;

Sahimi and Imdakm, 1988); this means the calculation is carried out by generating several independent realizations of membrane pore space. Thus, resulting MD vapor flux and any other transport properties of interest are the average value of the results obtained over all these realizations. This averaging process is applied on the overall results reported in this study regardless of membrane pore space description, NW or CT model.

The vapor flux transport mechanism through membrane pores depends upon the mean free path, λ, which is defined as the average distance that a molecule travels between two successive collisions (Present, 1958):

$$\lambda = \frac{K_B T}{P\sqrt{2\pi\sigma^2}} \qquad (14.1)$$

where K_B is the Boltzmann constant, T the absolute temperature [K], P is the applied operating pressure [Pa], and σ is the collision diameter [m]. The value of λ and how it is interrelated to the membrane pore sizes plays an essential role in determining the type of transport mechanism(s) governing vapour flux through membrane pores. In both models, temperature (T) and pressure (P) in Equation (14.1) are replaced by the arithmetic average pore temperature (T_{pore}) and the arithmetic average pore pressure (P_{pore}). The vapour transport mechanism through each pore (see Figs. 14.1a–c and 14.2) is then dictated by the ratio of mean free path, λ, to pore radius. For example, if the ratio of λ to the pore radius, r_p, is less than unity, then the transport mechanism through this pore is the viscous (Poiseuille) type of flow. In this case, the flow rate through the pore, N_{vis}, is given by (Imdakm and Matsuura, 2004; 2005; Imdakm and Sahimi, 1991; Imdakm et al., 2007; Present, 1958; Sahimi and Imdakm, 1988):

$$N_{vis} = \frac{P_{pore}\pi r_p^4}{8RT_{pore}\mu\delta} M_w \Delta P \qquad (14.2)$$

where ΔP is the pressure difference across the membrane pore, P_{pore} is the arithmetic average pore pressure, T_{pore} the arithmetic average pore temperature, and r_p is the pore radius. However, when the transporting species has a mean free path, λ, larger than the membrane pore radius, r_p, then the mass transport mechanism through this pore is Knudsen diffusion type of flow. Accordingly, the flow rate through the pore, N_{Kn}, is given as:

$$N_{Kn} = -\frac{2}{3\delta}\left(\frac{8}{\pi RT_{pore}M_w}\right)^{1/2} r_p \Delta P \qquad (14.3)$$

where M_w is the molecular weight of the transporting species. It must be stated that other transport mechanisms, such as surface diffusion and ordinary diffusion, are not taken into consideration as stated in previous studies (Imdakm and Matsuura, 2004; 2005; Imdakm and Sahimi, 1991; Imdakm et al., 2007; Sahimi and Imdakm, 1988). In most MD processes, the ordinary molecular diffusion resistance is neglected because it is proportional to the partial pressure of air in the membrane pores and in all MD processes, only traces of air are actually present within the membrane pores, which makes its contribution to the transporting process very small if not nil compared to other transport mechanisms. Moreover, the surface diffusion is also neglected due to the fact that the surface diffusion area of the membrane matrix is very small compared to the pore area and for hydrophobic membranes as those used in MD, the "*affinity*" between water and the membrane material is very weak, which may allow us to neglect the contribution of transport by surface diffusion, especially for porous membranes with large pore sizes and high porosities (Lawson and Lloyd 1996; 1997).

With the assumption that all MD processes operate on, the principle of vapor-liquid equilibrium (VLE), partial vapor pressure can be calculated using VLE data. Therefore, the saturation vapor pressure at the membrane feed-side can be estimated by the Antoine equation as a function of the interfacial surface temperature; see Figures 14.1c and 14.2 for a detailed description of vapor flux at the membrane feed-side (entry pores). These entry pores, particularly when the membrane pore space is described by a network model, plays a very essential role in vapor flux calculations,

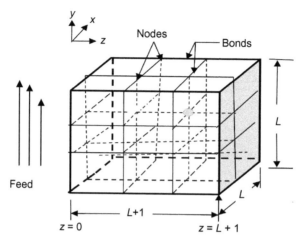

Figure 14.1a.   A schematic of three dimensional network (NW) model of interconnected bonds (pores) and nodes (sites).

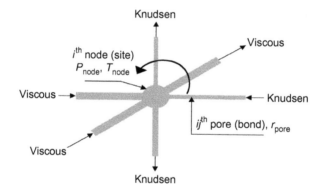

Figure 14.1b.   Possible vapor flux transport mechanism(s) through the ith node and connected bonds in network (NW) model.

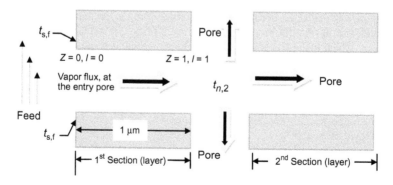

Figure 14.1c.   A schematic of MD process vapor flux through membrane pores (entry pores), when membrane pore space is described by three dimensional network model (NW).

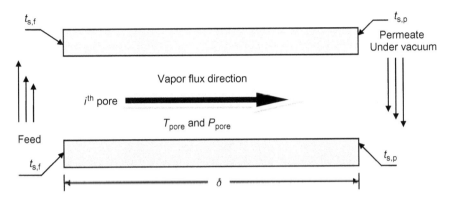

Figure 14.2.   A schematic of vapour flux of MD process through membrane pores when membrane pore space is described by capillary tube (CT) model.

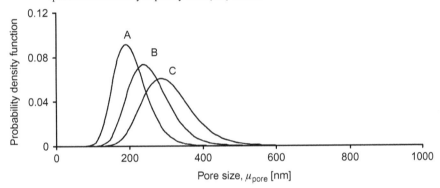

Figure 14.3.   Probability density function curves for membranes A ($\mu_{pore} = 200\,nm$), B ($\mu_{pore} = 250\,nm$) and C ($\mu_{pore} = 300\,nm$), with a constant $\sigma_{pore} = 1.25$ (Wang *et al.*, 2001).

for further details see Imdakm and Matsuura (2004; 2005) and Imdakm *et al.* (2007) and references therein. Figure 14.3 shows the geometrical pore sizes, standard deviations, and the corresponding probability density curves for three membranes, A, B, and C, which are employed for both NW and CT models. These pore size distributions are actually based on experimentally reported pore size distribution published in the literature (Wang *et al.*, 2001).

14.2.2   *Vapor flux calculation through membrane pore space*

As mentioned above, when the membrane pore space is described by a three-dimensional network model, the entire calculation procedure of mass and energy flux in any MD process, is an iterative procedure, thus, membrane-bulk interfacial surface temperatures and MD processes energy and vapour flux and other transport properties of interest are also calculated iteratively using proper iterative methods such as the method of successive substitutions (Imdakm and Matsuura, 2004; 2005; Imdakm *et al.*, 2007; Rice and Do, 1995) and references therein. When the membrane pore space is described by CT model, this iterative method is also applied. In this case, an initial guess of interfacial membrane surface temperatures, $t_{s,f}$, and $t_{s,p}$ is given to calculate membrane boundaries vapor pressure, $P_{s,f}$ and $P_{s,p}$ using the Antoine equation. Note that the applied permeate vapour pressure is lower than the saturation pressure ($P_{sat.}$). Since in this CT model, the membrane pore topology (connectivity) is totally ignored, the temperature and pressure distribution across the membrane cannot be calculated. Thus, inlet, outlet, and average vapor pressure and temperature are of the same values for all membrane pores. Accordingly, pore level vapor flux transport mechanism depends on the pore size, r_p only. When $\lambda < r_p$, then the flow rate of vapor through this pore, j_i, is given by viscous (Poiseuille) flow transport mechanism, Equation (14.2), while when $\lambda > r_p$, then the dominating vapour flux transport mechanism across this pore, is Knudsen diffusion, and j_i in this case is given by Equation (14.2). Note that N_{kn} and N_{vis} were replaced by j_i to carry out MD process vapor flux calculation (Imdakm and Matsuura, 2004; 2005; Imdakm *et al.*, 2007). Thus simulated process vapor flux, J, of the membrane is the sum of vapor flux of each pore calculated by Equation (14.2) or Equation (14.3), and it is given as: $J = \sum_{i=1}^{n} j_i$, where n is the sum of membrane pores in CT model, while in NW model, it is the total number of entry pores (see Figs. 14.1 and 14.2) divided by membrane boundary (feed or permeate) surface area, which is calculated from the pore size distribution and surface porosity, S_p. Note that heat energy and mass flux calculation procedure for both MD processes employed in this study have already been discussed elsewhere (Imdakm and Matsuura, 2004; Imdakm and Matsuura, 2005; Imdakm *et al.*, 2007; Present, 1958) and in references therein.

The Monte Carlo simulation model employed in this study is applied for MD process in which water is the only permeate. Accordingly, the physical properties of the vapor, such as vapor density (ρ), viscosity (μ), and thermal conductivity (k_v), are assumed to be of water vapor. The studies were carried out in this chapter, at feed temperature, $t_{b,f}$, in a range of 40–90°C, permeate bulk temperature, $t_{b,p}$, of 20°C (fixed), membrane boundaries heat transfer coefficients, h_f and h_p, in the range of 15,000 W m^{-2} K^{-1}, composite membrane surface porosity, S_p, of 3.5%, water vapor thermal conductivity, k_v, of 0.02 W m^{-1} K^{-1}, membrane solid phase (polymer) and thermal conductivity, k_s, of 0.04 W m^{-1} K^{-1}. As mentioned, membrane pore space is described by two distinct models, network (NW) model and capillary tube (CT). This data input is applied for all cases discussed in this chapter, unless they are specified differently, and the results reported are averaged over 20 realizations for each reported case.

14.3   SIMULATION RESULTS AND DISCUSSION OF DCMD PROCESS-VAPOR FLUX

14.3.1   *Effect of employed membrane pore space description*

Figure 14.4 shows a comparison of simulated DCMD process vapor flux vs. feed temperature, for membranes A and C, when the temperature polarization phenomenon was taken into consideration, i.e., when both h_f and h_p are of finite values of 15,000 W m^{-2} K^{-1} each, and the membrane pore space is described by NW and CT models. As can be seen, when the membrane pore space is

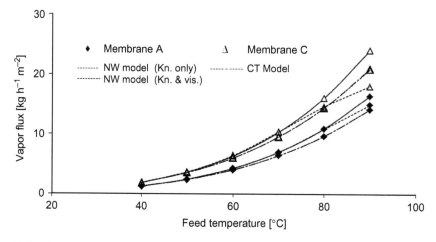

Figure 14.4.   Comparison of DCMD process vapor flux variation versus feed temperature for membrane A and C: (—) process is controlled by Knudsen only and (- - - -) process is controlled by Knudsen and viscous flow (NW model), and (- - -) process is controlled by Knudsen and viscous flow, but viscous effect was not observed (CT Model), $S_p = 3.5\%$ and h_f and $h_p = 15,000$ W m^{-2} K^{-1}.

described by CT model, the pore level vapor flux is mostly governed by Knudsen diffusion, the contribution of pore level viscous flux is of no significant value, and the resultant vapor flux is much less than when it is described by NW model.

Figure 14.5 shows membrane surface temperatures, $t_{s,f}$ and, $t_{s,p}$ vs. feed temperature when the temperature polarization is taken into consideration. As feed temperature increases membrane surface temperatures (feed and permeate) increase, resulting an increase in DCMD vapor flux. This increase in vapor flux continue up to 70°C after which the fraction of membrane pores governed by viscous flux increases resulting in significant decrease in temperature polarization coefficients, and vapor flux (see Figs. 14.4 and 14.6, and Imdakm and Matsuura, 2004; 2005 and

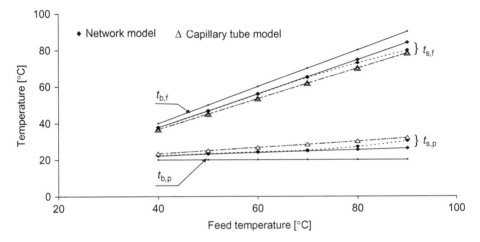

Figure 14.5. Variation of membrane surface temperature versus feed temperature for membrane C: (- - -) process is controlled by Knudsen and viscous flow and (—) process is controlled by Knudsen only (NW-model). (- - -) process is controlled by Knudsen and viscous flow (CT model), viscous effect not was observed. h_f and $h_p = 15,000 \, \text{W m}^{-2} \, \text{K}^{-1}$, and $S_p = 3.5\%$.

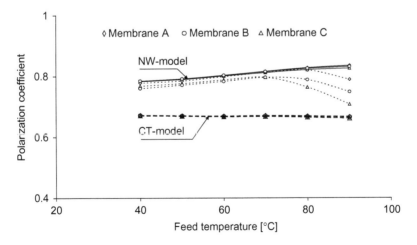

Figure 14.6. Comparison of simulated DCMD temperature polarization coefficient, τ, for membranes A, B, and C: (—) process is controlled by Knudsen diffusion only and (- - - -) process is controlled by Knudsen and viscous flow (NW-model). (- -) process is controlled by Knudsen only, and (- -) process is controlled by Knudsen and viscous flow (CT-model), viscous effect was not observed. h_f and $h_p = 15,000 \, \text{W m}^{-2} \, \text{K}^{-1}$ and $S_p = 3.5\%$.

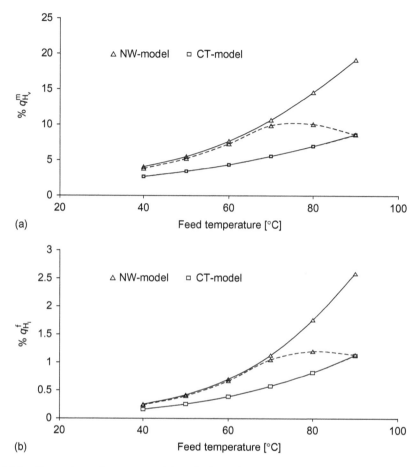

Figure 14.7.   Comparison of transport model predictions of percentage of energy transported by (a) liquid water enthalpy, $q_{H_1}^f$, and (b) saturated water vapor enthalpy, $q_{H_v}^m$, for membrane C: (——) process is controlled by Knudsen diffusion, and (- - - -) process is controlled by Knudsen and viscous flow. h_f and $h_p = 15,000\,\mathrm{W\,m^{-2}\,K^{-1}}$ and $S_p = 3.5\%$.

Imdakm *et al.*, 2007). When the membrane pore space is described by a capillary tube model, CT, the polarization coefficient is significantly lower than the NW model.

The observation in Figure 14.6 is further explained by Figure 14.7, where the percentage of energy that was consumed to evaporate liquid water at membrane bulk interface, $\% \, q_{H_v}^m$, is plotted vs. feed temperature. It is clear that the $\% \, q_{H_v}^m$ is much higher when membrane pore space is described by the NW model than when it is described by the CT model. In other words, this indicates that more energy could be consumed effectively to evaporate liquid water when membrane pore space is described by NW, thus pushing up the temperature polarization coefficient. Increasing membrane boundaries heat transfer coefficients, h_f and h_p, will increase, $t_{s,f}$ and decrease $t_{s,p}$, resulting in an increase process vapor flux and temperature polarization coefficient. This improvement of vapor flux and temperature polarization is more obvious when porous membrane pore space is described by NW model than when it is described by CT model.

14.3.2   *Effect of membrane pore size distribution*

In this study, we have investigated DCMD process performance when the pore size distribution is sufficiently narrow to be practically considered as a uniform pore size distribution. Let us first

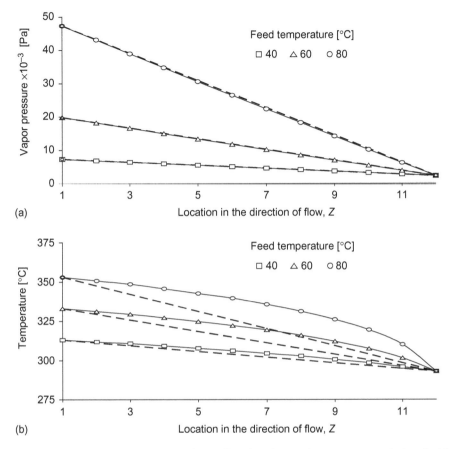

(a)

(b)

Figure 14.8.   Pressure (a) and temperature (b) profile when the membrane pore space is described by a network model (—), and when it is described by capillary tube model (– –), for membrane of uniform pore size distribution (120 nm), process vapor flux is dominated by Knudsen diffusion, h_f and $h_p \to \infty$ and $S_p = 3.5\%$.

consider the case when the temperature polarization is neglected. When the membrane pore space is described by an NW model of uniformly distributed pore radius, and process vapor flux is dominated by Knudsen flow only. Then, the resultant vapor pressure distribution is linear in the direction of flow (Z-direction only) as shown in Figure 14.8a. In other words, at any given value of Z, the pressure gradient is nil in the X and Y directions. This implies that, there is no flux among membrane pores in directions perpendicular to the direction of flow (Z). Note that the temperature distribution across the membrane (Z-direction) is not necessary linear, but it is also of same value at each network section, i.e., there is no temperature gradient perpendicular to the direction of fluid flow, Z (see Fig. 14.8b).

When membrane boundaries heat transfer coefficients are finite (h_f and $h_p =$ 15,000 W m^{-2} K^{-1}), the pore space model has a considerable effect on both vapor and temperature profile in the Z-direction, i.e., both vapor pressure and temperature are much lower when predicted by the CT model (see Fig. 14.9a,b), which results in much lower temperature polarization coefficient for the CT model as shown in Figure 14.10. This also makes the vapor flux predicted by the CT model lower than the NW model (Fig. 14.11).

Figure 14.12 shows vapor flux versus membrane thickness for both NW and CT model. As mentioned already, CT model predicts much lower flux than the NW model. The difference

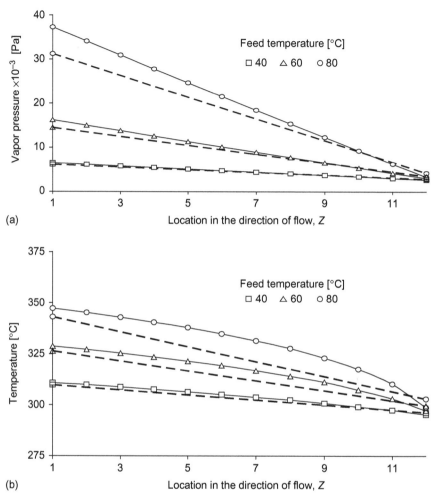

Figure 14.9.   Pressure (a) and temperature (b) profile when the membrane pore space is described by a network model (—) and when it is described by capillary tube model (--), for membrane of uniform pore size distribution (120 nm). Process is controlled by Knudsen diffusion, h_f and $h_p = 15{,}000\,\mathrm{W\,m^{-2}\,K^{-1}}$ and $S_p = 3.5\%$.

between the CT and NW model is much greater for the thinner membrane and as the thickness increases the difference decreases, both models approaching asymptotic values (Imdakm and Matsuura, 2004). The reason is that the heat conduction through the membrane is prevented when the membrane is thick and most of the heat is utilized for water evaporation for both CT and NW model. When the membrane is thin, heat is utilized more effectively for NW model than CT model.

14.4   SIMULATION RESULTS AND DISCUSSION – VMD PROCESS

14.4.1   *Effect of employed membrane pore space description*

The Monte Carlo simulation model developed in this study was also applied to describe VMD process performance. The following process variables are investigated because of their important

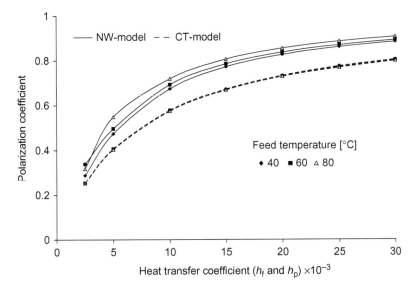

Figure 14.10. Comparison of temperature polarization coefficient, τ, versus membrane boundaries film heat transfer coefficients (h_f and h_p), for membrane of uniform pore size distribution (120 nm). Process is controlled by Knudsen diffusion only.

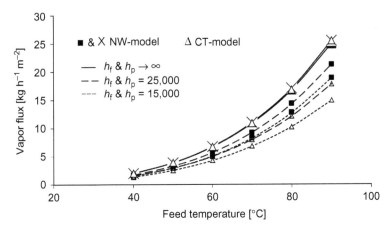

Figure 14.11. Simulated DCMD process vapor flux versus feed temperature, for membrane of uniform pore size distribution (120 nm): ■ and Δ process is controlled by Knudsen only, and □ process is controlled by Knudsen and viscous flow and h_f and $h_p \to \infty$. Viscous flux was not observed.

effects on the transporting process: (i) membrane pore size distribution, (ii) feed temperature, (iii) vacuum applied on membrane permeate side, and (iv) transport mechanism(s). These studies were carried out for three simulated membranes (A, B, and C, see Fig. 14.3) by applying both NW and CT models. As well, the average membrane temperature was assumed to be equal to respective membrane feed-side interfacial surface temperature, $t_{s,f}$, since there is no heat conduction involved in VMD. Figures 14.13, 14.14, and 14.15 show the simulated vapor flux for membranes A, B, and C, respectively, when the vapor flux through the membrane pores was assumed to be governed by Knudsen diffusion only. These figures show model productions are in excellent agreement with the theoretical expectations and in qualitative agreement with experimental data (Rice and Do,

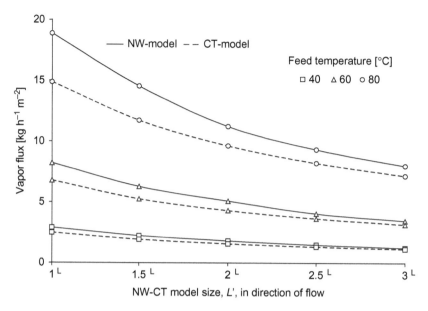

Figure 14.12.   Variation of DCMD process vapor flux versus simulated membrane size, L', increases in direction of flow for membrane of uniform pore size distribution (120 nm). Process is controlled by Knudsen diffusion only, h_f and $h_p = 15{,}000\,\mathrm{W\,m^{-2}\,K^{-1}}$.

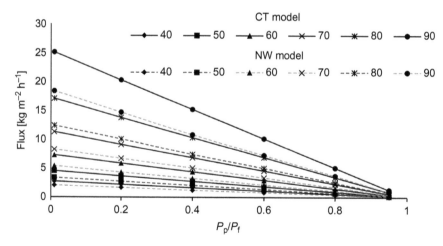

Figure 14.13.   Simulated VMD vapor flux for membrane A vs. ratio of the downstream pressure at different bulk feed temperatures [°C]. Process controlled by Knudsen diffusion transport mechanism only (solid line, CT model; broken line NW model).

1995) applied pressure on the permeate side, and this linearity does not alter with the increase of feed temperature and mean pore size from A to C regardless of adopted pore space model, either NW or CT. This is due to the vapor mechanism (Knudsen diffusion) imposed on the VMD process. In comparison, the simulation procedure was carried out under the same operating condition(s) when viscous (Poiseuille) flow was also included for membranes A and B. Figures 14.16, and 14.17 show that the simulated process vapour flux was influenced significantly by the increase of feed temperature; the flux became more notably non-linear for the NW model when feed

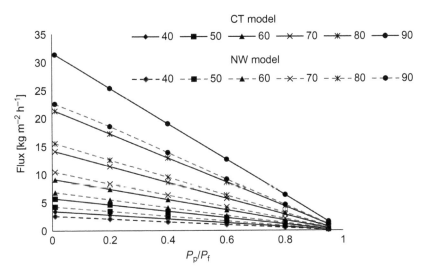

Figure 14.14. Simulated VMD vapor flux for membrane B vs. ratio of downstream pressure at different bulk feed temperatures [°C]. Process controlled by Knudsen diffusion transport mechanism only.

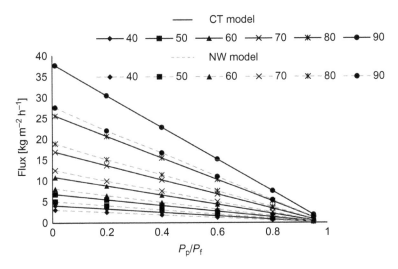

Figure 14.15. Simulated VMD vapour flux for membrane C vs. ratio of downstream pressure at different bulk feed temperatures [°C]. Process controlled by Knudsen diffusion transport mechanism only.

temperature becomes higher and the higher pressure is applied on the permeate side. The flux was much suppressed in the NW model due to the inclusion of viscous flow and the suppression is more significant for large pores (C) than the smaller pores (B). To explain the above results, the ratio of network pores in which viscous flow governs to the total number of number of network pores was plotted vs. P_p/P_f for membranes A to C (Fig. 14.18). The ratio increases with the increase of applied pressure on the permeate side and mean pore size. This was observed only when membrane pore space is described by the network, NW, model when membrane pore space is described by the CT model, it was not observed at all.

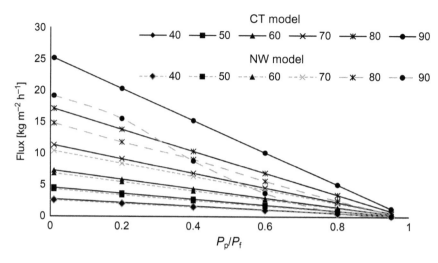

Figure 14.16.   Simulated VMD vapour flux for membrane B vs. ratio of downstream pressure at different bulk feed temperatures [°C]. Process controlled by Knudsen diffusion and viscous transport mechanisms (solid line; CT model; broken line NW model).

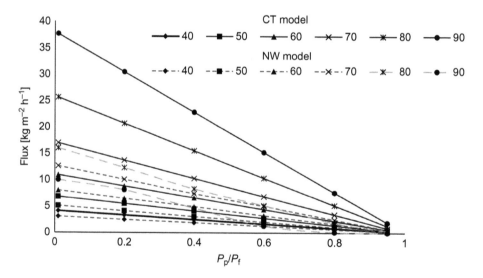

Figure 14.17.   Simulated VMD vapour flux for membrane C vs. ratio of downstream pressure at different bulk feed temperatures [°C]. Process controlled by Knudsen diffusion and viscous transport mechanisms (solid line, CT model; broken line NW model).

14.5   MONTE CARLO SIMULATION MODEL PREDICTION AND EXPERIMENTAL DATA

Finally, this Monte Carlo simulation model was tested against experimental data (see Table 14.1) obtained through personal communication with Dr. M. Khyat, see Appendix A.

As can be seen, the experimentally reported pore size distribution is very narrow (see Fig. 14.19), and therefore, resultant MC simulation model prediction(s) is almost identical regardless of membrane pore space description, CT, or NW model as clearly shown in Figure 14.20.

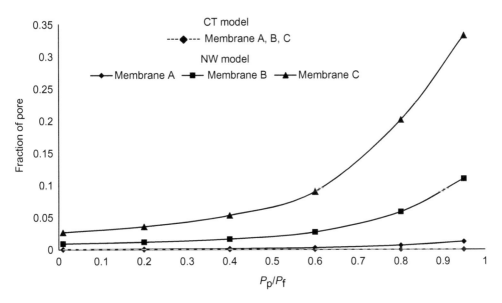

Figure 14.18.   Comparison of fraction of pores governed by viscous flow for membrane A, B and C: (- - - -) process is controlled by Knudsen only (CT-model), and (——) process is controlled by Knudsen and viscous flow mechanisms (NW model), operating temperature is 80°C.

Table 14.1.   Membrane characteristics: membrane thickness, δ; liquid entry pressure of water, LEP_{w}; void volume (volume porosity), ε_{v}; surface porosity, ε_{s}; mean pore size, μ_{p}; geometric standard deviation, σ_{p}; pore density, N. [Dr. M. Khyat personal communication].

Membrane	δ [μm]	LEP_{w} [MPa]	ε_{v} [%]	ε_{s} [%]	μ_{p} [nm]	σ_{p}	N [μm^{-2}]
TF200	55 ± 6	0.276 ± 0.009	69 ± 5	43.18	233.38	1.07	9.87
GVHP	118 ± 4	0.204 ± 0.003	70 ± 3	32.74	265.53	1.12	5.73

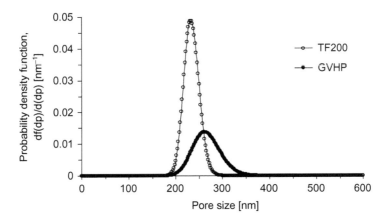

Figure 14.19.   Probability density curves of the membranes TF200 and GVHP.

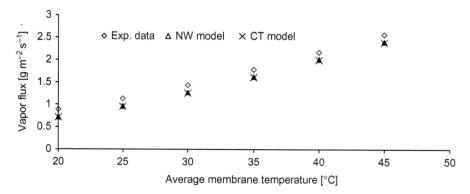

Figure 14.20.　Simulated and experimental DCMD vapor flux of TF200 membrane vs. average membrane temperature at stirring rate 500 rpm, and a bulk temperature difference (feed and permeate) of 10°C, see Appendix A for further details.

In general these MC model predictions are in excellent qualitative agreement with the reported experimental data.

14.6　SUMMARY AND CONCLUSION

In this chapter, we have discussed the effects of membrane pore space representation on simulated MD processes. For this purpose, a Monte Carlo simulation model developed in the previous studies (Imdakm and Matsuura, 2004; Present 1958) was summarized. This MC model is designed to be comprehensive in its approach and free of any adjustable parameters, and therefore, it is applicable to all MD processes published in the literature regardless of their different configurations. In any simulation of MD processes, a realistic membrane pore space description is very essential, and therefore, to illustrate its effects on the simulation of MD's transporting processes, membrane pore space is described by two distinct models, the first model is in the form of a three-dimensional network model (NW) of inter-connected sites (nodes) and bonds (pores), and the second model is the classical capillary tube model (CT) in which membrane pore inter-connectivity is totally ignored. In both models, the membrane pores are assumed to be cylinders of experimentally obtained pore size distribution. They may be also assumed to be, in some cases, of uniform distribution. When this MC simulation model is applied to describe MD processes of interest such as DCMD and VMD processes, the results obtained show clearly that inter-connectivity of the pore space is very essential and cannot be ignored in any realistic MD process simulation model. The only special case where the MC simulation does not depend on the inter-connectivity of the pore space is when membrane pores are assumed to be uniform, temperature polarization phenomenon can be ignored, and the dominating vapor flux transport mechanism in the membrane pores is linearly related to pressure drop across the membrane pores (e.g., Knudsen diffusion). This special case is evolved because the CT model is actually built on the assumption of linear pressure and temperature profile across the membrane pores. It has been found that MD processes vapor flux depends principally on membrane boundaries imposed operating condition(s), and the temperature distribution across the entire membrane pores, which is function of membrane pore space description, thus adding a constant adjusting parameter(s) to force models prediction fits experimental data, such as, the tortuosity factor, will not overcome the negligent of membrane pore space inter-connectivity and its effects on MD process vapor flux and related transport properties of interest, in addition to this it requires the availability of experimental data, which is not the case in these MD simulation models.

NOMENCLATURE

C	membrane distillation coefficient [kg m^{-2} s^{-1} Pa^{-1}]
r_{pore}	pore radius [m]
h_f and h_p	heat transfer coefficient at feed and permeate side [W m^{-2} K^{-1}]
$H_l\{T\}$	liquid water enthalpy at temperature T [kJ kg^{-1}]
$H_v\{T\}$	saturated vapor water enthalpy at temperature T [kJ kg^{-1}]
J	vapor flux [kg m^{-2} h^{-1}]
l	pore length [μm]
L	network size [number of nodes (sites)]
k_m	composite membrane effective thermal conductivity [W m^{-1} K^{1}]
k_s	membrane solid phase (polymer) thermal conductivity [W m^{-1} K^{-1}]
k_v	membrane vapor phase (water vapor) thermal conductivity [W m^{-1} K^{-1}]
P_f	feed side membrane pressure [Pa]
P_p	permeate side membrane pressure [Pa]
P_{pore}	average pore pressure [Pa]
$P_{b,f}$ and $P_{b,p}$	vapor pressure at feed and permeate side [Pa]
$P_{s,f}$ and $P_{s,p}$	vapor pressure at membrane-feed and membrane-permeate interface [Pa]
$q^m_{cond.}$	conduction heat transfer rate across membrane solid phase and vapor phase in direct contact with the feed [kW m^{-2}]
q^f_{Hv}	heat transfer due to liquid flux across membrane – feed boundary layer [kW m^{-2}]
q^m_{Hv}	heat transfer due to vapor flowing through membrane pores [kW m^{-2}]
S_p	membrane surface porosity
$t_{b,f}$ and $t_{b,p}$	bulk temperatures at feed and permeate sides [°C], respectively
$t_{s,f}$ and $t_{s,p}$	interfacial membrane-surface temperatures at feed and permeate sides [°C], respectively
T	temperature [K]
T_p	average pore temperature [K]
$T_{s,f}$	VMD process interfacial membrane-surface temperature at feed side [K]
X, Y and Z	axial coordinates

Greek symbols

μ_{pore}	mean pore size of the pores [m]
σ_{pore}	geometrical standard deviation of the pores
σ	collision diameter [m]
δ	membrane thickness [m]
λ	molecules mean free path [m]
τ	temperature polarization coefficient

REFERENCES

Bandini, S., Saavedra, A. & Sarti, G.C. (1997) Vacuum membrane distillation: experiments and modeling. *AICHE Journal*, 43, 398–408.

Calabro, V., Drioli, E. & Matera, F. (1991) Membrane distillation in textile wastewater treatment. *Desalination*, 83, 209–224.

Calabro, V., Jiao, B.L. & Drioli, E. (1994) Theoretical and experimental study on membrane distillation in the concentration of orange juice. *Industrial & Engineering Chemistry Research*, 33, 1803–1808.

Imdakm, A.O. & Matsuura, Y. (2004) A Monte Carlo simulation model for membrane distillation processes: direct contact (MD). *Journal of Membrane Science*, 237, 51–59.

Imdakm, A.O. & Matsuura, T. (2005) Simulation of heat and mass transfer in direct contact membrane distillation (MD): the effect of membrane physical properties. *Journal of Membrane Science*, 262, 117–128.

Imdakm, A.O. & Sahimi, M. (1991) Computer simulation of particles transport processes in flow through porous media. *Chemical Engineering Science*, 46, 1977–1993.

Imdakm, A.O., Khayet, M. & Matsuura, T. (2007). A Monte Carlo simulation model for vacuum membrane distillation process. *Journal of Membrane Science*, 306, 341–348.

Lagana, F., Barbieri, G. & Drioli, E. (2000) Direct contact membrane distillation: modeling and concentration experiment. *Journal of Membrane Science*, 166, 1–11.

Lawson, K.W. & Lloyd, D.R. (1996) Membrane distillation. II. Direct contact MD. *Journal of Membrane Science*, 120, 123–133.

Lawson, K.W. & Lloyd, D.R. (1997) Review membrane distillation. *Journal of Membrane Science*, 124, 1–25.

Mason, E.A. & Malinauskas, A.P. (1983) *Transport in porous media: the dust-gas model*. Elsevier, Amsterdam, The Netherlands.

Mason, E.A., Malinauskas, A.P. & Evans, R.B. (1967) Flow and diffusion of gases in porous media. *Journal of Chemical Physics*, 46, 3199–3207.

Phattaranawik, J. & Jiraratananon, R. (2001) Direct contact membrane distillation: effect of mass transfer on heat transfer. *Journal of Membrane Science*, 188, 137–143.

Phattaranawik, J., Jiraratananon, R. & Fane, A.G. (2003a) Heat transport and membrane distillation coefficients in direct contact membrane distillation. *Journal of Membrane Science*, 212, 177–193.

Phattaranawik, J., Jiraratananon, R. & Fane, A.G. (2003b) Effect of pore size distribution and air flux on mass transport in direct contact membrane distillation. *Journal of Membrane Science*, 215, 75–85.

Present, R.D. (1958) *Kinetic theory of gases*. McGraw-Hill, New York, NY.

Rice, R.G. & Do, D.D. (1995) *Applied mathematics and modeling for chemical engineers* (Appendix A). John Wiley & Sons Inc., New York, NY,

Sahimi, M. & Imdakm, A.O. (1988) The effect of morphological disorder on hydrodynamic dispersion in flow through porous media. *Journal of Physics* A: *Mathematical and General*, 21, 3833–3870.

Schofield, R.W., Fane, A.G. & Fell, C.J.D. (1987) Heat and mass transfer in membrane distillation. *Journal of Membrane Science*, 33, 299–313.

Schofield, R.W., Fane, A.G., Fell, C.J.D. & Macoun, R. (1990a). Factors affecting flux in membrane distillation. *Desalination*, 77, 279–294.

Schofield, R.W., Fane, A.G. & Fell, C.J.D. (1990b) Gas and vapor transport through microporous membranes. II. Membrane distillation. *Journal of Membrane Science*, 53, 173–185.

Wang, Q.A., Matsuura, T., Feng, C.Y., Weir, M.R., Detellier, C. & Van Mao, L. (2001). The sepiolite membrane for ultrafiltration. *Journal of Membrane Science*, 184 (30), 153–163.

Wu, Y., Kong, Y., Liu, J., Zhang, J. & Xu, J. (1991) An experimental study on membrane distillation: crystallization for treating wastewater in taurine production. *Desalination*, 80, 235–242.

APPENDIX A

Membrane: TF200-experimental data
Average pore size: $\mu_p = 233.38\,\text{nm}$
Geometric standard deviation: $\sigma_p = 1.07$
Surface porosity: $\varepsilon_s = 43.18\%$
Void volume (porosity): $\varepsilon_v = 68.73 \pm 5\%$
Thickness: $\delta = 55 \pm 6\,\mu\text{m}$
Pore tortuosity: $\tau = 1.59$
PTFE thermal conductivity: $k_s = 0.22\,\text{W}\,\text{m}^{-1}\,\text{K}^{-1}$
Experimental results of 500 rpm

t_f [°C] feed side	t_p [°C] permeate side	J [g m^{-2} s^{-1}]	h_f [W m^{-2} K^{-1}] feed	h_p [W m^{-2} K^{-1}] permeate
25	15	0.8854	3401.39	2618.07
30	20	1.1336	3523.97	2772.68
35	25	1.4381	3607.65	2924.90
40	30	1.7859	3632.25	3056.00
45	35	2.1721	3583.27	3145.50
50	40	2.5663	3456.58	3173.59

CHAPTER 15

Simulation of drying for multilayer membranes

Zawati Harun, Tze Ching Ong, David Gethin & Ahmad Fauzi Ismail

15.1 INTRODUCTION

Theoretically, a membrane can be divided into several layers as shown in Figure 15.1.

The first top layer having a fine pore structure is used as a filter for separation processes (Li, 2007). The next three layers which are the intermediate layers and the bottom layer are fabricated in the form of a porous structure to support the membrane itself and to facilitate the permeation process (Li, 2007). Obviously, the top separation layer with a fine porous structure possesses a hygroscopic property compared to the intermediate and bottom layers that have to be porous and non-hygroscopic. All layers have their own characteristic and structure and shrink at different rates when the membrane is dried.

Drying is one of the stages of ceramic membrane preparation that is undertaken between precursor formation and sintering. Although drying is one of the most energy intensive industrial processes (Mujumdar, 2006), it is also one of the least understood as it involves a lot of variables that change concurrently as the phases and stages change (Perré et al., 2007; Rattanadecho et al., 2007). Further complications arise if drying involves materials that comprise multilayers (Harun and Gethin, 2008; Harun et al., 2008; 2012; Ismail et al., 2012). In particular, the top hygroscopic layer of the ceramic membrane involves bound water that is strongly attached to the capillary wall. The amount of the water is quite difficult to measure experimentally due to the stronger capillarity suction and the slower permeation than in the non-hygroscopic material. Typically, drying of the membrane of a layered structure not only involves the compatible issue of drying parameters but the different materials that possess different properties also strongly influence the consistency of shrinkage. This often leads to defect formation which is associated with leakage due to cracks induced during the drying and firing processes. Warping and cracking occur due to non-uniform stresses arising from the pressure gradient caused by the flow of liquid during the constant rate period (CRP) of shrinkage, the macroscopic pressure gradient of escaping gases during falling rate period (FRP) and different thermal expansion of the ceramic material due to the temperature gradient (Ring, 1996). The flow of liquid, the macroscopic pressure gradient, and the temperature gradient are controlled by the drying rate, which is ultimately controlled by the external conditions (Ring, 1996). Therefore, understanding the stages and variables that evolve with time throughout the drying sequence is essential to eliminate all the problems associated with membrane manufacture.

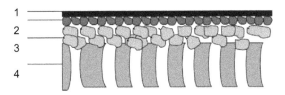

Figure 15.1. Schematic representation of an asymmetric membrane (extracted from Li, 2007).

In general, material that has undergone drying will induce several compression and strain changes in the porous network, which results in formation of a shrinkage mechanism (Perré et al., 2007). Shrinkage mechanism can be defined as dimensional changes due to the elimination of water in a porous structure, which will generate the non-uniform stress distribution, either from the inner structure and outer surface layer. This non-uniform stress is attributed mainly to drastic or fluctuated changes of moisture gradient (Kowalski and Mierzwa, 2013; Pourcel et al., 2007) or network capillary pressure that has a high tendency to build cracks. When drying is at a constant rate period, the surface of the green body is always wet by the flow of liquid to the surface, resulting in the rearrangement of particles in the green body, which is caused by the compressive capillary pressure. This eventually leads to shrinkage and deformation of the solid as particles come into contact to maintain an expelling of liquid flow to the surface. However, the dried product is free from stress at this moment as there is no crack formation being observed in this period (Colina and Roux, 2000). When drying reaches a decreasing drying rate period, the liquid-vapor interface starts to recede into the pores and capillary flow will continue as long as there is a continuous pathway from the liquid front of the green body surface. At this moment, the particles are in close contact or touching each other, which leads to minimal or no shrinkage happening beyond this period of drying. However, the critical point of saturation when entering FRP is the vital point for the crack initiation (Nascimento et al., 2005). There is much lower shrinkage in FRP but the fracture phenomenon becomes the most intensive in this period of drying process due to inhomogeneous moisture distribution in the solid matrix. The differential in strain corresponding to the surface that tends to contract faster than the interior causes non-uniform moisture and pressure gradient. Subsequently, this leads to crack formation. It appears that shrinkage of drying materials, which is also a characteristic property of the dried material and cracking, is more likely to happen in the low permeability material that offers higher resistance for moisture transfer that causes a nonhomogeneous moisture gradient between the upper and bottom layer shrinkage inside the solid matrix (Scherer, 1990). In this present study, a mathematical model of the drying process with coupled mass, heat and gas transfer for ceramic materials has been developed. This model will emphasize on the transport mechanism of liquid by capillary action, vapor and gas by diffusion and bulk air convective flow. The introduction of bound water over the later period (FRP2) of drying was incorporated by referring to the earlier work (Zhang et al., 1999). The material slab used has been divided into two layers, hygroscopic and non-hygroscopic layers, to represent the layered structure of the membrane.

15.2   MATHEMATICAL MODELING

15.2.1   Theoretical formulation for phase transport

The conservation law of the heat and mass balance equations applied here for the drying process is based on the work by Zawati et al. 2008 and others (Harun et al., 2012; Zhang et al., 1999). In the governing equations, the liquid phase is water in which vapor and air are dissolved and transported by diffusion, whilst the gas phase consists of the binary mixture of vapor and dry air which are also transported in the gas phase by diffusion. The transport of liquid by the capillarity effect is also included. Both mass and heat transport are taken into account. For the hygroscopic material, bound water is considered in the transport mechanism especially in the FRP2 (Defraeye et al., 2012; Zhang et al., 1999) stage. Based on these physical principles, the governing equations are expressed in terms of three primary variables; water pressure, P_l, temperature, T and gas pressure, P_g. The water mass conservation equation is formulated as follows for both non hygroscopic (15.1a) and hygroscopic material (15.1b):

$$\frac{\partial(\phi\rho_l S_l)}{\partial t} + \frac{\partial(\phi\rho_v S_g)}{\partial t} = -\nabla \cdot (\rho_l V_l) - \nabla \cdot (\rho_v V_v) - \nabla \cdot (\rho_v V_g) \tag{15.1a}$$

$$\frac{\partial(\phi\rho_l S_l)}{\partial t} + \frac{\partial(\phi\rho_v S_g)}{\partial t} = -\nabla \cdot (\rho_l V_l) - \nabla \cdot (\rho_v V_v) - \nabla \cdot (\rho_v V_g) - \nabla \cdot (\rho_l V_b) \tag{15.1b}$$

The water velocity, V_1 and gas velocity, V_g can be easily derived from Darcy's law (Harun and Gethin, 2008; Harun *et al.*, 2012):

$$V_1 = -\frac{Kk_1}{\mu_1}[\nabla(P_1 + Z)] \tag{15.2}$$

$$V_g = -\frac{Kk_g}{\mu_g}[\nabla(P_g)] \tag{15.3}$$

where Z is the vertical elevation from a datum. The vapor velocity by diffusion is defined as:

$$V_v = \frac{-D_{atm}\nu\alpha\theta_a}{\rho_v} \cdot \nabla\rho_v \tag{15.4}$$

where D_{atm}, ν, α, θ_a are the molecular diffusivity of water vapor through dry air, mass-flow factor, tortuosity factor and volumetric air term, respectively. Rearranging the above equation according to the measured variables gives:

$$V_v = \frac{-D_{atm}\nu\alpha\theta_a}{\rho_v}\left\{\rho_o\frac{\partial h}{\partial P_1}\nabla P_1 + (\rho_o + h\beta)\frac{\partial h}{\partial T}\nabla T + \rho_o\frac{\partial h}{\partial P_g}\nabla P_g\right\} \tag{15.5}$$

When the water content reaches the maximum irreducible level, the bound water flux is taken into consideration in Equation (15.1b). The bound water velocity was derived from Zhang *et al.* (1999) as:

$$V_b = -D_b\nabla\theta_b \tag{15.6}$$

where D_b, θ_b are the bound water conductivity and volume fraction of bound water respectively. Rearranging the above equation according to the measured variables gives:

$$V_b = -D_b\phi\left(\frac{\partial S_1}{\partial P_1}\nabla P_1 + \frac{\partial S_1}{\partial T}\nabla T + \frac{\partial S_1}{\partial P_g}\nabla P_g\right) \tag{15.7}$$

Application of mass balance to the flow of dry air within the pores of the material body dictates that the time derivative of the dry air content is equal to the spatial derivative of the dry air flux:

$$\frac{\partial(\phi\rho_a S_g)}{\partial t} = -\nabla(\rho_a V_g - \rho_v V_v) \tag{15.8}$$

The effects of conduction, latent heat and convection are considered in the energy equation as given as:

$$\frac{\partial((1-\phi)c_p\rho_s + \sum_{i=a,l,v}\phi s_i\rho_i c_i)}{\partial t} = \nabla(\lambda\nabla T) - \nabla(\rho_v V_v + \rho_v V_g)L - \nabla\left(\sum_{t=a,l,v}(T - T_r)\rho_i c_{pi} V_i\right) \tag{15.9}$$

15.2.1.1 *Thermodynamic relationship*
The existence of a local equilibrium at any point within the porous matrix is assumed. Kelvin's law is applied to give the equation:

$$h = \exp\left(\frac{P_w - P_g}{\rho_1 R_v T}\right) \tag{15.10}$$

The vapor partial pressure can be defined as a function of local temperature and relative humidity as:

$$\rho_v = \rho_o h \tag{15.11}$$

where the saturation vapor pressure, ρ_o is estimated using the equation in Mayhew and Rogers (1976) with the saturated vapor density expressed as a function of temperature, which here has the form:

$$\frac{1}{\rho_o} = 194.4 \exp\{-0.06374(T - 273) + 0.1634 \times 10^{-3}(T - 273)^2\} \tag{15.12}$$

The degree of saturation, S is an experimentally determined function of capillary pressure and temperature:

$$S_1 = S_1(p_c, T) \tag{15.13}$$

Saturation is expressed in the equation below (van Genuchten, 1980):

$$S_1 = \frac{\theta - \theta_r}{\theta_s - \theta_r} = \left(\frac{1}{1 + (\alpha\varphi(T))^n}\right)^m \tag{15.14}$$

where the parameters α, n and m are dependent on the porous material properties and these influence the shape of the water retention curve. The permeability of water and gas are based on Muelem's model and are given by the formulation below as in Baroghel-Bouny et al. (1999):

$$k_1(S_1) = \begin{cases} \sqrt{S_1}(1 - (1 - S_1^{1/m})^m)^2 & S_1 > S_{irr} \\ 0 & S_1 < S_{irr} \end{cases}$$

$$k_g(S_1) = \begin{cases} \sqrt{1 - S_1}(1 - S_1^{1/m})^m)^{2m} & S_1 < S_{cri(g)} \\ 0 & S_1 > S_{cri(g)} \end{cases} \tag{15.15}$$

15.2.2   Material data

As mentioned in Section 15.1, the thin ceramic membrane comprising a two-layer system is presented as shown in Figure 15.2.

In this study, the porous slab (0.003×0.004 m) was meshed using 12 quadratic serendipity elements and 51 nodes. These map the porous medium that is homogeneous, isotropic and composed of solid phase, water and vapor phase, gas phase and dry air phase. Nodes 1 to 7 are assumed to remain at the atmospheric condition, while the rest of the wall surfaces are assumed to be insulated and impermeable. The hygroscopic layer is located in elements 1, 5 and 9 while the rest of the elements represent the non-hygroscopic layer. The ambient temperature is 27°C for slow convective drying. Heat and mass coefficient are 1.5 W m^{-2} K^{-1} and 0.00175 m s^{-1}. The material properties of these two different porous slabs exposed to convective drying are listed in Table 15.1. The material properties for hygroscopic and non-hygroscopic layers are similar to alumina and clay ceramic profiles, respectively.

15.2.3   Boundary conditions

The drying environment was imposed at the top surface and the rest of the surfaces were assumed insulated and impermeable. Thus, the boundary condition for convective drying was applied to elements 1, 5 and 9 with reference to nodes 1 to 7. The general boundary condition that apply to convective mass and heat transfer are given as:

$$J_m = h_m(P_\infty^v - P_a^v) \tag{15.16}$$

$$J_T = h_T(T_f - T_\infty) \tag{15.17}$$

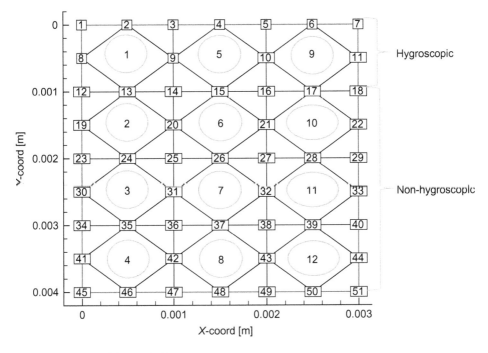

Figure 15.2.   Mesh of the slab body.

Table 15.1.   Physical properties of material body for hygroscopic and non-hygroscopic layer.

Layer	Symbol	Value	Unit	Reference
Hygroscopic	ρ_s	3970	kg m^{-3}	Witharana *et al.* (2012)
	Ø	0.1	–	Assumed by modeling and experimental data
	K	5.13×10^{-17}	m^2	Schmitz *et al.* (2012)
	C_p	775	J kg^{-1}K^{-1}	Callister and Rethwisch (2011)
	λ	39	W m^{-1}K^{-1}	Callister and Rethwisch (2011)
Non hygroscopic	ρ_s	2000	kg m^{-3}	Harun and Gethin (2008)
	Ø	0.4	–	Hall and Hoff (2012)
	K	2.9×10^{-15}	m^2	Hall and Hoff (2012)
	C_p	925	J kg^{-1}K^{-1}	Harun and Gethin (2008)
	λ	1.8	W m^{-1}K^{-1}	Harun and Gethin (2008)

15.2.4   *Solution of governing equations and numerical method*

The coupled heat and mass transfer equations described above, in 2-dimensions can be written into a matrix form as follows:

$$[\mathbf{C}(\Phi)\frac{\partial}{\partial t}\{\Phi\}] = \nabla([\mathbf{K}_{cx}(\nabla\Phi)]\mathbf{i}_x + [\mathbf{K}_{cy}(\nabla\Phi)]\mathbf{i}_y)\{\Phi\} + R(\nabla Z) \qquad (15.18)$$

where $\{\Phi\} = \{P_w, T, P_g\}$ is the column of unknowns; $[\mathbf{C}]$, $[\mathbf{K}_{cx}]$ and $[\mathbf{K}_{cy}]$ are 3×3 matrices. Each element of the matrix is a coefficient for the unknown $\{\Phi\}$; $\mathbf{i}_x$ and $\mathbf{i}_y$ are the unit direction vectors. In order to discretize this simplified second order non-linear coupled partial differential equation, the finite element method is used. The Galerkin method was used to minimize the residual error before the application of Greens theorem, to the dispersive term that includes second order derivatives; this simplified combined equation set can be expressed in the following form:

$$\mathbf{K}(\Phi)\Phi + \mathbf{C}(\Phi)\dot{\Phi} + \mathbf{J}(\Phi) = \{0\} \tag{15.19}$$

where,

$$\mathbf{K} = \begin{bmatrix} K_{11} & K_{12} & K_{13} & K_{14} \\ K_{21} & K_{22} & K_{23} & K_{24} \\ K_{31} & K_{32} & K_{33} & 0 \end{bmatrix} \qquad \mathbf{C} = \begin{bmatrix} C_{11} & C_{12} & C_{13} \\ C_{21} & C_{22} & C_{23} \\ C_{31} & C_{32} & C_{33} \end{bmatrix}$$

$$\mathbf{J} = \begin{Bmatrix} J_1 \\ J_2 \\ J_3 \end{Bmatrix} \qquad \Phi = \begin{Bmatrix} P_{ws} \\ T_s \\ P_{gs} \end{Bmatrix} \qquad \dot{\Phi} = \begin{Bmatrix} \dfrac{\partial P_{ws}}{\partial t} \\ \dfrac{\partial T_s}{\partial t} \\ \dfrac{\partial P_{gs}}{\partial t} \end{Bmatrix}$$

in which typical elements of the matrix are:

$$K_{ij} = \sum_{s=1}^{n} \int_{\Omega^e} K_{ij} \nabla N_r \nabla N_s d\Omega^e \quad (i,j = 1, 2, \ldots, 5)$$

$$C_{ij} = \sum_{s=1}^{n} \int_{\Omega^e} C_{ij} N_r N_s d\Omega^e$$

$$J_i = \int_{\Omega^e} K_{i4} \nabla N_r \nabla z d\Omega^e - \int_{\Gamma^e} N_r J_i d\Gamma^e$$

(Γ of the domain Ω).

The transient matrix and nonlinear second order differential equations above are then solved by using a fully implicit backward time stepping scheme along with a Picard iterative method which is taken into account for non-linearity.

15.3 RESULTS AND DISCUSSION

Figure 15.3 shows the saturation distributions of the proposed model and the saturation results from other models gained elsewhere. Initially, the proposed model which includes the hygroscopic equation and saturation conditions is compared with results gained from Harun and Gethin (2008) (non-hygroscopic system), Przesmycki and Sturmillo (1985) and Stanish et al. (1986) (hygroscopic system). Figure 15.3 shows the variation of average saturation computed by the proposed model. Obviously, the Harun and Gethin model and the proposed model show good agreement in the variations of saturation for the drying period up to the non-hygroscopic zone as reported by Harun and Gethin (2008). Presemycki and Sturmillo (1985) studied the drying of a brick and the result is in close agreement for saturation evolution when the drying period enters the hygroscopic zone. As for Stanish et al. (1986), the proposed model and Stanish et al. (1986) have similar saturation patterns, but the drying rates used by Stanish et al. (1986) was faster compared to the current model. Generally, the saturation curve consists of three periods. The CRP which is indicated by a straight line from point A to B, FRP1 as a gradual curve from point B to C and FRP2 that shows a declining slope from point C to D (refer to Fig. 15.3). An explanation of all these periods of drying are discussed in detail in most scientific texts on drying phenomena for both porous hygroscopic and non-hygroscopic materials (Harun and Gethin, 2008; Harun et al., 2008; 2012; Perré et al., 2007; Rattanadecho et al., 2007; Zhang et al., 1999).

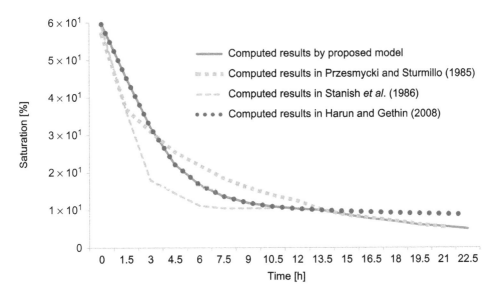

Figure 15.3.   Saturation curve during drying.

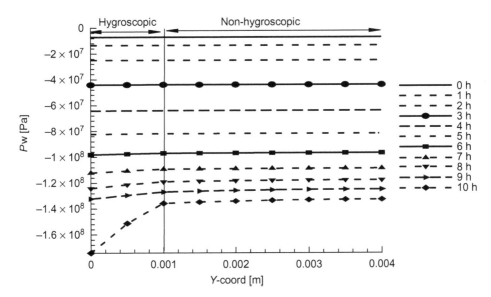

Figure 15.4.   Pore water pressure distribution as a function of depth and time.

Figure 15.4 shows the variation of pore water pressure and how it evolves with time and depth. Pore-water pressure at the initial stages of drying shows a constant condition as free water is dominant during the constant rate period (Harun and Gethin, 2008; Perré *et al.*, 2007; Ring, 1996) indicating free water continues to travel to the exposed surface by capillary action. The liquid migrates from regions with higher moisture concentration towards regions with lower moisture concentration as expressed by Darcy's law (Perré *et al.*, 2007). This CRP can last as long as the surface is supplied with liquid based on external factors and body properties (Perré *et al.*, 2007). When the solid porous body is wet, it has a capillary adhesion that holds particles together and it will weaken as liquid disappears during the drying (Ring, 1996). Eventually, the body may

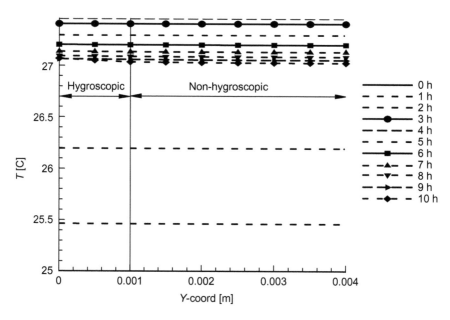

Figure 15.5.   Temperature distribution as a function of depth and time.

rupture during the CRP when there is a drastic change of pressure gradient in capillarity flow that will create nonhomogeneous shrinkage (Ring, 1996). When water saturation reaches the critical value that shows the early stage of falling rate period, the pore water pressure shows an increasing trend due to increase of capillary action. This slightly higher value of pore pressure water within the hygroscopic layer is due to greater capillary action of water that is strongly attached to the solid body and creates a greater capillary suction. As the drying proceeds towards the irreducible stage, the pore water pressure is represented by a bound water mechanism that is dominated by vapor diffusion. The bound water removal stage for the hygroscopic layer shows an abrupt increase of pore water pressure especially at the top surface that indicates a greater suction of bound water transfer near dryness.

The temperature variation over time and depth is shown in Figure 15.5. At the commencement of drying, temperature increases towards the ambient temperature as moisture removal is controlled by the external forces (Perré et al., 2007). Heat energy supplied from the surface towards the inner volume of the body will determine the drying rates (Perré et al., 2007). For multilayers comprising hygroscopic and non-hygroscopic systems, the temperature profiles do not show any obvious trend (Rattanadecho et al., 2007) as there are only slight differences in temperature changes within the two layers. This small change is impossible to plot using the big scale of temperature. The temperature profiles rise to a higher value than the environment temperature when drying reaches the falling rate period where gas diffusion takes places. The temperature inside the body reaches its higher values as this generates a greater diffusion process. With the increase of drying time, the temperature settles down towards the ambient temperature as no more energy is gained from the external environment. The drying state using the current conditions is within internal forces such as bound water movement or sorption diffusion (Zhang et al., 1999). The temperature profiles during drying are particularly important to study especially when dealing with very thin layers of brittle ceramic membrane. This is due to the fact that the lower stresses induced due to small temperature gradient (Ring, 1996) could generate a massive expansion in ceramic membrane structure. Thereby, controlling the temperature is essential in order to avoid excessive stress, which would warp or crack the solid.

The gas pressure evolution against time and depth is shown in Figure 15.6 where it remains nearly constant at atmospheric pressure as a result of free water capillary removal. As drying

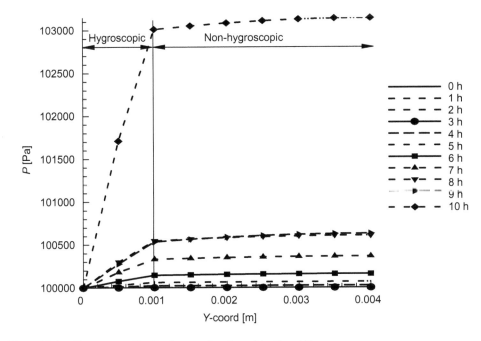

Figure 15.6.   Gas pressure distribution as a function of depth and time.

proceeds to the falling rate period, gas pressure starts to increase as pore liquid is replaced by gas. Diffusion of gases is important in this stage to enable continued mass transport towards the surface. Hence, this stage of drying is strongly affected by water vapor movement that occurs by diffusion in response to the vapor pressure gradient. The top hygroscopic layer shows an increasing trend in gas pressure due to the smaller pores, higher density and lower permeability value that restrains the gas movement to penetrate inside its body (Harun and Gethin, 2008). When it reaches the non-hygroscopic layer, the gas pressure remains nearly constant as gases are trapped inside the body. The gas pressure in the hygroscopic layer reaches its highest value as gases penetrate through the very fine capillaries (Ring, 1996) and it remains stable in the non-hygroscopic layer. As shown in Figure 15.6, during the FRP, stress will be induced due to the macroscopic pressure gradient of escaping gasses (Ring, 1996). Thereby, understanding the variation of gas pressure through simulation during drying will help to reduce defects in membranes.

Figure 15.7 shows the saturation distributions as a function of depth and time. This shows that moisture decreases very quickly at the early stage of drying. This movement of liquid is supplied by the capillary action as water continues to evaporate steadily. The surface of the material where heat is supplied to it is supposed to have a higher liquid extraction rate (Rattanadecho *et al.*, 2007). In contrast to drying of a single layer, the saturation distribution for a multilayer of hygroscopic materials shows only a small spatial variation as hygroscopic materials possess only a small porous network. The saturation reduction slows down when drying reaches the critical stage which is normally 0.3 for porous materials (Zhang *et al.*, 1999). Beyond this point vapor diffusion is the main supplier of mass transfer within the body. When the drying stage reaches the irreducible state which is at a relative saturation level of 0.09 (Zhang *et al.*, 1999), saturation is almost constant indicating that there is no water that can be removed as the material has been dried. As for the hygroscopic layer, bound water still remains within the body and transport continues through vapor diffusion until it reaches the equilibrium content at which point no more water can be removed (Zhang *et al.*, 1999).

Further observation on drying behavior was determined based on the potential of crack formation as the drying front continuously receded far into the membrane structure which was

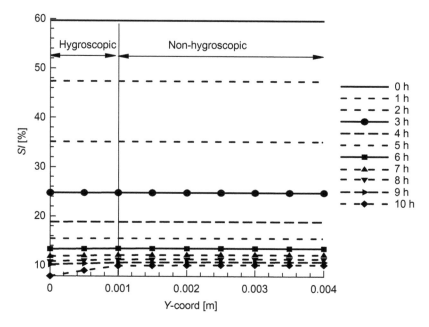

Figure 15.7.   Saturation distribution as a function of depth and time.

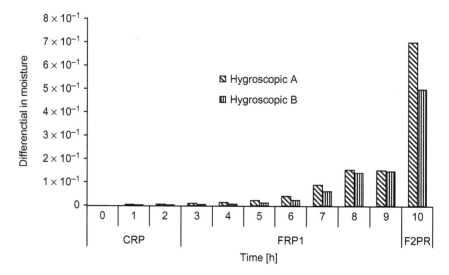

Figure 15.8.   Moisture gradient as a function of time.

accompanied by the existence of non-uniform stress due to the different moisture gradient levels. Thus two different material properties that represent the hygroscopic layer coupled with the same porous support structure are compared in terms of the difference in moisture gradient between drying front and inner wet structure body. Figure 15.8 depicts the moisture gradient evolution between a more hygroscopic characteristic (denoted as hygroscopic A) and less hygroscopic characteristic (denoted as hygroscopic B) in the top layer while both having similar bottom non hygroscopic material properties. Hence, the result and discussion of this section will only present the results of moisture gradient for the upper layer structure. As mentioned in the introduction, cracking is a characteristic of the dried porous matrix that is always associated with the non-homogenous moisture gradient within the solid matrix. As clearly illustrated in Figure 15.8 moisture gradient for hygroscopic A is higher than hygroscopic B at all times. Hence, it can be deduced that more hygroscopic behavior will offer more resistance of moisture transport or higher capillary suction which is attributed to lower permeability value and porosity. Consequently, drying of the upper surface layer and bottom layer exhibits non-uniform moisture content within the solid matrix which may cause higher possibilities of crack formation due to non-homogenous stress distribution especially in strongly hygroscopic materials. Since the gradient of moisture or capillary suction increases gradually with the drying period as increasing the capillary suction, crack formation is more likely to happen in hygroscopic material. Thus, drying from the top surface must be carefully controlled especially in hygroscopic material to avoid undesired defects at the end of the drying process.

15.4 CONCLUSION

The drying procedure of multilayer membranes is quite different from that of a single layer due to different characteristics of those membranes. The hygroscopic layer that inherits higher density, smaller permeability and smaller pores leads to higher pore water pressure and gas pressure. For hygroscopic materials, the transport of the bound water will also take place with the vapor pressure gradient in the gas phase as the driving force. Stresses due to pressure gradient arising from the liquid flow, macroscopic pressure gradient of escaping gases and temperature gradient can affect the drying speed. Material selection is also vitally essential as more hygroscopic characteristic contributed to higher non-uniform evaporation losses inside the porous matrix which increase the probabilities of defects due to crack formation. Based on these simulated results, it can be concluded that a better understanding of drying phenomena could be achieved, which will play an important role in predicting and determining the drying rate during the formation of the separation layer of the ceramic membrane.

ACKNOWLEDGEMENT

We are grateful for the KPTM scholarship for author 2.

NOMENCLATURE

μ	viscosity [N s m^{-2}]
C_{ij}	capacity coefficient [ad]
C_p	specific heat capacity [J mol^{-1} K^{-1}]
h	relative humidity [%]
h_m	mass transfer coefficient [m s^{-1}]
h_T	heat transfer coefficient [W m^{-2} K^{-1}]
J_m	mass transfer flux [kg m^{-2} s^{-1}]
J_T	heat transfer flux [W m^{-2}]
K	intrinsic permeability [m^2]
k	relative permeability [ad]
K_{ij}	kinetic coefficients [ad]
L	latent heat of vaporization [J kg^{-1}]
N_r	shape function of residual error [ad]
N_s	shape function of system variables [ad]
P	pressure [Pa]
R	gas constant [J mol^{-1} K^{-1}]
S	saturation [ad]

T	temperature [K]
t	time [s]
V	velocity [ms^{-1}]

GREEK SYMBOLS

Ø	porosity [ad]
θ	volumetric content [ad]
λ	thermal conductivity [W m^{-1} K^{-1}]
ρ	density [kg m^{-3}]
Ω^e	element domain [ad]

SUBSCRIPTS

∞	calculated
a, c, v, g, l, b	air, capillary, vapor, gas, liquid, bound water
cri	critical
F	final
irr	irreducible
r	residual
s	saturated

REFERENCES

Baroghel-Bouny, V., Mainguy, M., Lassabatere, T. & Coussy, O. (1999) Characterization and identification of equilibrium and transfer moisture properties for ordinary and high-performance cementitious materials. *Cement and Concrete Research*, 29 (8), 1225–1238.

Callister, W.D. & Rethwisch, D.G. (2011) *Materials science and engineering*. John Wiley & Sons, Inc., New York, NY.

Colina, H. & Roux, S. (2000) Experimental model of cracking induced by drying shrinkage. *The European Physical Journal* E, 194, 189–194.

Defraeye, T., Blocken, B. & Derome, D. (2012) Convective heat and mass transfer modelling at air-porous material interfaces: overview of existing methods and relevance. *Chemical Engineering Science*, 32, 49–58.

Hall, C. & Hoff, W.D. (2012) *Water transport in brick, stone and concrete*. Spon Press, London, UK.

Harun, Z. & Gethin, D.T. (2008) Drying simulation of ceramic shell build up process. *Modeling & Simulation, 2008. AICMS 08. Second Asia International Conference on Modeling & Simulation 2008, 13–15 May 2008, Kuala Lumpur, Malaysia*. IEEE. pp. 794–799.

Harun, Z., Gethin, D.T. & Lewis, R.W. (2008) Combined heat and mass transfer for drying ceramic (shell) body. *The International Journal of Multiphysics*, 2, 1–19.

Harun, Z., Nawi, N.M., Batcha, M.F. & Gethin, D.T. (2012) Modeling of layering ceramic shell mould. *Applied Mechanics and Materials*, 232, 548–552.

Ismail, N.F., Harun, Z. & Badarulzaman, N.A. (2012). A comparative study of double layers Al$_2$O$_3$/Al$_2$O$_3$ and Al$_2$O$_3$/SiO$_2$ prepared by microwave and natural drying. *International Journal of Integrated Engineering*, 4–1, 16–21.

Kowalski, S.J. & Mierzwa, D. (2013) Numerical analysis of drying kinetics for shrinkable products such as fruits and vegetables. *Journal of Food Engineering*, 114 (4), 522–529.

Li, K. (2007) *Ceramic membranes for separation and reaction*. John Wiley & Sons Ltd., Chichester, UK.

Mayhew, Y.R. & Rogers, G.F.C. (1976) *Thermodynamic and transport properties of fluids*. Blackwell, Oxford, UK.

Mujumdar, A.S. (ed.), (2006) *Handbook of industrial drying*. CRC Press, Boca Raton, FL.

Nascimento, J.J.S., Belo, F.A. & de Lima, A.G.B. (2015) Experimental drying of ceramics bricks including shrinkage. *Defect and Diffusion Forum*, 365, 106–111.

Perré, P., Remond, R. & Turner, I. (2007) Comprehensive drying models based on volume averaging: background, application and perspective. In: Tsotsas, E & Mujumdar, A.S. (eds.) *Modern drying technology*. Wiley VCH Verlag GmbH & Co. KGaA, Weinheim, Germany. pp. 1–12.

Pourcel, F., Jomaa, W., Puiggali, J.R. & Rouleau, L. (2007) Criterion for crack initiation during drying: alumina porous ceramic strength improvement. *Powder Technology*, 172 (2), 120–127.

Przesmycki, Z. & Strumillo, C. (1985) The mathematical modelling of drying process based on moisture transfer Mechanism. In: Mujumdar A.S. (ed.) *Drying '85*. Hemisphere, New York, NY. pp. 126–134.

Rattanadecho, P., Pakdee, W. & Stakulcharoen, J. (2007). Analysis of multiphase flow and heat transfer: pressure buildup in an unsaturated porous slab exposed to hot gas. *Drying Technology*, 26, 39–53.

Ring, T.A. (1996) *Fundamentals of ceramic powder processing and synthesis*. Academic Press, San Diego, CA.

Scherer, G.W. (1990) Theory of drying. *Journal of the American Ceramic Society*, 73 (1), 3–14.

Schmitz, A.V., Mutlu, Y.S., Glatt, E., Klein, S. & Nestler, B. (2012) Flow simulation through porous ceramics used as a throttle in an implantable infusion pump. *Biomedizinische Technik. Biomedical Engineering*, 57 (Suppl. 1), 277–280.

Stanish, M.A., Schajer, G.S. & Kayihan, F. (1986) A mathematical model of drying for hygroscopic porous media. *AIChE Journal*, 32 (8), 1301–1311.

van Genuchten, M. (1980) A closed-form equation for predicting the hydraulic conductivity of unsaturated soils. *Soil Science Society of America Journal*, 8, 892–898.

Witharana, S., Hodges, C., Xu, D., Lai, X. & Ding, Y. (2012) Aggregation and settling in aqueous polydisperse alumina nanoparticle suspensions. *Journal of Nanoparticle Research*, 14 (5), 851–862.

Zhang, Z., Yang, S. & Liu, D. (1999) Mechanism and mathematical model of heat and mass transfer during convective drying of porous materials. *Heat Transfer-Asian Research*, 28 (5), 337–351.

Part IV
Membranes for energy and environmental
applications (novel membrane developments
for gas separation)

CHAPTER 16

Effect of poly phenylene oxide (PPO) concentration on the morphological, thermal, crystalline and CO_2 permeation properties of as-synthesized flat sheet dense PPO polymeric membranes

Biruh Shimekit, Hilmi Mukhtar, Zakaria Man & Azmi Mohd Shariff

16.1 INTRODUCTION

Polymeric membranes are the most widely used industrial membranes for gas separation application due to their low cost and ease of processability compared to inorganic membranes. Early researches on gas separation applications using polymers focused on rubbery, semi-crystalline and some glassy polymers (Shu, 2007). The current industrial gas separation membrane materials are mostly derived from amorphous, thermoplastic polymers, of which poly (2,6-dimethyl-1,4-phenylene oxide) polymers; also commonly known as poly (2,6-dimethyl-1,4-phenylene ether) or usually written in short form as polyphenylene oxide (PPO) are well known as one of the most investigated polymeric materials for gas separation application. PPO is a linear amorphous thermoplastic polymer with a free phenolic hydroxyl on the head group of each chain. Structurally, PPO is made of phenylene rings linked together by ether linkages in the 1,4 or para-positions, with a methyl group attached to carbon atoms in the 2 and 6 positions (Chenar et al., 2006).

PPO has a glass transition temperature (T_g) ranging from 206 to 225°C depending on the molecular weight. The T_g of PPO increases with the molecular weight. The relatively high T_g of PPO is attributed to strong interactions between the polymeric chains (Chenar et al., 2006).

Its excellent mechanical, thermal and high impact strength have been attributed to its capacity to undergo rapid conformational transitions due to the free rotation of phenyl rings about the ether linkages in the polymer backbone. Due to the presence of phenyl rings, PPO is hydrophobic and has excellent resistance to water, acids, alcohols and bases (Mortazavi, 2004).

Among all glassy polymers, PPO shows one of the highest permeabilities to gases. This high permeability is attributed to the absence of polar groups in the main chain of PPO (Mortazavi, 2004; Sridhar et al., 2006). PPO also offers many advantages such as lower cost material, high thermal stability, widespread availability, and easy modification with functional groups. Hence, in the present study an attempt has been made to investigate the effect of the desired concentration of PPO on the morphological, thermal, crystalline and CO_2 permeation properties of as-synthesized flat sheet dense PPO membranes in the laboratory.

16.2 EXPERIMENTAL

16.2.1 Materials

PPO of 30,000 g mol^{-1} (high molecular weight) with intrinsic viscosity equal to 0.57 dL g^{-1} in chloroform at 25°C, 99.99% purity and density 1.06 g cm^{-3} was purchased from Sigma Aldrich® Co., Malaysia. Pure CO_2 was obtained from Gas Malaysia® Company, with purities at or above 99.99%. Analytical graded chloroform ($CHCl_3$) with 99.99% purity was also obtained from Sigma Aldrich® Co., Malaysia.

16.2.2   Membrane preparation

Prior to use, the poly (2,6-dimethyl-1,4-phenylene oxide), PPO powder was dried in a vacuum oven at 110°C for at least 12 h to remove the moisture. Flat sheet dense PPO membranes were prepared using the solution casting procedure where the oven dried PPO at desired loadings (5.5 wt%, 12.5 wt% and 20.5 wt%) were dissolved in 94.5, 87.5 and 79.5 wt% chloroform, respectively and stirred gently at 400 rpm for 1 h at or preferably below 60°C using magnetic stirrer to form a homogeneous solution. After mixing, the solution was left to stand for 12 h aimed at degassing the bubbles that might have been formed during mixing. Then the solution was poured and drawn by a casting knife on a flat, dry, smooth and dust-free glass plate. Afterwards, the obtained membrane in the form of thin film was carefully peeled off the glass plate spontaneously by the action of natural evaporation. While it was visually inspected for dust and pin holes, a razor blade was sometimes used to initiate delamination of the membranes at their edges. For cleaning and removal of moisture from the glass surfaces, acetone was used. The peeled off membranes were then kept in a vacuum oven for drying at 85°C for 12 h to remove the residual solvent and moisture prior to permeation test and were stored in a desiccator until used for further analysis.

16.2.3   Characterizations

16.2.3.1   Field emission scanning electron microscope (FESEM)
For morphological studies of the fabricated PPO membranes, field emission scanning electron microscope (FESEM) with model ZEISS SUPRA TM 55VP was used. In order to characterize membrane samples on the FESEM at the sample preparation stage, the membranes were cryo-genically fractured in liquid nitrogen by immersing them for several minutes to have clear cut of the cross sections and mounted on a circular stainless steel sample holder with electronically con-ductive double sided carbon adhesive tape. The samples were sputter-coated by gold/palladium using Polaron Range SC7640 to create a conductive coating environment for enhanced qual-ity of image. For most analyses of the FESEM, the micrographs were investigated using 5 kV accelerating voltage with magnification power (5000–100,000).

16.2.3.2   Thermogravimetric analyzer (TGA)
The thermal stability of the PPO membranes was examined using a Perkin Elmer® Instruments model Pyris 1 TGA (thermogravimetric analyzer) from 25–725°C at a heating rate of 10°C min^{-1} with inert nitrogen gas flushed at 20 mL min^{-1} in order to remove all corrosive gas involved in the degradation and to avoid thermal oxidative degradation. The actual analysis is performed by gradually raising the temperature of the sample and plotting the weight loss percentage against temperature. A 10 mg sample with the lower limit being set by the sensitivity of the balance and the upper limit by the size of the sample holder was checked. The T_g of the samples were determined as the midpoint temperature of the transition region in the second heating cycle.

16.2.3.3   Differential scanning calorimetry (DSC)
Differential scanning calorimetry (Shimadzu DSC60) was used to determine the glass transition temperatures (T_g) of the developed PPO membranes. Small sections of the PPO membranes were cut, weighed and placed into aluminum DSC pans. Samples were heated from 30 to 250°C at a rate of 10°C min^{-1} in N_2 atmosphere, and then the samples were cooled down to 30°C. The first cycles were carried out to remove the thermal history. The samples were heated again to 250°C with the same procedure for the second scan. The second scan thermograms were used to determine the glass transition temperature of the PPO membranes.

16.2.3.4   X-ray diffraction analysis (XRD)
The X-ray diffraction analysis of membranes was made by the XRD (Bruker A&S D8 advanced diffractometer) ranging from 2–60° 2θ at scanning speed of 1.2°C min^{-1} to determine the inter-segmental chain spacing (d-spacing). XRD spectra are plotted as intensity of XRD spectra (Cps)

against diffraction angle (2θ). As a general trend, the X-ray diffraction scans of amorphous polymers are typically dominated by a broad peak associated with the center to center chain distance of d-spacing.

16.2.3.5 *Response surface methodology (RSM)*

Response surface methodology (RSM) was employed to optimize the permeance of CO_2 in PPO homogeneous membranes. The results of experimental design were analyzed using MiniTab 14.13 statistical software to predict the response of the dependent response variable. The response surface is normally represented graphically, where the contour plots are often drawn to visualize the shape of the response surface. When the process is close to optimum, the second order model that incorporates curvature is usually required to approximate the response:

$$y = B_0 + \sum_{i=1}^{k} B_i x_i + \sum_{i=1}^{k} B_{ii} x_i^2 + \sum \sum_{i<j} B_{ij} x_i x_j + \varepsilon \tag{16.1}$$

where y is the predicted response, B_0 is the offset term, B_i is the linear effect, B_{ii} is the squared effect and B_{ij} represents the interaction effect. The optimized condition was obtained from contour plots graphically and also by solving the polynomial regression of determination R^2 and statistical significance with probability level of 0.5% was employed.

In the present study, two level factors were chosen, the factors included various PPO concentration (5.5, 12.5 and 20.5 wt%) and applied feed pressures (0.2, 0.6 and 1.0 MPa) and the permeance of CO_2 was the response. Screening of independent factors affecting the permeation of CO_2 was carried out with blocks of experiments.

16.2.3.6 *Gas permeation*

The CO_2 permeation flux of the as-synthesized PPO membranes was measured using a gas permeation testing unit. In such a system, the amount of CO_2 that permeates across the membrane per unit area (flux) was measured by constant volume-variable pressure method. This is maintained by the desired feed and permeate pressure together with constant volume with a known thickness and area of the membrane (14.5 cm^2) on permeate phase.

Each experiment was conducted at feed pressure of 0.2, 0.4, 0.6, 0.8 and 1.0 MPa and the permeate stream was at atmospheric pressure. The flow rate of the permeate stream was measured by the calibrated bubble flow meter. In this experiment, the time required to reach a desired volume of gas in the permeate stream was observed and recorded. The permeation of CO_2 through the membranes was measured twice at steady state condition. For ease of comparison and accuracy of results (i.e. to avoid any uncertainty due to slight fluctuation of the actual measurement of thickness of membranes), pressure normalized flux or permeance (*GPU*) was selected as the unit of measurement.

16.3 RESULTS AND DISCUSSION

16.3.1 *Effect of concentration of PPO polymer on the morphological properties of as-synthesized flat sheet dense PPO membranes*

The effect of concentration of PPO on the morphologies of the fabricated dense flat sheet PPO membranes was analyzed using the FESEM. Figure 16.1a–c shows the FESEM micrographs for the cross section view of the 5.5, 12.5 and 20.5 wt% PPO flat sheet dense membranes with their average thickness ranging between 3.5 and 5.5 µm.

Figure 16.1a–c clearly shows that there is no significant change in the morphology of the PPO membranes when increasing the concentration of the PPO polymer from 5.5 to 12.5 wt%. This is due to the homogeneity of the dense membranes that arise from good interaction of the PPO polymer and the chloroform at lower loadings of PPO. However, increasing PPO concentration

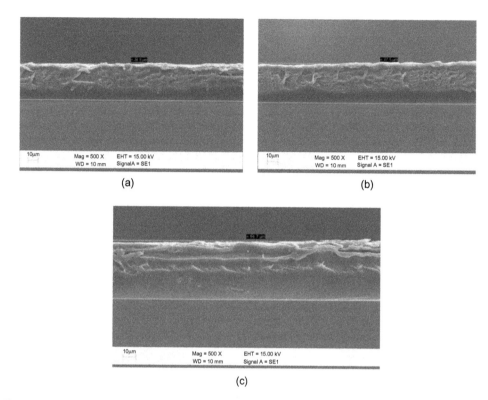

Figure 16.1.   FESEM of the cross sectional view of flat sheet PPO dense membranes of concentration: (a) 5.5 wt% (b) 12.5 wt% and (c) 20.5 wt% PPO.

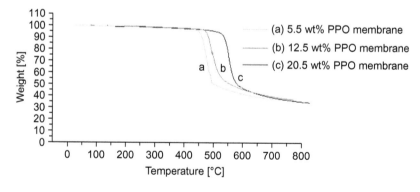

Figure 16.2.   Thermogravimetric (TGA) analyses of varied concentration of PPO membranes.

to 20.5 wt% has relatively increased the polymer chain entanglement (rigidity) than in the lower wt% PPO concentration.

16.3.2   *Effect of PPO concentration on the thermal properties of as-synthesized flat sheet dense PPO membranes*

Figure 16.2 describes the TGA profiles of 5.5, 12.5 and 20.5 wt% concentration of PPO membranes.

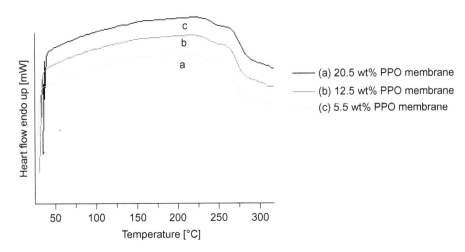

Figure 16.3.   DSC graphs of PPO dense membranes at varied concentration.

Figure 16.2 clearly shows that the weight losses of the PPO membranes were more significant for the lower wt% PPO polymer (such as 5.5 wt%) as there were high amounts of solvent present in the preparation stages indicating that at a given temperature, the applied heat removes more moisture and solvent from the lower wt% PPO polymers than the higher wt% PPO membranes (such as 20.5 wt%).

The TGA analysis of all PPO samples (Fig. 16.2) show that there was a maximum of 5.25% ± 0.1 loss of weight that was recorded till 400°C indicating that only the chloroform solvent (whose boiling point is 61.2°C) has been removed in the range. As the temperature increase from 400 to 500°C, the weight loss percent became significant (in the range of 51% to 54% for 5.5 wt% and 20.5 wt% PPO membranes, respectively).

For the range higher than 550°C, an average weight loss percentages of 8.14% ± 0.1, 8.06% ± 0.1 and 6.52% ± 0.1 were recorded for the 5.5, 12.5 and 20.5 wt% PPO membranes, respectively.

The maximum thermal degradation points for the tested PPO membranes were obtained by taking the derivatives of change in consecutive weight losses to the heating rate and plotting against temperature. By doing so, it was found that the thermal stability of the PPO membrane increases with an increase in the concentration of the PPO polymer, indicating that the rigidity (inhibition of polymer chain mobility) of the PPO polymer has been increasing through increasing the concentration of the critical polymer chain entanglement.

The highest thermal stability (463.25 ± 0.1°C) is attained for the 20.5 wt% PPO membrane, which exhibits higher decomposition temperature than the 5.5 and 12.5 wt% PPO membranes whose decomposition temperatures were 459.72 ± 0.1 and 461.23 ± 0.1°C respectively. The obtained decomposition temperature of the typical 20.5 wt% PPO membranes is comparable to the value of 456°C as reported by previous researchers such as O'Reilly (1977) and the value of 464°C also reported by Tran and Kruczek (2007).

The corresponding values for the glass transition temperature (T_g) of the PPO membranes (5.5, 12.5 and 20.5 wt%) have been obtained from the DSC curves are reported in Figure 16.3.

Figure 16.3 shows that the overall heat flow increased linearly with temperature until it reached a sharp change in the slope of the heat flow at 217.38, 218.26 and 219.14°C, for 5.5, 12.5 and 20.5 wt% PPO membranes, respectively. The next step change in the heat flow occurred at 219.33, 220.56 and 221.49°C again for the 5.5, 12.5 and 20.5 wt% PPO membranes, respectively. Hence, the midpoint temperature that defines the glass transition temperature of the aforementioned dense PPO membranes as calculated to be 218.36, 219.41 and 220.32°C for 5.5, 12.5 and 20.5 wt%

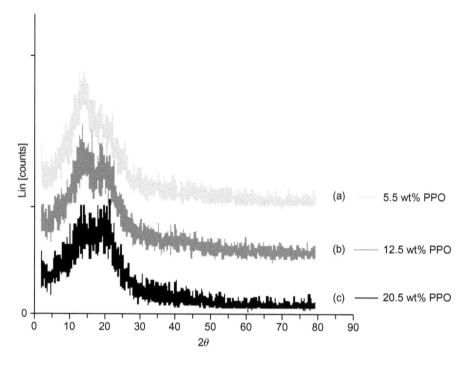

Figure 16.4.   XRD analysis of varied concentration of as-synthesized PPO membranes.

PPO membranes, respectively. Interestingly, the results obtained from the DSC were in a good agreement with previous researches (Cong et al., 2007; Guan et al., 2005; Yang et al., 2006).

Generally, the T_g value of the dense PPO membranes increases with an increase in the concentration of PPO in the casting solution confirming that the thermal stability of the PPO membranes is improved.

16.3.3   Effect of PPO concentration on the crystalline behavior of the as-synthesized flat sheet dense PPO membranes

Figure 16.4 depicts the X-ray diffraction (XRD) of varied concentrations of the as-synthesized PPO membranes.

Figure 16.4 shows that for the 5.5 wt% PPO, its major peak occurs at $2\theta = 14.2°$, which corresponds to a d-spacing of 6.23Å. For 12.5 wt% PPO, the major peak occurs at positions ($2\theta = 13.75°$ and $2\theta = 14.05°$) which corresponds to a d-spacing of 6.44 Å and 6.3 Å respectively. For the 20.5 wt% PPO, the major peak occurs at $2\theta = 21.5°$ which corresponds to a d-spacing of 4.13 Å. Generally, the obtained results are in a good agreement with the previous studies (Kim and Lee, 2001; Story and Koros, 1992).

As can clearly be observed from the analysis, the PPO membranes have the characteristics of amorphous to semi-crystalline behavior. It is also shown that the peak shift to a lower d-spacing (as shown for a higher concentrations such as 20.5 wt% PPO) suggest that the PPO polymer chains in the dense PPO membranes are arranged more orderly and tightly packed.

16.3.4 CO_2 permeation and parameters optimization

The typical CO_2 permeation data (measured using the gas permeation unit at constant volume and varied feed pressure) together with the RSM results for the tested PPO membranes whose PPO concentration ranges (5.5–20.5 wt%) are presented in Table 16.1.

Table 16.1. Comparison of the permeance of CO_2 using Equation (16.2) with that of the experiment in the PPO membranes.

PtType	Blocks	PPO [wt%]	Pressure [MPa]	Permeance [GPU]*	
				Experimental	Predicted
0	1	12.5	0.6	7.09	7.30
1	1	20.5	1.0	6.62	7.14
1	1	5.5	1.0	3.40	3.34
1	1	5.5	0.2	5.13	5.18
0	1	12.5	0.6	7.11	7.30
0	1	12.5	0.6	7.06	7.30
0	1	12.5	0.6	7.10	7.30
1	1	5.5	1.0	3.43	3.34
0	1	12.5	0.6	7.12	7.30
0	2	12.5	0.6	7.10	7.30
0	2	12.5	0.6	7.07	7.30
0	2	12.5	0.6	7.08	7.30
0	2	12.5	0.6	7.10	7.30
1	2	5.5	0.2	5.16	5.18
1	2	20.5	0.2	5.50	6.50
1	2	20.5	0.2	5.50	6.50
0	2	12.5	0.6	7.09	7.30
1	2	20.5	1.0	6.55	7.14

*GPU: pressure normalized flux or permeance; 1 GPU $= 0.33$ mol m^{-2} s^{-1} Pa^{-1}.

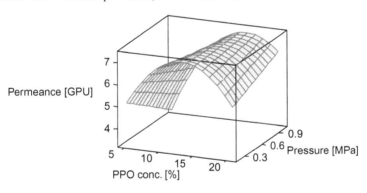

Figure 16.5. Surface plot of permeance of CO_2 in PPO homogeneous membrane versus pressure, PPO concentration.

By employing the RSM via the regression analysis, the measured CO_2 permeation data (Table 16.1) were fitted to the second order polynomial of the Equation (16.1). Hence, the quadratic regression model for permeation of typical CO_2 gas is given in Equation (16.2).

$$CO_2 \text{Permeance} = 2.34 + 0.81 \times PPO - 0.33 \times Pressure - 0.03 \times PPO^2$$
$$+ 0.02 \times PPO \times Pressure \tag{16.2}$$

The predicted permeance of CO_2 using (Equation (16.2)) is compared with the measured CO_2 permeation data (Table 16.1), using relatively few combinations of variables to determine the response function.

Further, in order to examine the effect of factors (feed pressure and PPO concentration) towards the CO_2 permeation, 3D surface graphical representation was used and presented in Figure 16.5.

As shown from Figure 16.5 and from the response surface methodology (RSM) that optimized the experimental conditions for the maximum permeation of CO_2 based on the present experiments, involving the concentration of PPO (5.5–20.5 wt%) and applied feed pressure (0.2–1.0 MPa), it was found that the optimal permeation of CO_2 is estimated at 7.26 GPU, the estimated optimum value of PPO concentration is 12.5 wt% and that of feed pressure is 0.25 MPa at 0.99 desirability. Hence, the 12.5 wt% PPO concentration is selected to be an optimal concentration.

16.4   CONCLUSION

Flat sheet dense PPO membranes of desired concentration (5.5, 12.5 and 20.5 wt%) were successfully fabricated in the laboratory. Morphological, thermal, crystalline and CO_2 permeation characteristic of the as-synthesized PPO membranes were also successfully analyzed. Cross-sectional views of the morphologies of the as-synthesized PPO membranes showed that increasing the concentration of the PPO polymer from 5.5 to 20.5 wt% demonstrates distinct morphology with possible polymer rigidification effect. The thermal stability of the PPO membranes also increased from (218.36 to $220.32 \pm 0.1^{\circ}$C) while increasing the concentration of the PPO polymer from (5.5 20.5 wt%). The X-ray diffraction analysis confirmed that the tested PPO membranes had the characteristics of amorphous behavior. The response surface methodology (RSM) demonstrated that for the feed pressure of (0.2–1.0) MPa, the optimal permeation of CO_2 was estimated to be 7.26 GPU at the 12.5 wt% PPO concentration and feed pressure of 0.25 MPa.

ACKNOWLEDGEMENT

The authors are grateful for the assistance and support provided by Universiti Teknologi PETRONAS.

REFERENCES

Chenar, M.P., Soltanieh, M., Matsuura, T., Tabe-Mohammadi, A. & Feng, C. (2006) Removal of hydrogen sulfide from methane using commercial polyphenylene oxide and cardo-type polyimide hollow fiber membranes. *Separation and Purification Technology*, 51, 359–366.

Cong, H., Hu, X., Radosz, M. & Shen, Y. (2007) Brominated poly(2,6-diphenyl-1,4-phenylene oxide) and its SiO_2 nanocomposite membranes for gas separation. *Industrial and Engineering Chemistry Research*, 46, 2567–2575.

Guan, R., Gong, C., Lu, D., Zou, H. & Lu, W. (2005) Development and characterization of homogeneous membranes prepared from sulfonated poly(phenylene oxide). *Journal of Applied Polymer Science*, 98, 1244–1250.

Kim, J.H. & Lee, Y.M. (2001) Gas permeation properties of poly(amide-6-b-ethylene oxide)-silica hybrid membranes. *Journal of Membrane Science*, 193, 209–225.

Mortazavi, S. (2004) *Development of polyphenylene oxide and modified polyphenylene oxide membranes for dehydration of methane*. PhD Thesis, University of Ottawa, Ottawa, ON, Canada.

O'Reilly, J.M. (1977) Conformational specific heat of polymers. *Journal of Applied Physics*, 48, 4043–4048.

Shu, S. (2007). *Engineering the performance of mixed matrix membranes for gas separation*. PhD Thesis, Georgia Institute of Technology, Atlanta, GA.

Sridhar, S., Smitha, B., Ramakrishna, M. & Aminabhavi, T.M. (2006) Modified poly(phenylene oxide) membranes for the separation of carbon dioxide from methane. *Journal of Membrane Science*, 280, 202–209.

Story, B.J. & Koros, W.J. (1992) Sorption and transport of CO_2 and CH_4 in chemically modified poly(phenylene oxide). *Journal of Membrane Science*, 67, 191–210.

Tran, A. & Kruczek, B. (2007) Development and characterization of homopolymers and copolymers from the family of polyphenylene oxides. *Journal of Applied Polymer Science*, 106, 2140–2148.

Yang, S., Gong, C., Guan, R., Zou, H. & Dai, H. (2006) Sulfonated poly(phenylene oxide) membranes as promising materials for new proton exchange membranes. *Polymers Advanced Technologies*, 17, 360–365.

CHAPTER 17

Critical concentration of PEI in NMP as a criterion for the preparation of asymmetric membranes for gas separation

Ahmad Arabi Shamsabadi, Masoud Bahrami Babaheidari & Kaveh Majdian

17.1 INTRODUCTION

Separation of gases is one of the essential processes for many industries such as refineries and petrochemical complexes. There are primarily four conventional techniques for gas separation including absorption, adsorption, cryogenic distillation and membrane process. The membrane separation process offers a number of advantages in terms of no phase change, no usage of sorbent materials, small footprint, low maintenance requirements, low cost, high process flexibility and high energy efficiency (Takht Ravanchi et al., 2009). Also membranes offer an attractive potential technology to offset traditional bulky and often less environmentally friendly means of separation (Koros and Fleming, 1993). Gas separation by polymer membranes is a proven technology that has found a wide range of industrial applications (Staudt-Bickel and Koros, 1999). Polymer membranes are widely used in separation processes as they have acceptable film forming capability, toughness, flexibility and low costs. They have gained an important place in chemical technology and are applicable to various separation processes of many chemicals and gases (Seo et al., 2006).

In the olefin plant of Marun Petrochemical Company, hydrogen is used as a green fuel for cracking furnaces and hydrogenation process and in poly-olefins plants it is used to terminate polymer reaction. Because of the inadequate purity of the methane to use as balance gas in the ethylene oxide plant of Marun Petrochemical Co., the stream after turbo-expanders, containing hydrogen and methane, is sent to flare. Hydrogen recovery from this tail gas is an economical and environmental policy of this company. It can help to recycle hydrogen as a valuable resource and to purify methane for using, instead of using nitrogen to increase the loading of the ethylene oxide plant by reducing the recycle gas compressor duty.

Commercially available polyetherimide (PEI) has several important advantages as a membrane material. This polymer has good chemical and thermal stability. The studies on gas permeation in the PEI dense film reveal that PEI exhibits impressively high selectivity for many important gas pairs. In 1987, Peinemann (1987) prepared PEI flat-sheet asymmetric membranes for CO_2/CH_4 and reported selectivity of about 30–40. Kneifel and Peinemann (1992) reported their results in preparing porous and dense PEI hollow fiber membranes for gas separation and they obtained selectivity of about 170 for He/N_2 separation. Wang et al. (1998) reported the results of PEI hollow fiber membranes for separation of N_2 from He, H_2, CO_2, CH_4 and Ar. Kurdi and Tremblay (2003) prepared PEI/metal complex asymmetric hollow fiber membranes for O_2/N_2 separation and obtained selectivity of 2–7. Ren and Deng (2010) prepared polyetherimide membrane and investigated the influence of various non-solvents on membrane morphology.

Low concentration of the polymer solution results in a porous membrane, which is not suitable for gas separation. Preparation of the polymer membranes with different concentrations is a method for optimizing the effect of the polymer concentration but it is expensive and time-consuming. Chung and Kafchinskin (1997) reported that for each couple of polymer and solvent there is a concentration at which molecule chains have good linkage and therefore the viscosity can change quickly by the polymer concentration.

17.2   EXPERIMENTAL

17.2.1   *Materials*

Polyetherimide (PEI) was obtained from Sigma-Aldrich, USA, in pellet form. Ethanol, methanol, isopropanol, *n*-hexane and anhydrous 1-methyl-2-pyrrolidinone (EMPLURA®, 99.5%, water <0.1%) were supplied from Merck, Germany.

Demineralized water was supplied from Marun Petrochemical Company. Polydimethylsiloxane (PDMS and curing catalyst) was bought from Z-mark Co., Italy. The pure gases including CH_4 (99.9%) and H_2 (99.9%) were supplied from Technical gas services. These gases were used in permeation measurement experiments.

17.2.2   *Characterizations*

Asymmetric membranes were prepared with 28 wt% PEI in NMP solvent by phase inversion method. The PDMS coating solution was prepared by dissolving the PDMS polymer resin and the cross-linking agent with a weight ratio of 10:1 in *n*-hexane to obtain a homogeneous solution. A conventional set up was used for gas permeance measurement with highly accurate transmitters and a PLC. The Mettler Toledo digital density meter and a Koehler viscometer (Model HKV3000) were used for density and viscosity measurements.

17.3   RESULTS AND DISCUSSION

17.3.1   *The influence of the polymer concentration on the solution density and viscosity*

Density of the polymer solution is needed for the calculation of thermodynamics and transport properties of the polymer solution. Results showed for PEI in NMP at a given temperature, the solution density increased linearly by increasing the polymer mass fraction. Also, as the temperature increased, the density decreased linearly, as well. The viscosity of the polymer solution depends on concentration and size (i.e., molecular weight) of the dissolved polymer, used solvent and temperature. Viscosity techniques are very popular because they are experimentally simple. By measuring viscosity, studying of the polymer solution properties is possible. Increasing of the polymer concentration can promote the solution viscosity where ln μ vs. X is linear where μ is the dynamic viscosity of the polymer solution (cp) and X is polymer concentration [wt%]. As temperature increases, the viscosity decreases. At higher temperatures, the molecular movement of the polymer is easier due to the following reasons:

- Decrease of the solvent viscosity at higher temperatures.
- Decrease of the inter-chain liaisons.
- Increase in the polymer solubility.

All these factors contribute to decrease the viscosity as the temperature increases.

17.3.2   *Critical concentration determination*

Above and below of the critical concentration, the entanglement of the macromolecular chains changes dramatically. Figure 17.1 shows the effect of the polymer mass fraction on the solution viscosity at 25°C. It was found that the solution viscosity increased as the polymer concentration of the solution was raised. As shown in Figure 17.1, the viscosity data showed a drastic slope change in a range of the polymer concentration between 20 and 30 wt%. The intersection of the tangents at these two concentrations determines the critical concentration of about 24 wt%, where a significant change in the degree of chain entanglement occurs (Chung and Kafchinskin, 1997). Therefore, the membranes which are cast from solutions with a polymer concentration higher than 24 wt% have a high entangled conformation, tighter intermolecular chain displacement and

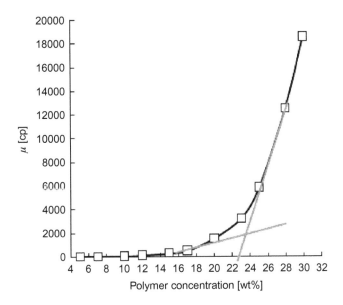

Figure 17.1.   The effect of PEI mass fraction in NMP solvent on viscosity at 25°C.

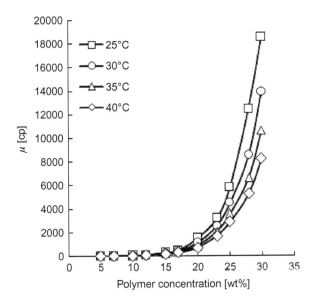

Figure 17.2.   The viscosity of PEI solutions in NMP solvent at various temperatures.

less defective outer skin layers than those cast from the solutions with a polymer concentration lower than 24 wt% (Ismail and Lai, 2003).

Because of the importance of temperature in the preparation of gas separation membrane, the range of 25–40°C was selected for study of the influence of temperature on the solution viscosity. Figure 17.2 represents the influence of temperature on the PEI solutions viscosity in NMP. It was found that the solutions viscosity decreased by increasing temperature. The critical concentration of PEI in NMP was not changed significantly at different temperatures.

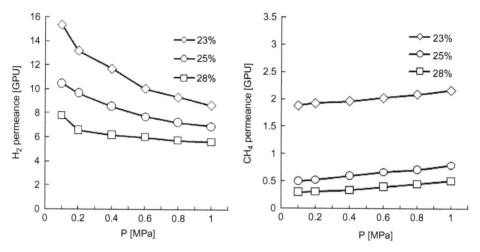

Figure 17.3 (right) and 17.4 (left). H_2 and CH_4 permeance in PEI membrane with different concentrations at different pressures and 25°C.

17.3.3   *The effect of the polymer concentration*

Polymer concentration has been identified as one of the most influential parameters on the membrane performance and morphology. Increasing the polymer concentration in the dope solution leads to a thicker selective layer and higher selectivity. Increasing the polymer concentration leads to stronger interaction between polymer/solvent, and polymer/non-solvent. These effects can promote the coagulation rate for dense selective layer and slower the precipitation rate for the sub-layer. Therefore, increasing the polymer concentration leads to an asymmetric membrane with a dense selective layer and a closed cell substrate suitable for gas separation.

Polymeric membranes were prepared with three different concentrations of PEI, lower and upper critical concentration, (23, 25 and 28 wt%). The prepared membranes were coated by 15 wt% PDMS in *n*-hexane solution. Figures 17.3 and 17.4 show H_2 and CH_4 permeance through the PEI membranes with different concentrations as a function of pressure at 25°C, respectively.

The results show that the permeance of both H_2 and CH_4 declined with increasing the polymer concentration due to the fact that membrane preparation from a dilute polymer solution results in a thinner and sometimes porous skin layer, leading to a higher permeability but low degree of selectivity for gas separation. Figure 17.5 shows the cross sectional SEMs of the membranes fabricated from different concentrations of PEI. These membranes exhibited a typical asymmetric structure with developed macrovoids and a thick dense top layer.

As shown in the cross sectional SEMs, the prepared membranes from casting solutions with higher polymer concentration exhibited less finger-like macro voids and a more obvious porous substructure. On the other hand, the membranes produced from the casting solutions with lower polymer concentrations showed lower overall thickness that has led to higher permeability and lower selectivity.

Figure 17.6 indicates the selectivity of H_2/CH_4 for different concentrations of PEI at different pressures and 25°C. The selectivity of H_2/CH_4 was enhanced by increasing the polymer concentration. By increasing the polymer concentration, it was observed that the skin layer thickness is increased and the surface porosity decreased.

17.3.4   *Effect of non-solvent type*

The formation of asymmetric polymer membranes by immersion precipitation process is based on the phenomenon of liquid-liquid phase separation. This phase separation is an important

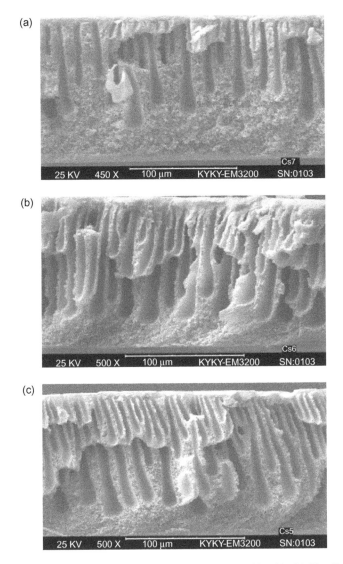

Figure 17.5.   Cross sectional SEM of un-coated membrane with (a) 28 wt%, (b) 25 wt% and (c) 23 wt% concentrations.

feature of the membrane formation process that occurs in the polymer solution after immersion in a non-solvent bath, with a great influence on membrane structure. The used non-solvent in the phase inversion method can affect the membrane behavior and morphology. Research on different coagulants in the PEI/NMP/non-solvent phase inversion process provides information about a good non-solvent for this process. The precipitation time is the most important parameter to determine membrane structure. The precipitation time was defined to describe the phase inversion process qualitatively. This time was recorded as the time period from when the wet cast film was immersed into the coagulation bath, to when the film separated completely from the glass plate. Water, water/2-propanol (4:1), methanol, ethanol and 2-propanol were use as non-solvents. Figure 17.7 represents the effect of difference in solubility parameters of solvent and no-solvent on precipitation time.

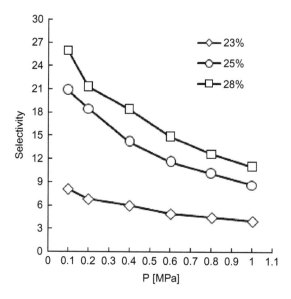

Figure 17.6.   Selectivity of H_2/CH_4 for different concentrations of PEI at different pressures and 25°C.

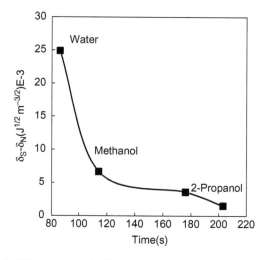

Figure 17.7.   Effect of difference in solubility parameters of solvent-no-solvent on precipitation time at 25°C.

Comparison between solubility parameters of non-solvents and solvent shows that lower difference between the solubility parameters enhances precipitation time. In general, the fast coagulation rate (low precipitation time) results in a formation of large finger-like macrovoids and cavity-like structures, whereas the slow coagulation rate results in a porous sponge-like structure (Deshmukh and Li, 1998).

Figure 17.8 presents SEM of fabricated membrane with different non-solvents. As shown in SEMs using water as non-solvent resulted in a finger-like membrane structure and for isopropanol sponge-like was formed; from water to isopropanol macro voids reduces and sponge-like structure was formed as expected according to Figure 17.7.

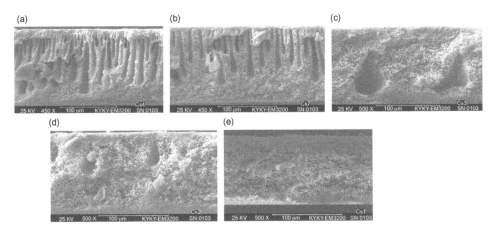

Figure 17.8.   Cross sectional SEM of fabricated membrane (28 wt%) with different non-solvents from left to right (water, water/isopropanol (4:1), methanol, ethanol and isopropanol).

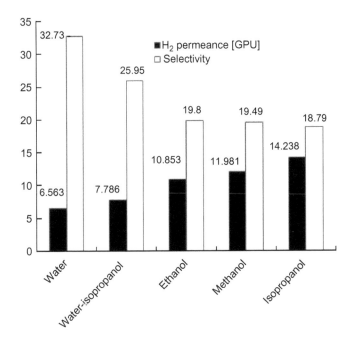

Figure 17.9.   Effect of different non-solvents on selectivity and permeability of pure H_2 at 0.1 MPa and 25°C.

Figure 17.9 shows selectivity and pure hydrogen permeance for different non-solvents at 0.1 MPa and 25°C. In the case of water, because of the lowest precipitation time finger-like macrovoids and cavity-like structures will be resulted and the highest selectivity and the lowest permeance was obtained. With increasing precipitation time more porous structures are formed and permeance was enhanced and selectivity was declined.

17.4   CONCLUSION

Critical concentration of the PEI solutions in NMP solvent was determined by viscometric method as 24 wt%. The influence of temperature on critical concentration was investigated and no significant effect was observed. The study of the polymer mass fraction and temperature effects on the viscosity and density of the PEI solution in NMP showed that as the temperature increased, both the viscosity and density decreased. In addition, by increasing the polymer mass fraction the viscosity and density of the polymer solution were enhanced. Finally, three asymmetric PDMS/PEI membranes with different concentrations of PEI were prepared and characterized for H_2/CH_4 separation. The results showed that the selectivity increased for the polymer concentrations in the dope solution higher than the critical concentration which led to a thick selective layer that resulted in preparation of more suitable membranes for gas separation. The use of water as non-solvent resulted in denser membrane due to its lower precipitation time (higher difference between solubility parameter of solvent and non-solvent). Using isopropanol resulted in membranes with sponge-like structure.

ACKNOWLEDGEMENT

We are grateful for the financial support provided by Marun Petrochemical Company and Petroleum University of Technology. The authors thank Dr. Ali Kargari for his assistance during this research.

REFERENCES

Chung, T.S. & Kafchinskin, E.R. (1997) The effects of spinning conditions on asymmetric 6FDA/6FDAM polyimide hollow fibers for air separation. *Journal of Applied Polymer Science*, 65, 1555–1569.

Deshmukh, S.P. & Li, K. (1998) Effect of ethanol composition in water coagulation bath on morphology of PVDF hollow fiber membranes. *Journal of Membrane Science*, 150, 75–85.

Ismail, A.F. & Lai, P.U. (2003) Effects of phase inversion and rheological factors on formation of defect-free and ultrathin-skinned asymmetric polysulfone membranes for gas separation. *Separation and Purification Technology*, 33, 127–143.

Kneifel, K. & Peinemann, K.V. (1992) Preparation of hollow fiber membranes from polyetherimide for gas separation. *Journal of Membrane Science*, 65, 295–307.

Koros, W.J. & Fleming, G.K. (1993) Membrane-based gas separation. *Journal of Membrane Science*, 83, 1–80.

Kurdi, J. & Tremblay, A.Y. (2003) Improvement in polyetherimide gas separation membranes through the incorporation of nanostructured metal complexes. *Polymer*, 44, 4533–4540.

Peinemann, K.V. (1987) Method for producing an integral asymmetric membrane and the resultant membrane. US Patent 4,673,418.

Ren, J.Z. & Deng, M. (2010) Morphology transition of asymmetric polyetherimide flat sheet membranes with different thickness by wet phase-inversion process. *Separation and Purification Technology*, 74, 119–129.

Seo, Y., Kim, S. & Hong, S.U. (2006) Highly selective polymeric membranes for gas separation. *Polymer*, 47, 4501–4504.

Staudt-Bickel, C. & Koros, W.J. (1999) Improvement of CO_2/CH_4 separation characteristics of polyimides by chemical crosslinking. *Journal of Membrane Science*, 155, 145–154.

Takht Ravanchi, M., Kaghazchi, T. & Kargari, A. (2009) Application of membrane separation processes in petrochemical industry: a review. *Desalination*, 235, 199–244.

Wang, D., Li, K. & Teo, W.K. (1998) Preparation and characterization of polyetherimide asymmetric hollow fiber membranes for gas separation. *Journal of Membrane Science*, 138, 193–201.

CHAPTER 18

Powder preparation effect on oxygen permeation flux of hollow fiber LSCF6428 ceramic perovskite membrane

Haikal Mustafa, Ammar Mohd Akhir, Sharifah Aishah Syed Abdul Kadir, Ahmad Fauzi Ismail, Pei Sean Goh, Be Cheer Ng & Mohd Sohaimi Abdullah

18.1 INTRODUCTION

Separation of oxygen from air for industrial applications is a growing operation that produces nearly 100 million tons of oxygen each year. The demand for oxygen will tremendously increase in the near future because virtually all large-scale clean energy technologies will require oxygen as one of the operation feeds (Emsley, 2001). The current industrial standard process to produce pure oxygen from air is using cryogenic distillation or pressure swing adsorption (PSA). The cryogenic distillation process is known to be complex, expensive and energy intensive. The latter practice, PSA involves the separation of air at ambient temperature using molecular sieve adsorbents to trap nitrogen in order to produce oxygen with purities of 90 to 95%. The inability to achieve higher oxygen purity has reduced the probability of PSA to be used in future clean energy endeavors (Hashim et al., 2011). It is obvious that both methods have their disadvantages.

Based on the finding of Teraoka et al. (1985) oxygen separation from air through perovskite dense ceramic membrane $La_{1-x}Sr_xCo_{1-y}Fe_yO_{3-\delta}$ showed potential to become an alternative method to produce pure O_2. Perovskite-type membrane with a structure of ABO_3 contains transition metals at the B site that show high electrical conductivity. The partial substitution of A site cations by other metal cations with lower valencies resulted in the formation of oxygen vacancies and appearance of ionic conduction. This oxygen sorptive property coupled with the electronic conductive properties as discussed earlier has suggested the possibility of using the defect perovskite-type oxides as an oxygen permeating membrane which can work without any need of electrodes and external electric circuit (Teraoka et al., 1985).

Previous research (Luyten et al., 2000; Tan et al., 2008a; Teraoka et al., 1988; Xu and Thomson, 1998) also indicated that perovskite dense membranes can provide a reliable and efficient option to produce pure oxygen.

In order to achieve high oxygen flux through the perovskite dense membrane, several factors must be taken into consideration. Firstly, the powder synthesis routes in which the route chosen must be able to produce powder with a single phase perovskite structure, maintain correct stoichiometry and exhibit high surface area (Richardson et al., 2003). Secondly, the membrane configuration should also exhibit the least oxygen permeation resistance and has appreciable structural strength under different air flow rate and pressure (Tablet et al., 2005). Lastly, it is imperative that the perovskite chemical composition selected is able to show good chemical stability under a reducing environment and at elevated temperature (Xu and Thomson, 1998). In this research, by using laboratory prepared and commercially procured LSCF6428 powders, 2 types of hollow fiber membranes were fabricated, characterized and evaluated. The purpose of this investigation is to study the relationship between the LSCF6428 powder preparation method and the oxygen permeation flux of the fabricated hollow fiber membranes.

LSCF6428 powders prepared through 3 methods; solid state reaction, reactive grinding and co-precipitation were used for comparison due to their ability in producing powders that meet the criteria of single phase perovskite structure, achieving the desired stoichiometry compound and exhibiting relatively high average surface area value around $5\,m^2\,g^{-1}$ (Richardson *et al.*, 2003; Xu and Thomson, 1999; Zawadzki *et al.*, 2011). Furthermore, these 3 methods of preparing the perovskite powder with desired product properties have been well established and the procedures involved are well documented. However, in this study, when compared to commercially procured LSCF6428 powder, the laboratory prepared powder was certainly less inferior by having a lower average surface area value than the commercially procured powder.

Membrane was fabricated in hollow fiber configuration due to several reasons. In most previous studies, disc-shaped membranes with only limited membrane area ($<5\,cm^2$) were usually used because they can be easily fabricated using conventional static-pressing methods (Yaremchenko *et al.*, 1999; Zeng *et al.*, 1998). Although a multiple planar stack can be adopted to enlarge the membrane area to a plant scale, many problems such as high temperature sealing, connection, and pressure resistance have to be faced (Li *et al.*, 1999). Tubular perovskite membranes were developed to reduce the engineering difficulties, especially the problem associated with high-temperature sealing (Steele, 1996; Zeng *et al.*, 1998). Their main disadvantages are small surface area/volume ratio and the high membrane thickness, which would in turn lead to low oxygen permeation fluxes, and these problems make them unfavorable in practical applications. On the other hand, hollow-fiber ceramic membranes with an asymmetric structure possess much larger membrane area per unit volume for oxygen permeation activities. By adopting long hollow fibers and keeping the two sealing ends away from the high-temperature zone, the problem of high-temperature sealing no longer exists. Furthermore, because of the asymmetric structure, the membrane's resistance to oxygen permeation is substantially reduced compared to that of symmetric membranes prepared by conventional methods. In addition, the integrated porous layers on either side or both sides of the membrane also provide much larger gas-membrane interfaces for oxygen-exchange reactions, leading to an enhancement of surface oxygen-exchange kinetics and thus to an improved oxygen permeation flux (Tan *et al.*, 2004).

LSCF6428 composition was selected for this research due to its abilities that exhibit chemical stability at reducing environment and elevated temperature. A research by Xu and Thomson (1998) found that LSCF6428 composition showed no phase changes when the LSCF powder was heated at 850°C in helium or argon for 8 h or nitrogen for 12 h. The partial substitution of the A site (La^{3+}) with Sr^{2+} in LSCF can maintain its chemical stability in inert gas at 850°C. It was also reported that the membrane is able to retain its mechanical and chemical stability in air at operating temperatures up to 960°C. In another research conducted by Tan *et al.* (2010), the extended period stability experiments have recorded excellent stability of LSCF6428 membranes system which was operated for more than 1167 h at around 960°C to produce 99.4% purity of oxygen concentration. Obviously, LSCF6428 composition has advantages to be operated in a reducing environment, high temperature conditions and for a prolonged period of operation.

18.2   EXPERIMENTAL

18.2.1   *LSCF6428 powders*

The commercially prepared LSCF6428 powder was purchased from Inframat Advanced Materials, United States of America. In order to acquire laboratory prepared LSCF6428 powder using the solid state reaction method, stoichiometric amounts of La_2O_3, Fe_2O_3, Co_2O_3, $SrCO_3$ powders were mixed and milled with water using a planetary ball miller at 300 rpm for duration of 24 h. The mixture was then dried in an oven for 24 h at 80°C. Finally the sample was calcined in a furnace at 900°C for 5 h. The reactive grinding route involved milling a stoichiometric mixture of La_2O_3, $SrCO_3$, Co_2O_3 and Fe_2O_3 powders with ethanol using a ball mill for period of 48 h. The powders were then calcined for 5 h at 900°C. The last method, the co-precipitation method, involved dissolving a stoichiometric mixture of $La(NO_3)_3 \cdot 6H_2O$, $Sr(NO_3)_2$, $Fe(NO_3)_3 \cdot 9H_2O$ and

$Co(NO_3)_3 \cdot 9H_2O$ in deionized water (0.2 M solution). An equal volume of precipitant solution was prepared of higher molarity, ensuring the precipitating ion was in excess throughout the experiment such as 5 M potassium hydroxide (KOH). Whilst stirring the precipitant solution vigorously, the individual nitrate solution was gradually added. The resulting slurries were vacuum filtered prior to being washed with deionized water and dried in an oven at $110°C$ for 24 h. A mortar and pestle were used to mix together the dried powders. Samples of the powders were characterized to determine their properties using X-ray diffraction (XRD) analysis, scanning electron microscopic (SEM) imaging and Brunauer, Emmett, Teller (BET) analysis. These values were then compared to the commercial purchased powders properties, and only one of the laboratory powders that had the best values was chosen to proceed with the membrane fabrication stage along with the purchased powder.

18.2.2 Hollow fiber membranes

Previous research work by Tan *et al.* (2004) was used as reference for the fabrication process. In this process, polyethersulfone (PESf) was used as the polymer solution, poly(vinylpyrrolidone) (PVP) as the plasticizer, *N*-methyl-2-pyrrolidone as the binder. Firstly, a calculated quantity of PESf with PVP additive was dissolved in the weighed NMP solvent in a 250 mL (wide-neck bottle). The weighed amount of LSCF6428 powder was then added gradually under stirring at around 300 rpm to ensure the uniform dispersion of the powder in the polymer solution. The stirring was carried out continuously for 48 h before the spinning process. After degassing at room temperature for 30 min, the starting solution was then transferred to a stainless steel reservoir and pressurized to 0.10 MPa using nitrogen. A spinneret with an orifice diameter and inner diameter of 2.0 and 1.0 mm, respectively, was used to obtain the hollow fiber precursors. Distilled water and tap water were used as the internal and external coagulants, respectively. The formed hollow fiber precursors were immersed in a water bath for more than 24 h to complete the solidification process. Finally, the resulting hollow fiber precursors were heated in a furnace at $600°C$ for 2 h to remove the organic polymer binder and were then sintered at high temperatures $1300°C$ for 4 h to allow the fusion and bonding to occur. Through these procedures, two types of hollow fiber membranes were fabricated, i.e. C-LSCF (commercial powder) and L-LSCF (laboratory powder). The membranes were then characterized through XRD analysis, SEM imaging and 3 point bending test.

18.2.3 Oxygen permeation experiments

The experiments were also conducted based on previous work that was reported by Tan *et al.* (2004; 2005) as point of reference. In order to proceed with oxygen permeation testing, the hollow fiber membrane module was first prepared by using two types of quartz tube of different lengths. A ceramic cement paste, Sauereisen Aluseal Adhesive Cement No. 2 purchased from Ellsworth Adhesives, Malaysia was used for the sealing purpose of the module. The exposure length of the hollow fiber membranes used for the experiment was set at 2 cm due to its low mechanical strength and brittleness characteristic which inhibits longer exposure length. After the membrane module was set up, it was left to cure for 24 h at room temperature. The final stage of the rig set up involved the connection of all the gas tubes to the membrane module. Two sets of module that consist of each C-LSCF and L-LSCF membrane, respectively, were constructed for the oxygen permeation experiments. In the experiments, both modules were subjected to 3 different temperatures (800, 850 and $900°C$). In order to achieve proper pressure drive, the feed oxygen gas was set at 0.11 MPa and nitrogen gas was set at 0.10 MPa, respectively. When the furnace reached the desired temperature, these gases were released for a period of time to attain a stable condition. Afterward, the sample gas at the module outlet was collected using a sample bag and oxygen concentration was tested using gas chromatography equipment. The oxygen permeation flux was then calculated using the established formulas. (Tan and Li, 2002).

Table 18.1.   LSCF6428 powders surface values from BET analysis [$m^2 \, g^{-1}$].

Method	1st Sample	2nd Sample	3rd Sample	Average
Solid-state reaction	7.08	7.09	7.08	7.08
Reactive grinding	6.69	6.67	6.70	6.69
Co-precipitation	5.77	5.63	5.75	5.72
Commercial	8.61	8.59	8.62	8.61

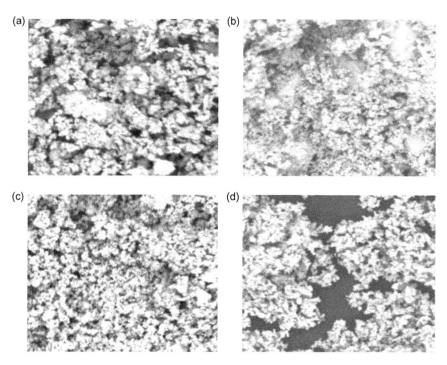

Figure 18.1.   SEM images of LSCF6428 powder types: (a) commercial, (b) solid state reaction method, (c) reactive grinding and (d) co-precipitation.

18.3   RESULTS AND DISCUSSION

18.3.1   *LSCF6428 powders preparation*

Table 18.1 shows the surface area value of each LSCF powders. C-LSCF powder exhibited the highest surface area followed by solid-state reaction (SS), reactive grinding, (RG) and co-precipitation (CP) powder. This trend of descending surface area values has also been observed and reported by Richardson *et al.* (2003) and Zawadzki *et al.* (2011). Powder produced through solid state reaction possessed the highest surface area value of 7.08 $m^2 \, g^{-1}$ among the 3 laboratory methods. This value is consistent with the previous work which said that the SS method's ability to attain such value was due to its powder preparation by grinding physically the raw materials (Richardson *et al.*, 2003). SEM images of the powders showed relatively small particle size and uniform structure except powder produced via the co-precipitation method. Figure 18.1d clearly shows in the powder formation. Such formation is expected since this method involves agglomeration of particles to form a new powder phase. Based on the XRD analysis results shown in Figure 18.2, powders produced through SS and RG gave a single phase perovskite

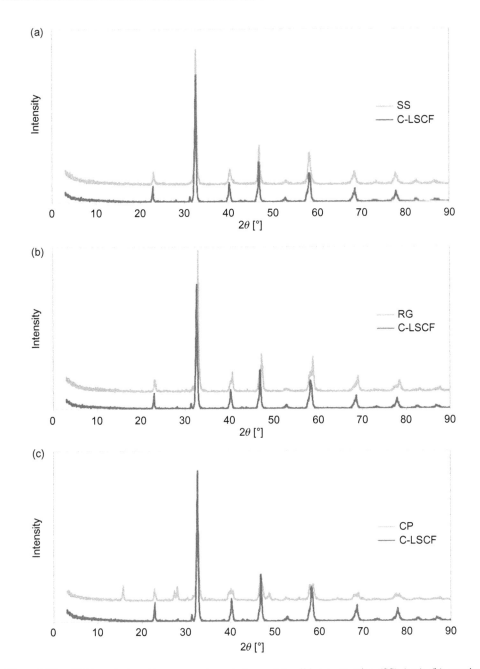

Figure 18.2.   XRD analysis pattern comparisons between (a) solid-state reaction (SS) (top), (b) reactive grinding (RG) (top) and (c) co-precipitation (CP) (top) and C-LSCF powder (bottom).

pattern while the powder fabricated via the CP route showed some impurities building up. As reported by Taheri *et al.* (2010), the occurrence of impurities in the co-precipitation produced powder may be due to the loss of strontium hydroxide during washing of the co-precipitated gel. From the preparation process aspect, even though reactive grinding applied the same principle as solid state reaction which is physical compounding, it required a longer period of milling, making it an unfavorable method to produce large quantity LSCF6428 powder. Also the usage of ethanol

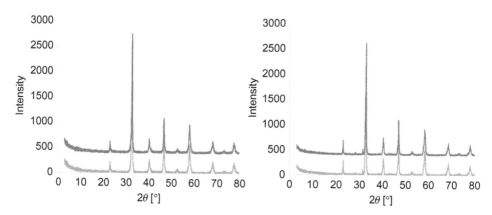

Figure 18.3.   XRD analysis pattern of (a) laboratory prepared and (b) commercially procured between LSCF6428 powder (bottom) and LSCF6428 membrane (top).

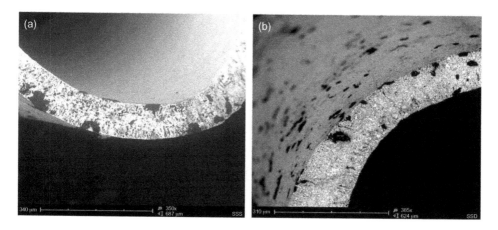

Figure 18.4.   SEM images of the fabricated LSCF6428 membrane: (a) L-LSCF and (b) C-LSCF.

in its procedure posed some safety issues such as combustion during the milling process. Powder produced through co-precipitation showed impurities building up in its composition, non-uniform powder particle size and relatively low surface area value. In addition, co-precipitation method involved numerous steps in its process. Therefore, this method has been identified as a poor choice to produce a high volume of LSCF6428 powder, given that the powder prepared through solid state reaction has achieved the desired requirements; single phase perovskite structure, small and uniform particle size and high surface area value. Its straight forward approach that only involves a few parameters makes this method a suitable alternative technique to produce perovskite LSCF6428 in large quantity. Therefore, it was chosen to proceed in the membrane fabrication process.

18.3.2   *Hollow fiber membranes fabrication*

The results of XRD analysis and SEM imaging are shown in Figures 18.3 and 18.4, respectively. From XRD analysis results, it was found that the membrane maintained its single phase perovskite structure albeit with increased intensity as compared to its powder due to enlargement of the crystal in the membrane resulting from the sintering process at high temperature. From the SEM images, both membranes exhibited fingerlike structures that formed near both the inner and

Table 18.2.   Oxygen permeation fluxes collected from C-LSCF Module and L-LSCF Module at various operating temperatures.

Temperature [°C]	C-LSCF Module [cm^{-3} cm^{-2} min^{-1}]				L-LSCF Module [cm^{-3} cm^{-2} min^{-1}]			
	Sample 1	Sample 2	Sample 3	Average	Sample 1	Sample 2	Sample 3	Average
900	0.1643	0.1637	0.1647	0.1642	0.1651	0.1643	0.1662	0.1652
850	0.0934	0.0922	0.0940	0.0932	0.0937	0.0949	0.0941	0.0942
800	0.0616	0.0627	0.0614	0.0619	0.0621	0.0629	0.0631	0.0627

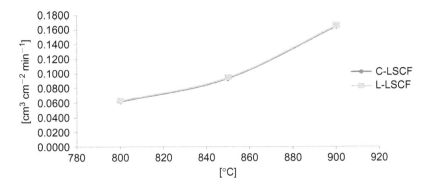

Figure 18.5.   Oxygen permeation flux comparison between L-LSCF membrane module and C-LSCF membrane module.

outer walls, and a sponge-like layer appeared at the center of the fiber membrane. The formation of such asymmetric structures can be attributed to the rapid precipitation occurring at both the inner and outer walls that are close to the coagulants, resulting in short finger-like pores, but the low precipitation at the center of the fiber resulted in the formation of a sponge-like structure (Tan *et al.*, 2004). However, these features are less significant on the C-LSCF membrane due to the smaller average particle size of its powder (high surface area value), which makes the powder closely packed during the membrane fabrication process, hence resulting in better fusion during the sintering stage (Liu *et al.*, 2001). The 3 point bending test provided the mechanical strength measurement of 28 MPa for L-LSCF membrane and 30 MPa for C- LSCF membrane but both values are less than the acceptable application value of 45 MPa (Liu *et al.*, 2001). The low mechanical strength possessed by the membranes might limit their application in extensive and prolonged periods of oxygen permeation. However, such values were expected since the outer and inner diameter values of the fabricated membranes were considerably less compared to membranes used in other previous research (Tan *et al.*, 2004; Tan *et al.*, 2008b). Liu *et al.* (2006) suggested that the mechanical value can be improved without any changes to the diameter of the membrane by increasing the sintering temperature or prolonging the duration of the sintering stage.

18.3.3   *Oxygen permeation experiments*

The oxygen permeation results are given in Table 18.2 while Figure 18.5 illustrates the performance comparison between the two types of membrane. Based on the results obtained, three important findings were revealed. Firstly, both membranes showed best oxygen permeation flux at an operating temperature of 900°C with C-LSCF and L-LSCF module giving 0.1642 and 0.1652 cm^3 cm^{-2} min^{-1}, respectively. These values clearly do not satisfy the industrial application requirement for oxygen production of 3.5 cm^3 cm^{-2} min^{-1} (Steele, 1996). Furthermore, the values were far below 0.7343 cm^3 cm^{-2} min^{-1}, which has been previously reported by Tan *et al.* (2004). It is worth mentioning that, the experiment methodologies of the current research were referred to that reported by Tan *et al.* Such low values were attributed to the application of shorter length hollow fiber membrane in this research. Secondly, the oxygen permeation fluxes exhibited by both membranes were similar at all 3 temperature settings. Based on previous studies (Li *et al.*, 1999; Zeng *et al.*, 1998), the assumption was made that, by using LSCF powders with higher surface area value, the membrane module of C-LSCF should provide a higher value of oxygen permeation flux than the L-LSCF module. However, in those studies, the powder surface area values were related to its ability to be compacted into a disk-shaped membrane. One of the main problems encountered in disk membrane application is that the membrane should be as thin as possible to reduce the oxygen permeation resistance without compromising its mechanical properties, since mechanical strength is essential for prolonged usage. Higher surface area value, which powders of smaller particle size exhibit, allows the powder to be fabricated into thin and dense disk shaped membrane. It seems that the type of membrane configuration chosen for this research; which is hollow fiber design has played a more significant role in determining the value of oxygen permeation flux through the membrane. As shown by previous studies by Tablet *et al.* (2005) and Liu and Gavalas (2005), which had made direct comparison between disk-shaped membrane and hollow fiber membrane in terms of oxygen permeation, it has been clearly affirmed that the hollow fiber membrane configuration demonstrated considerably higher flux value, due to its asymmetric structure which provides less resistance towards the oxygen permeation and also exhibits larger membrane area per unit volume for oxygen interaction with the membrane surface. Lastly, the values calculated also showed significant increment with higher operation temperature value. This trend has been explained by previous research studies (Tan and Li, 2002; Tan *et al.*, 2004) which proposed that operating temperature is one of the major factors in determining the oxygen permeation flux due to the order-disorder transition of oxygen vacancy. The oxygen vacancies tend to disorder at elevated temperatures, maximizing their configurational entropy, thus becoming highly mobile, which favors greater oxygen permeation flux through the membrane (Kruidhof *et al.*, 1993).

Based on the experiments conducted and results collected, there is no significant information that suggested that different types of powder preparation method will affect oxygen permeation flux of the resultant hollow fiber membrane. Nevertheless, as suggested by the previous research of Richardson *et al.* (2003), the method chosen must have the ability to create LSCF powder with the desired stoichiometric composition, small size and uniform powder particle, and relatively high surface area value. These characteristics are required to ensure the successful fabrication of gas tight hollow fiber membrane. A single phase perovskite structure is required to allow the permeation of oxygen ions through the membrane. Application of small and uniform powder particle size will result in greater dispersion of the particle powder in polymer and solvent during the dope preparation stage in the hollow fiber fabrication process. By having a well mixed dope solution, it will ensure that the solution will react properly with the internal and external coagulant during the spinning process to produce the asymmetric structure in the finished hollow fiber (Tan *et al.*, 2004). In terms of the difference in surface area, as evidenced from the SEM images and 3 point bending results from this research, a higher average surface area and smaller particle size have resulted in close packing of the powder during the sintering process, followed by improved fusion of the membrane throughout the sintering process. This behavior leads to enhanced mechanical properties of the membrane, hence resulting in improved structural strength.

18.4   CONCLUSION

Both membrane modules, L-LSCF and C-LSCF exhibited similar oxygen permeation flux at each operating temperature setting. This indicates that hollow fiber membrane derived from LSCF6428 powder that was prepared through the solid state reaction method showed the same oxygen permeation flux as hollow fiber membrane derived from commercially procured LSCF6428 powder. Given that the powder derived from solid state reaction has a lower surface area value than its counterpart, it was hypothetically assumed that the latter would provide better oxygen permeation results. However, this correlation was only applicable to oxygen permeation experiments that involved disk shape membrane design. In conclusion, based from the results obtained, the powder preparation method does not directly affect the oxygen permeation flux of hollow fiber membrane. However, it does affect the fabricated membrane properties such as its structural strength. Due to its asymmetric structure that provides the least oxygen permeation resistance and which offers a large surface area for oxygen ion interaction, the membrane configuration of hollow fiber design seems to be a more important factor in influencing the oxygen permeation flux.

ACKNOWLEDGEMENT

We are grateful for the research facilities provided by Advanced Membrane Research Technology Centre of Universiti Teknologi Malaysia and the research funding received through the Fundamental Research Grant Scheme given by the Malaysian Ministry of Higher Education.

REFERENCES

Emsley, J. (2001) *Nature's building blocks: an A-Z guide to the elements.* Oxford University Press, Oxford, UK.

Hashim, S.S., Mohamed, A.R. & Bhatia, S. (2011) Oxygen separation from air using ceramic-based membrane technology for sustainable fuel production and power generation. *Renewable and Sustainable Energy Reviews*, 15, 1284–1293.

Kruidhof, H., Bouwmeester, H.J.M., v. Doorn, R.H.E. & Burggraaf, A.J. (1993) Influence of order-disorder transitions on oxygen permeability through selected nonstoichiometric perovskite-type oxides. *Solid State Ionics*, 63–65, 816–822.

Li, S., Jin, W., Xu, N. & Shi, J. (1999) Synthesis and oxygen permeation properties of $La_{0.2}Sr_{0.8}Co_{0.2}Fe_{0.8}O_{3-\delta}$ membranes. *Solid State Ionics*, 124, 161–170.

Liu, S. & Gavalas, G.R. (2005) Oxygen selective ceramic hollow fiber membranes. *Journal of Membrane Science*, 246, 103–108.

Liu, S., Tan, X., Li, K. & Hughes, R. (2001) Preparation and characterisation of $SrCe_{0.95}Yb_{0.05}O_{2.975}$ hollow fiber membranes. *Journal of Membrane Science*, 193, 249–260.

Liu, Y., Tan, X. & Li, K. (2006) $SrCe_{0.95}Yb_{0.05}O_{3-\delta}$ (SCYb) hollow fiber membrane: preparation, characterization and performance. *Journal of Membrane Science*, 283, 380–385.

Luyten, J., Buekenhoudt, A., Adriansens, W., Cooymans, J., Weyten, H., Servaes, F. & Leysen, R. (2000) Preparation of $LaSrCoFeO_{3-\delta}$ membranes. *Solid State Ionics*, 135, 637–642, 2000.

Richardson, R.A., Ormerod, R.M. & Cotton, J.W. (2003) Influence of synthesis route on the powder properties of a perovskite-type oxide. *Ionics*, 9, 77–82.

Steele, B.C.H. (1996) Ceramic ion conducting membranes. *Current Opinion in Solid State and Materials Science*, 1, 684–691.

Tablet, C., Grubert, G., Wang, H., Schiestel, T., Schroeder, M., Langanke, B. & Caro, J. (2005) Oxygen permeation study of perovskite hollow fiber membranes. *Catalysis Today*, 104, 126–130.

Taheri, Z., Nazari, K., Seyed-Matin, N., Safekordi, A., Ghanbari, B., Zarrinpashne, S. & Ahmadi, R. (2010) Comparison of oxygen permeation through some perovskite membranes synthesized with EDTNAD. *Reaction Kinetics, Mechanisms and Catalysis*, 100, 459–469.

Tan, X. & Li, K. (2002) Modeling of air separation in a LSCF hollow-fiber membrane module. *AIChE Journal*, 48, 1469–1477.

Tan, X., Liu, Y. & Li, K. (2004) Preparation of LSCF ceramic hollow-fiber membranes for oxygen production by a phase-inversion/sintering technique. *Industrial & Engineering Chemistry Research*, 44, 61–66.

Tan, X., Liu, Y. & Li, K. (2005) Mixed conducting ceramic hollow-fiber membranes for air separation. *AIChE Journal*, 51, 1991–2000.

Tan, X., Wang, Z., Liu, H. & Liu, S. (2008a) Enhancement of oxygen permeation through $La_{0.6}Sr_{0.4}Co_{0.2}Fe_{0.8}O_{3-\alpha}$ hollow fiber membranes by surface modifications. *Journal of Membrane Science*, 324, 128–135.

Tan, X., Pang, Z. & Li, K. (2008b) Oxygen production using $La_{0.6}Sr_{0.4}Co_{0.2}Fe_{0.8}O_{3-\alpha}$ (LSCF) perovskite hollow fiber membrane modules. *Journal of Membrane Science*, 310, 550–556.

Tan, X., Wang, Z., Meng, B., Meng, X. & Li, K. (2010) Pilot-scale production of oxygen from air using perovskite hollow fiber membranes. *Journal of Membrane Science*, 352, 189–196.

Teraoka, Y., Zhang, H.M., Okamoto, K. & Yamazoe, N. (1988) Mixed ionic-electronic conductivity of $La_{1-x}Sr_xCo_{1-y}Fe_yO_{3-\delta}$ perovskite-type oxides. *Materials Research Bulletin*, 23, 51–58.

Teraoka, Y., Zhang, H.M., Furukawa, S. & Yamazoe, N. (1985) Oxygen permeation through perovskite-type oxides. *Chemistry Letters*, 14, 1743–1746.

Xu, S.J. & Thomson, W.J. (1998) Stability of $La_{0.6}Sr_{0.4}Co_{0.2}Fe_{0.8}O_{3-\delta}$ perovskite membranes in reducing and nonreducing environments. *Industrial & Engineering Chemistry Research*, 37, 1290–1299.

Xu, S.J. & Thomson, W.J. (1999) Oxygen permeation rates through ion-conducting perovskite membranes. *Chemical Engineering Science*, 54, 3839–3850.

Yaremchenko, A.A., Kharton, V.V., Viskup, A.P., Naumovich, E.N., Tikhonovich, V.N. & Lapchuk, N.M. (1999) Mixed electronic and ionic conductivity of $LaCo(M)O_3$ (M = Ga, Cr, Fe or Ni): V. Oxygen permeability of Mg-doped $La(Ga, Co)O_8O_{3-\delta}$ perovskites. *Solid State Ionics*, 120, 65–74.

Zawadzki, M.A., Grabowska, H. & Trawczyński, J. (2011) Effect of synthesis method of LSCF perovskite on its catalytic properties for phenol methylation. *Solid State Ionics*, 181, 1131–1139.

Zeng, Y., Lin, Y.S. & Swartz, S.L. (1998) Perovskite-type ceramic membrane: synthesis, oxygen permeation and membrane reactor performance for oxidative coupling of methane. *Journal of Membrane Science*, 150, 87–98.

CHAPTER 19

Preparation of porous PVDF/montmorillonite hollow fiber mixed matrix membrane contactor via phase inversion method to capture CO_2 in membrane contactor

M. Rezaei-DashtArzhandi, Ahmad Fauzi Ismail, Seyed A. Hashemifard, Gholamreza Bakeri & Takeshi Matsuura

19.1 INTRODUCTION

Threatening levels of anthropogenic greenhouse gases, particularly carbon dioxide (CO_2), pose great environmental, economic and operational impacts to the earth. Hence, the reduction of CO_2 content in gas streams and its emission to the atmosphere seems inevitable (Ismail and Mansourizadeh, 2010). Membrane contactors, which are designed to absorb CO_2, are a newly devised apparatus with a solid, microporous and hydrophobic membrane. Membrane contactors have drawn considerable attention due to their superior characteristics compared to conventional methods such as high surface area, no flooding, loading, entrainment and modularity. Nevertheless, the membrane contactors suffer from some drawbacks. For instance, the membrane itself during the process causes an additional resistance to the overall mass transport that becomes more significant when a fraction of membrane pores is filled with liquid absorbent (Hashemifard et al., 2011). Therefore, there is a need to fabricate membranes with low mass transfer resistance (high porosity) and high wetting resistance (high hydrophobicity). A number of methods have been proposed to enhance the membrane hydrophobicity and reduce the wettability, which dispersing small fillers throughout a polymer called mixed matrix membranes (MMMs) have presented an interesting approach. Moreover, the organic/inorganic combination is proved to enhance thermal, chemical and mechanical stability of polymeric membranes (Drioli et al., 2006; Sridhar et al., 2007).

Recently, inorganic clay-nanoparticles were developed as fillers or additives in polymer matrix, since the addition of very low clay content (less than 10%) can improve several properties of polymeric membranes (Bakeri et al., 2010; Drioli et al., 2006; Sridhar et al., 2007).

In the study, hydrophobic montmorillonite (MMT) was used as filler for the fabrication of porous PVDF hollow fiber MMM via wet phase inversion method for CO_2 absorption in contactor. MMT type clay possesses a high aspect ratio with a layered structure. It has a structure consisting of two silicate tetrahedral sheets with an edge-shared octahedral sheet of either aluminum or magnesium hydroxide. The effects of MMT nano-clay on the membrane morphology, permeability, hydrophobicity and CO_2 absorption performance were investigated. Since both PVDF polymer matrix and MMT particles are hydrophobic, higher surface hydrophobicity, wetting resistance and performance can be expected.

19.2 EXPERIMENTAL

19.2.1 Materials

Commercial PVDF polymer pellets (Kynar®740) were supplied by Arkema Inc., PA, USA. A hydrophobic MMT and Nanomer 1.30TC surface modified by octadecylamine (25–30 wt%) was purchased from Fluka, and 1-methyl-2-pyrrolidone (NMP >99.5%) was used as solvent without further purification. Lithium chloride (LiCl, $\geq$99%) (Sigma-Aldrich®) was used as pore-former additive in the polymer dopes. Methanol (GR grade, 99.9%) was purchased from MERCK and used for post-treating the prepared membranes. Tap water was used as the coagulant in the spinning process. Distilled water was used as a liquid absorbent in the CO_2 absorption experiments.

19.2.2   Fabrication of asymmetric porous hollow fiber MMMs

The selected clay particles, PVDF polymer pallets and LiCl were vacuum dried in an oven for 48 h at 65 $\pm$ 2°C to remove moisture content. Different MMT contents (1.0, 3.0 and 5.0 wt% of PVDF) were dispersed in the solvent (NMP) and sonicated for 1 h at 40°C. LiCl of 2.5 wt% was added, then 18 wt% PVDF was gradually added to the mixture and to ensure the complete dissolution of the polymer and nano-filler, the mixture was stirred overnight. The prepared spinning dope solution was degassed. The hollow fiber spinning details through the wet phase inversion can be found elsewhere (Bakeri *et al.*, 2011). The prepared membranes then were dried for further characterizations.

19.2.3   Field emission scanning electron microscopy (FESEM) examination

A field emission scanning electron microscope (FESEM) (JEOL JSM-6701F) was used to observe the morphology of the spun PVDF hollow fiber membranes and to investigate the extent of the adhesion between the filler particles and the polymer matrix by the standard methods. The dried hollow fiber samples were cut into small pieces of approximately 5 cm length to be fractured in nitrogen for observation of the cross-section. The FESEM micrographs of cross-section, internal surface and an external surface of the hollow fiber MMMs were taken at various magnifications.

19.2.4   Gas permeation test and contact angle measurement

A hollow fiber with an effective length of 10 cm was glued with epoxy glue at one end and the other end was potted to a stainless steel fitting. The purified nitrogen (N_2) gas was supplied to the shell side and the pressure increased at an interval of 0.05 MPa. The permeation rate of gas coming out from the lumen side was measured by the soap-bubble flow meter.

The hollow fibers were dried at 60 $\pm$ 2°C for 12 h and contact angles were measured on the upper surface of the membranes by the sessile drop technique using a goniometer (model G1, Krüss GmbH, Hamburg, Germany).

19.2.5   Gas absorption test

The CO_2 absorption experiment was carried out to evaluate the performance of the fabricated membranes. Ten hollow fibers with the effective length of 17.5 cm were packed in a contactor module. Distilled water and pure CO_2 flowed in the lumen and shell sides respectively. The CO_2 concentration in the liquid outlet at various flow rates was measured by titration using 0.02 M sodium hydroxide (NaOH) solution and phenolphthalein to determine the CO_2 flux.

19.3   RESULTS AND DISCUSSION

19.3.1   Morphology study

Dispersion of hydrophobic nano-fillers into PVDF polymer matrix, the extent of adhesion between them and the morphology of prepared MMMs were investigated by FESEM images. The cross-sectional, inner surface and outer surface micrographs in different clay loading are given in Figure 19.1. It can be seen from cross sectional FESEM images that the membranes have a thin skin layer and the finger-like structure in the substrate extends from the outer surface and meets

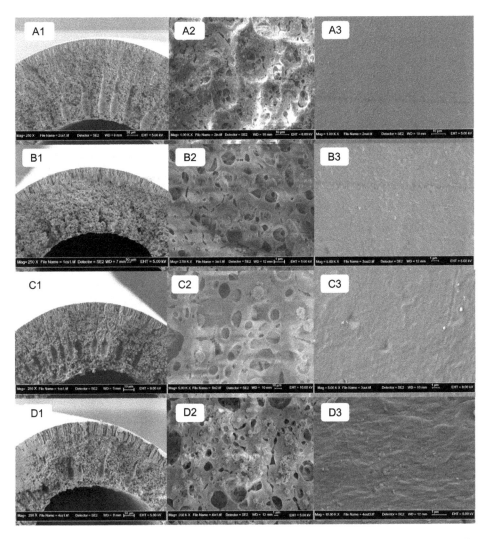

Figure 19.1.   Field emission scanning electron microscopy (FESEM) morphology of the PVDF hollow fiber membranes: (A) without MMT; (B) with 1 wt% MMT; (C) 3 wt% MMT; (D) 5 wt% MMT; (1) cross section; (2) inner surface and (3) outer surface.

with the sponge-like structure in the middle of the cross-section. It is well known that the solvent-coagulant exchange rate during phase inversion and the thermodynamic stability of the polymer solution control the structure of asymmetric membranes. As for the thermodynamic stability, incorporation of MMT clay filler into PVDF polymer matrix decreased thermodynamic stability of spinning solutions and accelerated the solvent/coagulant demixing rate. Consequently, MMMs with many finger-like macrovoids of smaller diameters underneath a very thin skin layer are formed. Regarding the inner surface micrographs (Fig. 19.2), an inner skinless surface with open microporous structure are observed due to the high concentration of solvent in internal coagulant and slow solidification process. In addition, no pores are visible on the outer membrane surfaces at magnification of 50k, confirming the small pore size of the outer membrane surface, which will be verified later in the gas permeation test.

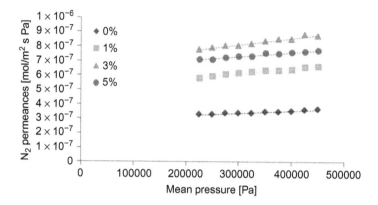

Figure 19.2.   Measured N_2 permeance of the PVDF hollow fiber MMMs as a function of mean pressure.

Table 19.1.   Characteristics of the prepared PVDF hollow fiber MMMs.

Membrane	Plain PVDF	1% MMT	3% MMT	5% MMT
Permeance of N_2 gas at 0.7 MPag	3.7×10^{-7}	6.6×10^{-7}	8.7×10^{-7}	7.7×10^{-7}
Effective surface porosity (ε/L_p) [m^{-1}]	87	124	170	237
Mean pore size, $r_{P,m}$ [nm]	26	32	31	21
Contact angle (θ)	80°	84°	88°	99°

19.3.2   *Gas permeation and contact angle measurement*

The results for the gas permeation experiments are shown in Figure 19.2. The mean pore size and effective surface porosity were further obtained, and the results are shown in Table 19.1. Figure 19.2 and Table 19.1 show that the gas permeance increased by increasing MMT loading up to 3 wt% and then went down at 5 wt%. The effective surface porosity, on the other hand, increased continuously with the nano-clay filler loading (Table 19.1). These observations, particularly those of the effective surface porosity, are in accordance with the increase of finger-like pores beneath the outer surface (see Fig. 19.1). It can be attributed to the good dispersion of layered clay into the polymer matrix which caused lower thermodynamic stability and faster solidification of polymer solution. Generally, incorporation of molecular sieve type nano-fillers into the polymer matrix can enhance the membrane permeability. A slight decrease of N_2 permeance for 5 wt% MMM compared to 3 wt% MMT loading is believed to be due to the detected decrease of membrane pore size.

The contact angles of the outer hollow fiber membrane surfaces also are included in Table 19.1. The contact angle of pristine PVDF membrane increased by MMT incorporation from 80° to 99° of 5 wt% MMT MMM. Most likely the contact angle increase was due to the loading of highly hydrophobic MMT nano-clay and the presence of hydrophobic component on the membrane surface.

Conclusively, it can be predicted that the fabricated MMMs with higher surface hydrophobicity and permeance can exhibit higher absorption performance.

19.3.3   *CO_2 absorption test*

Physical CO_2 absorption test was carried out supplying water absorbent in the lumen side and pure CO_2 gas in the shell side in a counter-current direction. The average absorption fluxes of CO_2 at different loadings of nano-clay as a function of the absorbent flow rates were obtained

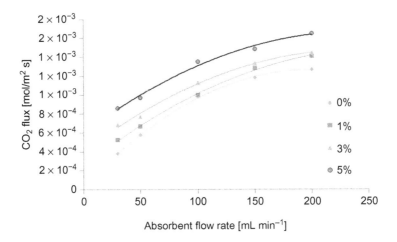

Figure 19.3. CO_2 absorption performance of the hollow fiber MMMs (pure CO_2-distilled water system).

and the results are illustrated in Figure 19.3. The flux of prepared membranes increased with an increase in the liquid flow rate due to the reduction of liquid side mass transfer resistance. The absorption flux also increased by increasing MMT content in the polymer matrix and 5 wt% MMT MMM exhibited a maximum absorption flux. This was attributed to the highest surface hydrophobicity and small pore size of 5 wt% MMT MMM compared to the other membranes which restricted the intrusion of absorbent into the membrane pores. The highest CO_2 absorption flux of 1.59×10^{-3} mol m^{-2} s^{-1} for embedded 5 wt% MMT MMM at the flow rate of 200 mL min^{-1} was approximately 21% higher than the flux of PVDF membrane without filler. The obtained flux of the synthesized membrane was superior to several in-house made and commercial membranes. The physical CO_2 absorption in a contactor using commercial PTFE (Sumitomo Electric Fine Polymer) and PP (Mitsubishi Rayon) hollow fiber membranes was conducted by Xu *et al.* (2008). The contactor test revealed higher absorption flux of PTFE than PP membrane; however both commercial membranes exhibited significantly lower absorption performance than embedded 5 wt% MMT MMM in this work. Rahbari-Sisakht *et al.* (2012) blended surface modified macromolecules (SMM) to PVDF spinning solution to improve hydrophobicity and CO_2 absorption performance. The highest performance of 5 wt% MMT MMM of 1.59×10^{-3} mol m^{-2} s^{-1} was approximately 26% higher than the surface modified membrane by SMM at the same flow rate which could be attributed to achieved high wetting resistance of 5 wt% MMT MMM.

Therefore, it can be concluded that membrane with 5 wt% loading of MMT is the best in terms of absorption rate, stability (least wetting) and hydrophobic clay particles can act as a superior additive to the polymer matrix to fabricate membranes with improved structure and enhanced surface hydrophobicity.

19.4   CONCLUSION

Porous hollow fiber membranes were prepared by the wet phase inversion method with nano-clay particles and LiCl additives. The incorporation of the clay filler into PVDF polymer matrix enhanced the formation of finger-like macrovoids. Results of gas permeation test showed that the membranes possess very small mean pore sizes and high surface porosities. As well, the surface hydrophobicity increased with the clay filler incorporation. The CO_2 absorption flux increased when MMT nano-filler was incorporated and showed a maximum when the nano-filler loading was 5 wt%.

ACKNOWLEDGEMENT

The author is thankful to the Ministry of Education for the Research University Grant vote number of Q.J130000.3009.00M02 and Research Management Centre (RMC), UTM for research management activities.

REFERENCES

Bakeri, G., Ismail, A.F., Shariaty-Niassar, M. & Matsuura, T. (2010) Effect of polymer concentration on the structure and performance of polyetherimide hollow fiber membranes. *Journal of Membrane Science*, 363, 103–111.

Bakeri, G., Ismail, A.F., Rana, D. & Matsuura, T. (2011) The effect of phase inversion promoters on the structure and performance of polyetherimide hollow fiber membrane using in gas-liquid contacting process. *Journal of Membrane Science*, 383, 159–169.

Drioli, E., Criscuoli, A. & Curcio, E. (eds.) (2006) Membrane contactors: fundamentals, applications and potentialities. *Membrane Science and Technology*, Volume 11. Elsevier Science, Amsterdam, The Netherlands.

Hashemifard, S.A., Ismail, A.F. & Matsuura, T. (2011) Effects of montmorillonite nano-clay fillers on PEI mixed matrix membrane for CO_2 removal. *Chemical Engineering Journal*, 170, 316–325.

Ismail, A.F. & Mansourizadeh, A. (2010) A comparative study on the structure and performance of porous polyvinylidene fluoride and polysulfone hollow fiber membranes for CO_2 absorption. *Journal of Membrane Science*, 365, 319–328.

Rahbari-Sisakht, M., Ismail, A.F. & Matsuura, T. (2012) A novel surface modified polyvinylidene fluoride hollow fiber membrane contactor for CO_2 absorption. *Journal of Membrane Science*, 415–416, 221–228.

Sridhar, S., Smitha, B. & Aminabhavi, M. (2007) Separation of carbon dioxide from natural gas mixtures through polymeric membranes – a review. *Separation and Purification Technology*, 36, 113–174.

Xu, A., Yang, A., Young, S., de Montigny, D. & Tontiwachwuthikul, P. (2008) Effect of internal coagulant on effectiveness of polyvinylidene fluoride membrane for carbon dioxide separation and absorption. *Journal of Membrane Science*, 311, 153–158.

Part V
*Membranes for environmental
and energy applications
(gas separation application)*

CHAPTER 20

An experimental investigation into the permeability and selectivity of PTFE membrane: a mixture of methane and carbon dioxide

Sina Gilassi & Nejat Rahmanian

20.1 INTRODUCTION

Research and technology innovations in the 1970s led to the significant commercial practice of gas separation by membranes that exists today. These advances involved developing membrane structures that could produce high fluxes and modules for packing a large amount of membrane area per unit volume (Murphy *et al.*, 2009). At present, the share of using a polymeric membrane in the capture of CO_2 is increasing and gradually the membrane technology is considered as the promising method in separation units, although the number of commercial membranes is not high. CO_2 capture from natural gas is one of the controversial topics that many researchers and engineers try to find the best method satisfying both high efficiency and low capital cost. In common, chemical physical absorption towers are applied to remove CO_2 from natural gas in order to prevent pipeline corrosion, even though the other component such as H_2S gives rise to operating problems. The obscure angle of a conventional unit is related to the high energy consumption while the absorbent needs to be purified by the regeneration units which implement the temperature as a unique manipulating parameter for separating amine groups. The great advantages of using the membrane in gas industry are the low capital cost, easy installation and maintenance so that for this simple reason, new membranes come to the market for different types of processes. Capture of CO_2 from natural gas accounts for one of the major difficulties so that the engineers try to employ membrane modules as to alter the process efficiency. However, there are only a limited number of membranes that can be used in real industry and the research still continues over this interesting topic (Burggraaf and Cot, 1996).

In this research, PTFE film is used which has specific chemical and physical properties such as resistance in high temperature, inertness, high hydrophobicity. The two important factors, which are taken into account for having better separation performance, are selectivity and permeability. According to Fick's law, the flux passing through the membrane is a function of the pressure gradient, thickness, diffusivity coefficient and solubility of the pure gas inside the membrane (Geankoplis, 2003):

$$N_A = \frac{DS(p_{A1} - p_{A2})}{l} \tag{20.1}$$

where p_{A1} and p_{A2} are the ambient pressures on upstream and downstream of a membrane, respectively. l is the membrane thickness. The product DS is called the permeability coefficient P. There is no doubt that an increase in permeability give rise to making a significant change in total flux. Some researchers (Brewis and Dahm, 2006) tried to alter this factor using chemical treatment and modification. Most of the more recent researches have focused on developing membrane materials with a better balance of selectivity and productivity (permeability) so that it seems the most likely route for expanding the use of this technology.

In this study, the permeability coefficients of pure and pair gas component CO_2 and CH_4 are determined by experimental work and the results are plotted separately to show the differences between the two measurements. In addition, the diffusivity and solubility factor are defined according to the gas transport equations. Moreover, the experimental work indicates the effect of operating conditions such as pressure and volumetric flow rate of feed on the permeate stream.

20.2   THEORY OF GAS TRANSPORT IN MEMBRANE

The movement of gas molecules through pore channels is described by three parameters, the diffusivity coefficient, solubility and permeability coefficient. The gas molecules are absorbed on one side of the membrane, diffuse through the membrane while the concentration of gas is changing through the membrane, and finally the molecules leave the other side of the membrane.

Fick's law governs the correlation between the above-mentioned theory and defines the diffusivity coefficient which is different depending upon the gas component and structure. Generally the diffusivity is a function of pressure, temperature and concentration. Moreover, both experimental work and theoretical equations can be applied to predict diffusivity of different gas mixtures (Bird et al., 2007). The solution-diffusion model describes the transport through the dense polymeric membrane which involves the first and second Fick's law (Pineri and Escoubes, 1993):

$$F = -D\frac{\partial C(x,t)}{\partial x} \tag{20.2}$$

$$\frac{\partial C(x,t)}{\partial t} = D\frac{\partial^2 C(x,t)}{\partial x^2} \tag{20.3}$$

As the gas concentration $C(x,t)$ within the membrane at point x and time t is related to the gas pressure $p(x,t)$, it can be expressed by the interpretation of Henry's law:

$$C(x,t) = Sp(x,t) \tag{20.4}$$

As the first Fick's law defines the gas flux in constant concentration, the volumetric flow rate of gas, Q, which passes the membrane, is determined by the second Fick's law when the stationary state is obtained (Pineri and Escoubes, 1993).

20.2.1   Determination of permeability, diffusivity coefficient

The time lag permeation experiment is an acceptable method to determine the permeability and diffusivity coefficient of gases in polymer films (Al-Ismaily et al., 2012). The pressurized gas is contacted with the polymeric membrane and the permeation stream, which leaves the low pressure side is recorded and expressed as a function of time. In this method, there are three stages, which consist of (i) the transient permeation, (ii) pseudo-steady-state permeation as the concentration difference throughout the membrane is constant, (iii) unsteady state permeation as the concentration of downstream is remarkably high (Chen et al., 2010). After passing across the first stage, the pseudo-steady state is described by the Equation (20.5) (Crank, 1975). This method was originally introduced by Daynes (Do, 2001):

$$Q = \frac{DC_1}{l}\left(t - \frac{l^2}{6D}\right) \tag{20.5}$$

where Q is the amount of gas that flowed through the membrane whose thickness and surface area are shown by l and A during specific time t and C_1 is the feed concentration. The difference in concentration and pressure profile leads to independent equations to determine the flux of gas F through the film (Pineri and Escoubes, 1993):

$$F = -D\frac{\Delta C}{l} = -P\frac{\Delta p}{l} \tag{20.6}$$

By plotting Q versus t, the steady-state line enables the calculation of both the permeability coefficient (P) from the slope and the time-lag θ by extrapolation on the time axis (Pineri and Escoubes, 1993):

$$P = \frac{dQ}{dt}\frac{1}{A}\frac{l}{\Delta p} \tag{20.7}$$

$$\beta = \frac{1}{A}\frac{l}{\Delta p} \tag{20.8}$$

$$\theta = \frac{l^2}{6D} \tag{20.9}$$

The determined time-lag value is the most common method for estimating the gas diffusion coefficient. Additionally, it is proportional to the reverse value of diffusion coefficient and the thickness of membrane. Then, the permeability coefficient is given by the product of diffusion coefficient and solubility:

$$P = DS \tag{20.10}$$

The ratio of the permeability of CO_2 and CH_4 determines the selectivity of CO_2/CH_4 gas mixture as follows:

$$\alpha = \frac{P_{CO_2}}{P_{CH_4}} \tag{20.11}$$

20.2.2 Fundamentals of diffusion and adsorption in porous media

The best way to select a good membrane is to deeply consider the structure of porous media in terms of mobility of molecules when a great number of molecules pass through the microscopic cross sectional area in different operating conditions. Generally, there has been a critical definition of different regimes which would form inside the pore channels. If a pure gas is composed of rigid, no attracting spherical molecules of diameter d and mass m, and the concentration of gas molecules is sufficiently small so that the average distance between the gas molecules is remarkably greater than their diameter d, the molecular velocity is given by:

$$\bar{u} = \sqrt{\frac{8KT}{\pi m}} \tag{20.12}$$

in which K is the Boltzmann constant (Bird *et al.*, 2007). Thus, the molecular velocity of two components CO_2 and CH_4 is different inside the pore channels. Then, the motion of molecules of gas mixture changes inside the pore channels and affects the average distance between two collisions that is given by:

$$\lambda = \frac{1}{\sqrt{2}\pi d^2 n} \tag{20.13}$$

where n represents the number density. According to the previous study on the viscosity of multi components gas mixture (Saksena and Saxena, 1965), the viscosity of gas molecules with the aim of comparison with the Newton's law can be written by:

$$\mu = \frac{1}{3}nm\bar{u}\lambda \tag{20.14}$$

20.2.3 Mass transport in membrane

The transport of gases in the membrane is challenging (Bird *et al.*, 2007). The Knudsen number (see Equation (20.15)) is the ratio of mean free path λ and diameter of pore channel d whose magnitude clarifies the mass transport mechanism that is divided as follows: (i) ordinary diffusion governed by Maxwell-Stefan equations; (ii) Knudsen diffusion; (iii) viscous flow defined by the Hagen-Poiseuille equation; (iv) diffusion-adsorption; (v) thermal transportation; and (vi) thermal diffusion. Basically, The Knudsen number clarifies the regime in which the molecules of gas mixture move inside the pore channels. In this study, PTFE membrane with pore size of 0.2 μm is used and the size of gas molecules of both CO_2 and CH_4 are smaller than the pore. Predictably, it is expected that the mechanism of motion inside the pore channels is testified by both viscous and diffusion regimes. In the viscous regime, the pressure drop occurs because of the interaction of molecules and the tortuosity of the membrane, whereas the creeping of molecules on the surface of the pore channel is considered in the diffusion-adsorption regime. The solution-diffusion model draws attention to the gas mixture solubility and the diffusivity factors which are interpreted with Darcy's law and Fick's law (Wijmans and Baker, 1995):

$$Kn = \frac{\lambda}{d} \tag{20.15}$$

20.2.4　Gas solubility vs. pressure and temperature

The concept between the gas solubility and temperature is very similar to the increase of vapor pressure with temperature so that increasing gas temperature causes an increase in kinetic energy. The higher kinetic energy causes the faster movement of molecules which finally break intermolecular bonds and escape from solution. Liquids and solids exhibit practically no change of solubility with changes in pressure. Based on the Henry's law, the solubility of a gas in a liquid is directly proportional to the partial pressure of that gas above the surface of the solution (Kotz et al., 2006). This law defines the solubility of gas inside the solid materials expressed by Equation (20.4). The partial pressure of each component is defined as follows:

$$P = xp^* \tag{20.16}$$

where p^* is the vapor pressure of pure component, and P is the partial pressure of the pure component (Incropera, 2011).

20.2.5　Temperature and pressure dependence of diffusivities

For the binary gas mixture at low pressure, D_{AB} is inversely proportional to the pressure and is directly proportional to temperature. It is also almost independent of the composition for a given gas pair (Bird et al., 2007). The following equation for estimating D_{AB} at low pressure has been developed from a combination of kinetic theory and corresponding-states arguments (Bird et al., 2007):

$$\frac{pD_{AB}}{(p_{cA}p_{cB})^{1/3}(T_{cA}T_{cB})^{5/12}(1/M_A + 1/M_B)^{1/2}} = a\left(\frac{T}{\sqrt{T_{cA}T_{cB}}}\right)^b \tag{20.17}$$

where D_{AB} is in $cm^2\,s^{-1}$, p is in atm, and T is in K. Analysis of experimental data gives the dimensionless constant $a = 2.745 \times 10^{-4}$ and $b = 1.823$ for nonpolar gas pairs. At high pressure, and in the liquid state, the behavior of D_{AB} is more complicated. The simplest and best understood situation is that of self-diffusion (inter-diffusion of labeled molecules of the same chemical species) (Bird et al., 2007)[5].

$$D_m = D_{AB}\frac{\varepsilon}{\tau} \tag{20.18}$$

where D_m is the diffusivity coefficient of membrane and ε and τ are porosity and tortuosity, respectively.

20.2.6 Gas permeation membrane processes

In membrane processes with two gas phases and a solid membrane, the equilibrium relation between the solid and gas phase is given by (Geankoplis, 2003):

$$H = \frac{S}{22.414} = \frac{C}{p} \tag{20.19}$$

where S is the solubility of A in m^3 STP $atm^{-1} m^{-3}$ solid, and H is the equilibrium relation in kg mol m^{-3} atm^{-1}. This is similar to Henry's law (Geankoplis, 2003).

20.3 EXPERIMENTAL

20.3.1 Materials

PTFE film/membrane with pore size of 0.2 μm, porosity over 60% was provided by Millipore Co. (Ireland). CH_4 and CO_2 were supplied by Linde Co. (Malaysia) and density of gas components in kg m^{-3} are 0.692 and 1.842, respectively. The temperature was 30°C and pressure was below 0.6 MPa.

20.3.2 Equipment

Both pure and mixed feed stream were supplied to the permeation cell to consider the difference between permeability in two various conditions. The equipment is called CO2SMU available at the Universiti Teknologi Petronas and was fabricated by Spectron Sdn. Bhd. This equipment is employed to record all the results, which consist of permeation, reject flow rate stream as well as the concentration with the aim of an online analyzer. This unit is run by programmable logic control and the operator can alter the manipulated parameters and the result is noted in the log file.

20.3.3 Methods

Two different procedures were selected to run the experiments. In the first procedure, the pure gas component was fed to the permeation cell with different volumetric flow rates, so that the flow rate of permeate and reject stream showed the capacity of separation in PTFE membrane. The upstream pressure was gradually increased by 0.6 MPa and in each pressure point, the permeate stream and time required to collect the volume were recorded. Five experiments were made under the same operating condition (pressure and temperature) for the purpose of validity. The above-mentioned method was performed for both CO_2 and CH_4 separately. In the second procedure, firstly, two gas components were mixed with the static mixer before the feed gas entered the permeation cell. Similarly, the flow rates of permeate stream were recorded and compared with the single component experiment. In addition, by using the fundamental gas separation and transport equations, all results were classified and plotted versus time.

20.4 RESULTS AND DISCUSSION

As the objective of this work is to find the correct diffusivity and permeability factor of PTFE, firstly pure CO_2 is fed to the permeation cell at temperature 30°C and pressure 0.6 MPag. The permeate and retentate stream are calculated by simple mass balance equation whereas the concentration of feed is constant.

Figure 20.1 shows the change in downstream volume versus time. According to the Equation (20.7), a line can be fitted to the data and the slope of this line is multiplied by the slope correction term, β, which is 0.00639, to define the permeability. In this experiment, the permeability factor for pure CO_2 was found to be 5.24 barrer.

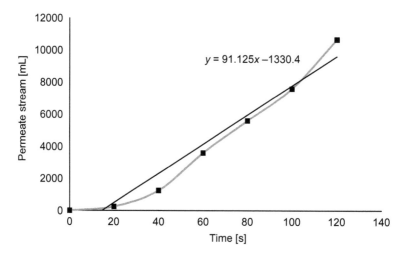

Figure 20.1.   Permeation flow rate of pure CO_2 versus time.

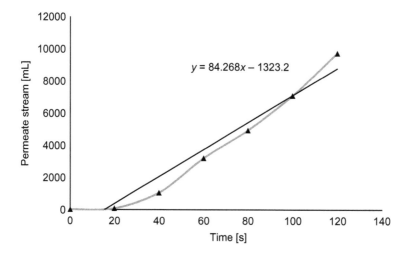

Figure 20.2.   Permeation flow rate of pure CH_4 versus time.

In the next step, pure CH_4 was fed to the permeation cell and data of permeate and retentate streams were recorded. Similarly, the feed concentration was constant and the result was plotted in Figure 20.2. Using the same methodology as above the permeability for pure CH_4 is found to be equal to 4.63 barrer.

In the next step, before the feed enters the permeation cell, a static mixer is applied to mix two gas components completely while the total gas pressure was kept at 0.6 MPag. The permeate stream is introduced to the gas analyzer to measure the concentration of both components. The results are shown in Table 20.1.

Figure 20.3 shows the result of CH_4 permeation volume versus time. By fitting the linear first order equation to this curve, the permeability is equal to 3.24 barrer.

Similar to Figure 20.1, the volume of CO_2 that was collected on the permeate side is plotted versus time in Figure 20.4. The permeability is equal to 4.97 barrer.

Table 20.1. Flow rate (slpm) and composition of feed and permeate for the CH_4/CO_2 mixture.

Feed flow rate [slpm][a]	CH_4 feed concentration [%wt]	CO_2 feed concentration [%wt]	Permeate flow rate [slpm]	CH_4 permeate concentration [%wt]	CO_2 permeate concentration [%wt]
26	90	10	23.4	81.20	18.7
26	70	30	23.8	69.1	30.9
26	50	50	24.1	48.3	51.7
26	30	70	24.8	29.4	70.6
26	10	90	25.1	9.1	90.9

[a] slpm: standard liters per minute

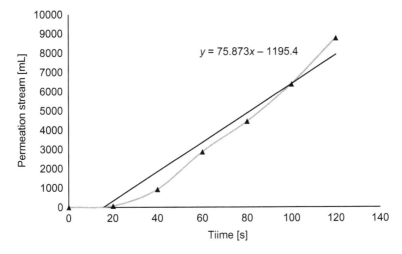

Figure 20.3. CH_4 permeate volume versus time for the mixed gas experiment ($CO_2 = 30$ wt% at 0.4 MPa).

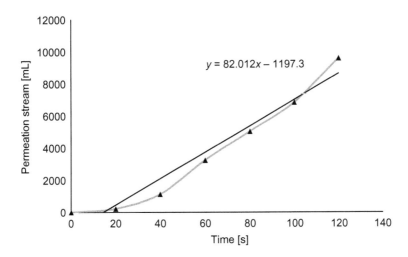

Figure 20.4. CO_2 volume in permeate versus time for the mixed gas experiment ($CO_2 = 30$ wt% at 0.4 MPa).

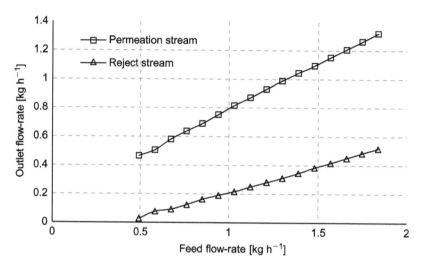

Figure 20.5.   Outlet flow-rate versus feed flow-rate of CH_4/CO_2 mixture.

Figure 20.5 shows the permeation stream flow rate versus feed flow rate. By increasing the feed flow rate, the permeation flow rate increases. On the other hand, increasing the concentration of feed component affects the permeation flow rate, similarly, this effect is developed for the retention flow rate (Ismail *et al.*, 2005).

20.5   CONCLUSION

Two experimental procedures in the calculation of the permeability factor were conducted so that pure and mixed gas were fed into the permeation cell and the result concerning permeate and reject flow rate, were recorded and plotted. According to the definition of selectivity, the ratio of more permeable component CO_2 to that of the less permeable component was determined so that the selectivity factor was different in the two experiments. In the first experiment for the pure gas component, $\alpha = 1.18$ and in the second experiment for mixture gas, $\alpha = 1.07$, which is 5 percent lower than the first experiment result. This measurement procedure can be carried out for other polymeric membranes with higher selectivity and the difference between experimental selectivity becomes more distinctive. The role of operating conditions such as pressure was significant, and the effects on change in the diffusivity coefficient were traced. In the first experiment, the pressure of the pure gas component on the surface of the interface cannot be replaced by partial pressure to calculate the solubility factor. So in conclusion, the permeability varies with the diffusivity coefficient changed by the temperature and, the gas velocity concerning to the motion of gas molecules inside the membrane pores and channels. Therefore, the first experiment is discarded by considering the equation of gas transport and equilibrium rules. The determination of the permeability factor is valid in the case that the feed consists of two mixed components.

ACKNOWLEDGEMENT

Dr. N. Rahmanian would like to thank Ministry of Science, Technology and Innovation of Malaysia for their financial support (Project No: 06-02-SF0178).

NOMENCLATURE

C mole concentration [mol m^{-3}]
D mass diffusivity [m^2 s^{-1}]
F molar flux of species A [mol s^{-1} m^{-2}]
H equilibrium relation [kg mol m^{-3} atm^{-1}]
l thickness of membrane [mm]
M molecular weight [g mol^{-1}]
N_A mass flux of species A [kg s^{-1} m^{-2}]
P^* vapor pressure [Pa]
P_{CO_2} permeability of CO_2 [barrer]
P_{CH_4} permeability of CH_4 [barrer]
P_{A1} pressure of component A_1 [Pa]
P_{A2} pressure of component A_2 [Pa]
Q volumetric flow rate of gas [m^3 s^{-1}]
t time [s]
S solubility [m^3 (STP) atm^{-1} m^{-3}]

Greek symbols

α selectivity factor
β slop correction term [m^{-1} Pa^{-1}]
θ time-lag

REFERENCES

Al-Ismaily, M., Wijmans, J. & Kruczek, B. (2012) A shortcut method for faster determination of permeability coefficient from time lag experiments. *Journal of Membrane Science*, 423, 165–174.

Bird, R.B., Stewart, E.E. & Lightfoot, E.N. (2007) *Transport phenomena*. John Wiley & Sons, Hoboken, NJ.

Brewis, D. & Dahm, R. (2006) *Adhesion to fluoropolymers*. Smithers Rapra Publishing, Shrewsbury, UK.

Burggraaf, A. & Cot, L. (eds.) (1996) *Fundamentals of inorganic membrane science and technology*. Elsevier Science, Amsterdam, The Netherlands.

Chen, Y., Zhang, Y. & Feng, X. (2010) An improved approach for determining permeability and diffusivity relevant to controlled release. *Chemical Engineering Science*, 65 (22), 5921–5928.

Crank, J. (1975) *The mathematics of diffusion*. Clarendon Press, Oxford, UK.

Do, D.D. (2001) *Adsorption analysis: equilibria and kinetics*. Imperial College Press, London, UK.

Geankoplis, C. (2003) *Transport processes and separation process principles (includes unit operations)*. Prentice Hall Press, Upper Saddle River, NJ.

Incropera, F.P. (2011) *Fundamentals of heat and mass transfer*. John Wiley & Sons, Hoboken, NJ.

Ismail, A.F., Kusworo, T.D., Mustafa, A. & Hasbullah, H. (2005) Understanding the solution-diffusion mechanism in gas separation membrane for engineering student. *Proceedings of the 2005 Regional Conference on Engineering Education*. pp. 155–159.

Kotz, J., Treichel, P. & Weaver, G. (2006) *Chemistry and chemical reactivity*. Cengage Learning, Boston, MA.

Murphy, T.M., Offord, G.T. & Paul, D.R. (2009) Fundamentals of membrane gas separation. In: Drioli, E. & Giorno, L. (eds.) *Membrane operations: innovative separations and transformations*. Wiley-VCH Verlag GmbH & Co. KGaA, Weinheim, Germany. pp. 63–82.

Pineri, L.R.E.G.M. & Escoubes, M. (1993) Polyaniline membrane for gas separation and ESR experiments. *3rd International Conference on Effective Membrane Processes – New Perspectives*. p. 323.

Saksena, M. & Saxena, S. (1965) Viscosity of multicomponent gas mixture. *Physica A*, 31, 18–25.

Wijmans, J. & Baker, R. (1995) The solution-diffusion model: a review. *Journal of Membrane Science*, 107 (1), 1–21.

CHAPTER 21

The effect of piperazine with N-methyldiethanolamine in emulsion liquid membrane for carbon dioxide removal

Norfadilah Dolmat & Khairul S.N. Kamarudin

21.1 INTRODUCTION

Carbon dioxide (CO_2) is one of the acid gases that are released from various industrial processes, contributing to global warming (Teramoto *et al.*, 2000). CO_2 in natural gas can also cause equipment plugging due to dry-ice formation, catalyst poisoning and reduction of the heating value (Aroonwilas and Tontiwachwuthikul, 1997). There are several methods that can be used to remove CO_2. Chemical and physical absorption, membrane separation, adsorption and cryogenic separation are those methods that have been used for CO_2 removal (Mandal and Bandyopadhyay, 2006; Moon and Shim, 2006). Meanwhile, absorption using alkanoamines such as primary, secondary and tertiary amines is a common technique in separation industries nowadays (Wong and Bioletti, 2002). However, amine that is in contact with metal surfaces leads to corrosion and damages the equipment (Kosseim *et al.*, 1984).

Liquid membrane technology is one of the alternative techniques that can be used to remove CO_2 efficiently. Liquid membrane consists of an organic phase that separates the two aqueous phases which are the feed and receiving phases. The separation occurs when the solute permeates through the liquid phase from the feed to receiving phase. Bulk liquid membrane (BLM), supported liquid membrane (SLM) and emulsion liquid membrane (ELM) are three different types of liquid membrane. Among these three types of liquid membrane, ELM seems to have a potential due to the large mass transfer area and simple operation (Chakraborty *et al.*, 2003). Indeed, ELM was successfully used in the food and dairy, chemical, cosmetic, pharmaceutical and biotechnology industries, in waste water treatment and in medical applications (Nath, 2008).

Absorption of carbon dioxide by using ELM technique can prevent corrosion problems (Li, 1968). This is due to the fact that, in the ELM technique, the aqueous phase containing amine in the form of small particles is dispersed in the organic phase. The organic layer prevents the amines from being directly in contact with the surface of the equipment. The formation of small particles provides a large interfacial area for absorption.

Although N-methyldiethanolamine (MDEA) is corrosive, it is commonly used in the separation process due to a high CO_2 loading capacity (1 mol CO_2 reacts with 1 mol of MDEA). However, MDEA is a slow reacting alkanoamines (Bishnoi, 2000). Therefore, the addition of activator or promoter such as piperazine (PZ) is necessary to increase the reaction between amine with CO_2. PZ contains two basic nitrogen atoms and can theoretically react with 2 mol of CO_2. Besides, the reaction rate constant of PZ with CO_2 is about one order of magnitude higher than that of MEA (Bishnoi and Rochelle, 2000). Therefore, PZ is used as a promoter with other amines for CO_2 capture process (Arunkumar and Bandyopadhyay, 2007; Bishnoi and Rochelle, 2002; Cullinane and Rochelle, 2004; Xu *et al.*, 1992; Zhang *et al.*, 2001). However, to separate CO_2 effectively, this process requires a stable emulsion. Hence, the objective of this chapter is to discuss the stability of emulsion and CO_2 absorption at various concentrations of MDEA and PZ. PZ has been used together with MDEA in order to increase the CO_2 absorption. Emulsion liquid membrane technique has been applied to prevent corrosion of metal surfaces caused by the MDEA and PZ.

21.2   EXPERIMENTAL

21.2.1   Materials

Emulsion liquid membrane (ELM) was formed by homogenizing aqueous and organic phases. The aqueous phase consists of MDEA (extractant), PZ (promoter) and NaOH solution, while the organic phase consists of SPAN 80 (surfactant) and kerosene (diluent). MDEA supplied by Qrec[TM] and PZ supplied by Merck were used as received. In addition, deionized water and sodium hydroxide pellets supplied by Ashland Incorporation were used in the preparation of the aqueous solution.

21.2.2   Aqueous phase preparation

100 mL of aqueous phase was prepared by adding a mixture of MDEA and PZ into aqueous 0.1 M NaOH solution. The solution was stirred for 15 min. Different samples were prepared by varying the amount of PZ and NaOH solution. Based on previous research, the MDEA solution was fixed at 8% volume (Bhatti and Kamarudin, 2010). Table 21.1 shows the volume of MDEA, PZ and NaOH for aqueous phase preparation. The stirring speed and temperature of the heating plate were fixed at 700 rpm and 30°C, respectively. Table 21.2 shows the volume of the aqueous phase based on MDEA/PZ ratio of 4:3. The amount of MDEA and PZ increases while NaOH solution decreases to observe the effect of MDEA/PZ on CO_2 removal.

21.2.3   Organic phase preparation

The 100 mL organic phase was prepared by adding 8 mL of Span-80 into the 92 mL of kerosene. The solution was heated for 15 min. The stirring speed and temperature of the heating plate was fixed at 700 rpm and 30°C, respectively. The same procedure was repeated for other samples.

21.2.4   ELM preparation

Figure 21.1 shows the homogenization of aqueous and organic phase using a high performance disperser Ultra Turrax® T25 with 18G mixing shaft. The organic phase mixture was placed in the beaker. Then, the aqueous phase mixture was added drop wise into the organic solution and homogenized for 5 minutes. The speed of the mixer was fixed at 10,000 rpm. Table 21.3 summarizes the formulation of ELM preparation.

Table 21.1.   Volume for aqueous phase preparation.

MDEA [mL]	PZ [mL]	0.1 M NaOH [mL]
8	0	92
8	2	90
8	3	89
8	4	88
8	5	87
8	6	86

Table 21.2.   Volume for aqueous phase preparation at different quantities of MDEA and PZ.

MDEA [mL]	PZ [mL]	0.1 M NaOH [mL]
12	9	79
16	12	72
20	15	65
24	18	58

Figure 21.1.   Homogenization of aqueous and organic phase using high performance disperser Ultra Turrax®
T25 with 18G mixing shaft.

Table 21.3.   Formulation of ELM preparation.

Aqueous phase [mL]			Organic phase [mL]	
MDEA	PZ	NaOH, 0.1 M	Kerosene	Span-80
8	0	92	92	8
8	2	90	92	8
8	3	89	92	8
8	4	88	92	8
8	5	87	92	8
8	6	86	92	8
12	9	79	92	8
16	12	72	92	8
20	15	65	92	8
24	18	58	92	8

21.2.5   ELM characterization and performance

21.2.5.1   Stability studies

To determine the stability of the emulsion, the samples were filled in the graduated test tubes and placed in the room at 25°C. After 24 h, the formation of layers was measured. Demulsification will result in the formation of 3 layers (organic phase, emulsion, aqueous phase) whereas sedimentation of emulsion will result in the formation of 2 layers. Sedimentation is a process of the particles to settle down to the bottom of the solution. It is the instability of the particle in the emulsion due to the size or density of particles. Thus, stability was calculated based on the following equation:

$$\%\text{Stability} = \frac{V_T - V_S}{V_T} \times 100\% \tag{21.1}$$

where
V_T = Total volume [mL]
V_S = Top layer volume [mL]

21.2.5.2   Carbon dioxide absorption analysis

A rotating disc contactor (RDC) column (Fig. 21.2) connected to the gas chromatography (GC) was used for the CO_2 absorption study. The column was filled with 200 mL of the prepared emulsion. The flow rate of carbon dioxide to the column was fixed at 20 LPM (liter per minute). The gas was allowed to flow into the column for 1 minute. The speed of the rotating disc was in a range of 450–500 rpm. The pressure of the RDC was also recorded. Gas chromatography (GC) was used to determine the amount of CO_2 in the inlet and out streams of the RDC since the amount of substance in a sample is proportional to the area under the peak of that substance. The percentage of removal is the amount of CO_2 absorbed by the ELM and was calculated based on the area under the CO_2 peak (Equation (21.2)):

$$\text{Percentage of removal } [\%] = \frac{\text{area } CO_{2\,in} - \text{area } CO_{2\,out}}{\text{area } CO_{2\,in}} \tag{21.2}$$

Figure 21.2.   Rotating disc contactor column.

21.3   RESULTS AND DISCUSSION

21.3.1   ELM characterization

21.3.1.1   Stability of emulsion

The separation of ELM was measured after the sample was left for 24 h in the graduated test tube at 25°C. The sample formed 2 layers, which indicates only sedimentation has occurred. The stability was calculated based on Equation (21.1). Figure 21.3 shows the emulsion stability before absorption at different MDEA/PZ compositions. A sample containing only MDEA has the highest stability (98%). After the addition of 2 mL of PZ into 8 mL of MDEA solution, the stability of the emulsion decreased from 98 to 89.9%. The stability of MDEA and PZ was influenced by the viscosity of the solution. When combining PZ with MDEA, the viscosity of the solution increased. In general, a high viscosity solution requires more energy during homogenization to form small size particles that are well dispersed in the emulsion. Since the emulsification speed and time were fixed at 10,000 rpm and 5 minutes respectively, the sedimentation started to occur. Thus,

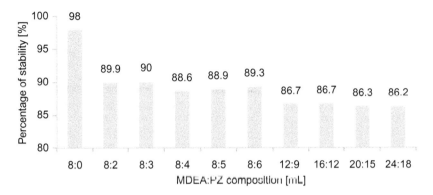

Figure 21.3.   The percentages of emulsion stability before CO_2 absorption.

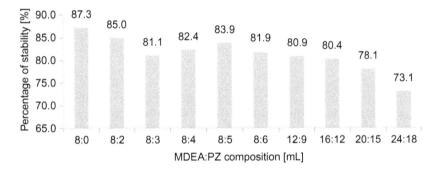

Figure 21.4.   The percentages of emulsion stability after CO_2 absorption.

sedimentation is enhanced and emulsion stability is decreased as the viscosity of solution increases due to the addition of PZ. The emulsion stability containing 8 mL MDEA and PZ (between 3 and 6 mL) were in a range of 88.6 to 90%. However, when the total quantity of MDEA with PZ was increased (based on 4:3 ratio), the emulsion stability was further reduced. The sample contained 24 mL MDEA with 18 mL PZ has the lowest percentage of emulsion stability (86.2%).

The emulsion stability was further reduced after CO_2 absorption due to the particle size increase. Figure 21.4 shows the percentage of emulsion stability after contacting with CO_2. The emulsion containing 8 mL MDEA retained the highest emulsion stability (87.3%) while 24 mL MDEA with 18 mL PZ had the lowest percentage of emulsion stability (73.1%). The emulsion stability for sample containing 8 mL MDEA with an increasing volume of PZ (3 mL to 6 mL) was in the range of 81.1 to 83.9%. As the quantity of MDEA and PZ increased (from 8:6 to 24:18), the emulsion stability decreased continuously.

21.3.1.2   pH analysis

In this study, the pH value of the emulsion before and after contacting with CO_2 was measured to observe its relationship with the stability of the emulsion and the percentage of CO_2 removal. The pH value of emulsion before and after contacting with CO_2 is shown in Figure 21.5 together with the composition of MDEA and PZ in the aqueous phase solution. The figure shows that before the emulsion contacted with CO_2, the sample containing 8 mL MDEA has the lowest pH value (10.83). After addition of PZ (8:2 and 8:3), the pH increased to 11.19 to 11.41. The increase of pH was due to high pK_a value of PZ ($pK_a = 10.33$). Despite this effect, the pH was maintained

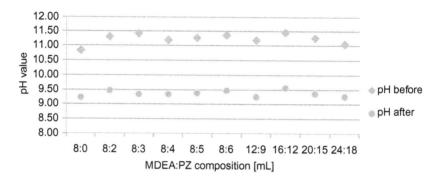

Figure 21.5.   pH value of emulsion before and after contacting with CO_2 at different composition of MDEA and PZ.

below 11.5 for all emulsions. After the absorption, the pH values decreased. This was due to the acidic nature of CO_2 gas.

21.3.2   *ELM performance*

21.3.2.1   *Carbon dioxide removal*
Figure 21.6 illustrates the percentage of CO_2 removal using different quantities of MDEA/PZ. The sample containing 8 mL MDEA and 6 mL PZ has the highest percentage of CO_2 removal (54.8%) while the sample containing 24 mL MDEA with 18 mL PZ showed the lowest percentage of CO_2 removal (34.1%). The percentage of CO_2 removal increased as the amount of PZ increased. It clearly shows that the addition of PZ into the solution increases the percentage of CO_2 removal as compared to use of MDEA only. For the reaction, MDEA has a CO_2 loading capacity of 1.0 mol CO_2 per mole of amine. However, when more PZ is added, the emulsion becomes less stable because of the viscosity increase, which will result in a decreased percentage of CO_2 removal. This trend is also found in Figure 21.6.

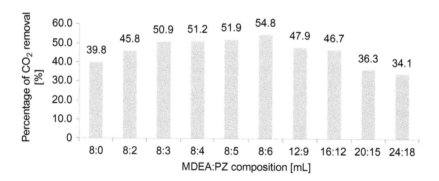

Figure 21.6.   The percentages of CO_2 removal.

21.3.3   *Formulation of MDEA and PZ for CO_2 removal*

Figures 21.7 and 21.8 show the relationship between the percentage of CO_2 removal and the percentage of the emulsion stability, before and after the CO_2 removal using different compositions of MDEA and PZ. From the figures, emulsion containing 8 mL MDEA has the highest percentage of stability before and after absorption, but the percentage of CO_2 removal was low. However, the addition of PZ into the emulsion decreased the percentage of stability of emulsion (before and

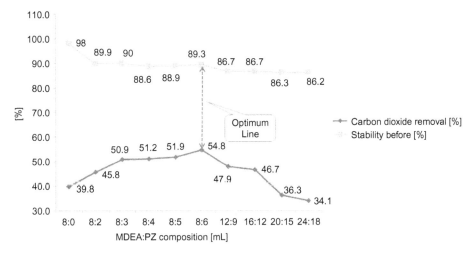

Figure 21.7.   The relationship between the percentages of CO_2 removal and percentage of emulsion stability before absorption.

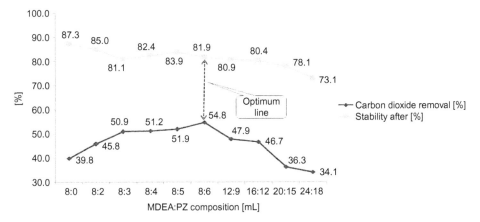

Figure 21.8.   The relationship between the percentages of CO_2 removal and percentage of emulsion stability after absorption.

after absorption) but increased the percentage of CO_2 removal. When the quantity of MDEA and PZ increased based on the ratio of 4:3, the stability of the emulsion before and after absorption was reduced. Similarly, the percentage of CO_2 removal was also reduced. Based on the formulation of emulsion in Table 21.3, the emulsion containing 8 mL MDEA and 6 mL PZ has the highest percentage of stability before and after absorption, and the highest CO_2 removal. Therefore, this study has shown that using a homogenization speed of 10,000 rpm and homogenization time of 5 minutes, emulsion formulation of 8 mL of Span 80, 92 mL of kerosene and 8 mL MDEA + 6 mL PZ has the highest CO_2 removal.

21.4   CONCLUSION

This study shows the effect of piperazine (PZ) addition to MDEA in the aqueous phase on stability of emulsion and percentage of CO_2 removal. The emulsion stability refers to sedimentation of the particles that leads to the formation of 2 layers. The stability of emulsion decreased when the

PZ was added to MDEA. This was due to high viscosity of solution that required more energy to homogenize the emulsion to produce small particles. In contrast, the addition of PZ to 8 mL of MDEA increased the percentage of CO_2 removal. However, with further increase of PZ, the emulsion stability started to decrease. Therefore, this study showed that emulsion formulation containing 8 mL of Span 80 and 92 mL of kerosene (organic phase), and 86 mL NaOH, 8 mL MDEA and 6 mL PZ (aqueous phase) was the best formulation for CO_2 removal in a rotating disc contactor column. The homogenization speed of 10,000 rpm and homogenization time of 5 minutes were able to produce a stable emulsion. This result provides new information about emulsion formulation and its capability to remove carbon dioxide.

ACKNOWLEDGEMENT

The authors are grateful to the Ministry of Science, Technology and Innovation (MOSTI) and Universiti Teknologi Malaysia (VotNo. 4S047) for supporting the research on carbon dioxide removal using emulsion liquid membrane technique.

REFERENCES

Aroonwilas, A. & Tontiwachwuthikul, P. (1997) High-efficiency structured packing for CO_2 separation using 2-amino-2-methyl-1-propanol (AMP). *Separation and Purification Technology*, 12, 67–79.

Arunkumar, S. & Bandyopadhyay, S.S. (2007) Kinetics and modeling of carbon dioxide absorption into aqueous solutions of piperazine. *Chemical Engineering Science*, 62, 7312–7319.

Bhatti, I. & Kamarudin, K.S.N. (2010) Removal of carbon dioxide using water-in-oil emulsion liquid membrane containing triethanolamine (TEA). *Journal Applied Science Research*, 6 (12), 2251–2256.

Bishnoi, S. (2000) *Carbon dioxide absorption and solution equilibrium in piperazine activated methyldiethanolamine*. PhD Thesis, The University Of Texas at Austin, Austin, TX.

Bishnoi, S. & Rochelle, G.T. (2000) Absorption of carbon dioxide into aqueous piperazine: reaction kinetics, mass transfer and solubility. *Chemical Engineering Science*, 55, 5531–5543.

Bishnoi, S. & Rochelle, G. (2002) Absorption of carbon dioxide in aqueous piperazine/methyldiethanolamine. *American Institute of Chemical Engineering Journal*, 48, 2788–2799.

Chakraborty, M., Bhattacharya, C. &. Datta, S. (2003) Studies on the applicability of artificial neural network (ANN) in emulsion liquid membranes. *Journal of Membrane Science*, 220 (1–2), 155–164.

Cullinane, J. & Rochelle, G. (2004) Carbon dioxide absorption with aqueous potassium carbonate promoted by piperazine. *Chemical Engineering Science*, 59, 3619–3630.

Kosseim, A.J., McCullough, J.G. & Butwell, K.F. (1984) Treating acid & sour gas: corrosion-inhibited amine guard ST process. *Chemical Engineering Progress*, 80 (10), 64–71.

Li, N.N. (1968). Separating hydrocarbons with liquid membranes. US Patent 3,410,794.

Mandal, B.P. & Bandyopadhyay, S.S. (2006) Absorption of carbon dioxide into aqueous blends of 2-amino-2-methyl-1-propanol and monoethanolamine. *Chemical Engineering Science*, 61, 5440–5447.

Moon, S.H. & Shim, J.W. (2006) A novel process for CO_2/CH_4 gas separation on activated carbon fibers-electric swing adsorption. *Journal of Colloid and Interface Science*, 298, 523–528.

Nath, K. (2008) *Membrane separation process*. Phi Learning Pvt. Ltd., New Delhi, India.

Teramoto, M., Sakaida, Y., Fu, S.S., Ohnishi, N. & Matsuyama, H. (2000) An attempt for the stabilization of supported liquid membrane. *Separation and Purification Technology*, 21 (1–2), 137–144.

Wong, S. & Bioletti, R. (2002) Carbon dioxide separation technologies. Alberta Research Council, Canada.

Xu, G.W., Zhang, C.F., Qin, S.J. & Wang, Y.W. (1992) Absorption of carbon dioxide into solutions of activated methyldiethanolamine. *Industrial Engineering Chemistry Research*, 31, 921–927.

Zhang, X., Zhang, C.F. & Xu, G.W. (2001). An experimental apparatus to mimic CO_2 removal and optimum concentration of MDEA aqueous solution. *Industrial Engineering Chemistry Research*, 40, 898–901.

CHAPTER 22

Separation of olefins from paraffins by membrane contactor – a review

Gholamreza Bakeri, Ali Kazemi Joujili, Mostafa Rahimnejad,
Ahmad Fauzi Ismail & Takeshi Matsuura

22.1 INTRODUCTION

Olefins are among the most important products in petrochemical plants. Generally, they are mixed with paraffins therefore it is necessary to use efficient technologies with lower energy consumption for difficult separation of olefins from paraffins. Because of high operating costs and capital investment for current commercially practiced separation processes, many researches have been done into better olefin/paraffin separation techniques.

An olefin/paraffin separation scheme using membrane based techniques offers great advantages such as reducing the operating cost and capital investment, as well as eliminating the operating problems in distillation towers such as flooding, entrainment, etc. (Faiz and Li, 2012). Most of the researches in this field have been devoted to gas separation membranes which have some problems such as low flux and selectivity that prohibit their application in petrochemical industries.

On the other hand, membrane contactor offers many advantages such as high flux and selectivity that are related to the separation mechanism in the contactor. As the membrane is porous and solute gas diffuses through the pores, the flux is much higher than a gas separation membrane where the gas should dissolve in the membrane and then diffuse through the dense skin layer. Furthermore, the selectivity in contactor applications depends on the absorbent that can be chosen to maximize the selectivity. Chilukuri et al. (2007) studied the separation of propylene from propane using membrane contactor with a capacity of 300 tons/year and achieved reasonable results. Therefore, the application of contactor with an efficient membrane for olefin/paraffin separation would be an attractive alternative to replace the current separation processes.

22.2 OLEFIN AND PARAFFIN

Nowadays, olefin based products (especially with low molecular weight such as C_2–C_5) encompass an important part of the petrochemical industries and people's lives and yield a wide range of products such as cosmetics, textile products, paints, tools etc. (Fallanza et al., 2012). Ethylene and propylene, with a purity of more than 99.9%, are the most important olefins that are used as feedstock in petrochemical plants for producing not only polyethylene and polypropylene but also many kinds of chemicals such as ethylene oxide, vinyl chloride, etc. (Ceresana, 2011; Hilyard, 2008). Since olefins are so important in petrochemical industries, their production and consumption capacity show increasing trends, e.g. in 1995, USA production capacity for ethylene was 21.3 million metric tons while in 1998, it increased to 23.6 million tons (Bessarabov, 1998; Burns and Koros, 2003).

22.2.1 Olefin production technologies

About 70% of the olefins are produced by the cracking method and the other 30% are made through catalytic cracking and paraffin dehydrogenation processes (Fallanza et al., 2011; Hilyard, 2008).

As cracking methods are not selective, new technologies with higher selectivity in reaction (such as methanol to propylene, MTP) were developed (Lurgi & NPC-RT, 2007).

22.2.2   *The importance of olefin/paraffin separation*

Since olefin and paraffin (with the same carbon number) have close physical properties such as boiling point, their separation is so hard and needs a distillation tower with many trays, which increases the operating costs and capital investment (Hirschfelder *et al.*, 1964; Perry and Green, 1984; Rajabzadeh *et al.*, 2010). On the other hand, in olefin production processes a mixture of olefin and paraffin is produced that needs efficient methods for separation. Furthermore, oil refinery plants produce a mixture of olefin/paraffin which, when used as a fuel, emits toxic materials (Safarik and Eldridge, 1998). Therefore, it is worthwhile to develop new technologies for separation of olefin from paraffin.

22.2.3   *Olefin/paraffin separation methods*

There are several methods for olefin/paraffin separation such as cryogenic distillation, extractive distillation, molecular sieve adsorption, absorption and membrane based processes (Chang *et al.*, 2002; Ghosh *et al.*, 1993). Among these methods, membrane technologies are the newest ones which present some advantages over the other methods such as modularity, lower operating cost etc.

22.2.4   *Membrane processes*

Membranes have been known to be used in separation processes for about 100 years (Lonsdale, 1982) but this technology was not used in industry for a long time. In the 1940s, membrane was used for separation of uranium isotopes on a commercial scale. With the development of asymmetric membranes by Loeb and Sourirajan (1963) in the early 1960s, membrane separation processes have been widely used in the dairy, food and beverage industries (Jiao *et al.*, 2004). Membrane technologies present some advantages such as lower energy consumption (Scholz and Lucas, 2003), modular design with smaller equipment (Porter, 1990; Scholz and Lucas, 2003), higher surface to volume ratio (Drioli *et al.*, 2006), more flexibility in operation (Drioli *et al.*, 2006) and less environmental pollution.

22.3   MEMBRANE CONTACTOR

Membrane contactor is one of the newest techniques for gas separation where mass transfer occurs between gas stream and liquid absorbent without dispersion of phases into each other. The streams flow on opposite sides of a porous membrane and the gas fills the pores of the membrane while the liquid is immobilized at the entrance of the pores, so the phases are in contact with each other at the mouth of the pores. In other words, unlike traditional separation equipment, there is no direct contact between phases (Baker, 2012). Membrane contactor has some advantages compared to other separation techniques (Bakeri *et al.*, 2011; Gabelman and Hwang, 1999; Nymeijer D.C. *et al.*, 2004; Rajabzadeh *et al.*, 2010) such as:

- The available contact area between phases remains nearly constant at different flow rates as it depends on the membrane characteristics.
- As there is no phase dispersion, no emulsion forms.
- Scale up/down or capacity change of contactor system is very straightforward as the system is modular.

- Higher efficiency is obtained in comparison to dispersive systems such as packed towers because of the higher contact area, as reported that the contact area in contactor systems is 4–30 times more than in packed towers (Simioni *et al.*, 2011).
- The separation process in membrane can be done with high selectivity since as strong an absorbent as possible can be selected.

As the energy consumption in contactor systems is low and the flux and selectivity are high, this technology offers higher efficiency as reported by Klassen *et al.* (2005). However, membrane contactor has some disadvantageous such as (Gabelman and Hwang, 1999):

- Membrane makes an extra resistance in mass transfer between phases. However, this resistance is compensated by the high contact area generated by membrane.
- Membranes may be subjected to fouling which is not easy to be controlled.
- Membrane life time is limited so the cost of periodic membrane replacement should be considered.

In summary, even though membrane contactor has some limitations, its benefits are large enough to use it on a commercial scale.

Most of the researches in the field of membrane contactors were devoted to CO_2 separation and few studies were done on olefin/paraffin separation. Qi and Cussler (1985a; 1985b) were the first who worked on the CO_2 capture using membrane contactor. Bessarabov *et al.* (1995) used non-porous membrane in two stages for ethane/ethylene separation. In recent years, a number of researches have been done on olefin/paraffin separation, most of which were for ethane/ethylene and propane/propylene separation Bessarabov *et al.* (1995).

In membrane contactor, olefin reacts with the absorbent and makes a complex with Ag and Cu ions that are used as absorbent. Then, in the next stage gas should be separated from the absorbent by breaking the complex through a change in temperature and pressure. Even though it seems that the separation process is performed in two stages and membrane increases the mass transfer resistance, the high interfacial area provided by the membrane enhances the overall performance of the contactor system.

22.4   MEMBRANE CHARACTERISTICS

Some membrane properties such as porosity and hydrophobicity affect the performance of membrane in contactor applications. Li *et al.* (1999) reported that the membrane mass transfer coefficient depends on membrane porosity, membrane tortuosity and membrane thickness:

$$k_m = \frac{D_e \varepsilon}{\tau l_m} \tag{22.1}$$

where l_m is membrane thickness, τ is membrane tortuosity, ε is membrane porosity and D_e is effective diffusivity that depends on membrane pore size (r) as presented by:

$$D_e = \frac{1}{\left(\frac{1}{D} + \frac{3}{2r}\sqrt{\frac{\pi M}{8RT}}\right)} \tag{22.2}$$

where M is the molecular weight of diffusing gas, D is bulk diffusivity, R is the universal gas constant and T is absolute temperature. The diffusivity of gas through membrane pores depends on pores size, i.e. for pore diameter less than 100 nm the Knudsen mechanism is the governing one, for diameters more than 100 nm the bulk diffusion is the governing one and between these two ranges, both mechanisms are dominant (Kumar *et al.*, 2003).

22.4.1 Membrane stability

The absorbent can alter the surface properties of membrane so the membrane material should be selected according to the type of absorbent. Mansourizadeh and Ismail (2009) reported the compatibility of different membranes and absorbents. Rajabzadeh et al. (2010) studied the stability of PVDF membrane in olefin/paraffin separation using $AgNO_3$ solution with a concentration of $200 \, mol \, m^{-3}$ and reported that the absorbent did not reduce the membrane stability.

22.5   MATHEMATICAL CORRELATIONS FOR MASS TRANSFER COEFFICIENTS

In membrane contactors, gas and liquid flow on opposite sides of a porous membrane where the solute gas(es) diffuses from the bulk of the gas to entrance of the pore, then through the pore to the other end of the pore and finally is absorbed by the liquid. Therefore, there are three mass transfer resistances in series which are:

- Mass transfer resistance in the gas boundary layer.
- Membrane mass transfer resistance.
- Mass transfer resistance in liquid boundary layer.

The overall mass transfer coefficient ($K_{overall}$) is given as Equation (22.3) (Fallanza et al., 2011; Feron and Jansen, 2002):

$$\frac{1}{K_{overall}} = \frac{d_o}{k_g d_i} + \frac{1}{E_a k_l H} + \frac{d_o}{k_m d_{lm}} \tag{22.3}$$

where d_i, d_o and d_{lm} are inner, outer and logarithmic mean diameters of membrane respectively, H is Henry's constant, k_l, k_m and k_g are liquid side, membrane (when the pores are gas filled) and gas side mass transfer coefficient, respectively, and E_a is the enhancement factor. The overall mass transfer coefficient ($K_{overall}$) can be obtained experimentally through the mass transfer flux as:

$$K_{overall} = \frac{Q_l(C_l^{out} - C_l^{in})}{A \Delta C_l^{ave}} \tag{22.4}$$

where Q_l is the liquid flow rate, C_l is the solute gas concentration in liquid, A is the contact area which is calculated on inner diameter of membrane if liquid flows through the lumen side of membrane and ΔC_l^{ave} is logarithmic mean of transmembrane concentration difference of solute gas in terms of liquid which is calculated by:

$$\Delta C_l^{ave} = \frac{(HC_g^{in} - C_l^{in}) - (HC_g^{out} - C_l^{out})}{\ln\left(\frac{HC_g^{in} - C_l^{in}}{HC_g^{out} - C_l^{out}}\right)} \tag{22.5}$$

where C_g is the concentration of solute gas in the gas stream and "in" and "out" refer to the entrance and exit of the liquid/gas stream from the membrane contactor. Fallanza et al. (2011) studied the absorption of propylene in membrane contactor with cross flow arrangement and reported Equation (22.6) for calculation of $K_{overall}$:

$$\frac{1}{K_{overall}} = 7.807 u_e^{-0.32} + 1557.5 \tag{22.6}$$

where u_e is the superficial velocity of liquid.

Table 22.1. The correlations for calculation of mass transfer coefficient in shell side of contactor.

Conditions	Correlation	Flow regime	Reference
$0.8 < Re < 20$; $0.4 < \varphi < 0.46$	$Sh = 2.15Re^{0.42}Sc^{0.33}$	Transverse	Zheng *et al.* (2005)
$0.1 < Re < 0.4$	$Sh = 0.41Re^{0.36}Sc^{0.33}$		Fouad and Bart (2007)
$0.05 < Re < 5$; $0.49 < \varphi < 0.53$	$Sh = 1.76Re^{0.82}Sc^{0.33}$		Schöner *et al.* (1998)
$0 < Re < 500$; $0.04 < \varphi < 0.4$	$Sh = \beta[d_e(1-\varphi)/L]Re^{0.6}Sc^{0.33}$ β: 5.8 for hydrophobic and 6.1 for hydrophilic fibers	Parallel	Prasad and Sirkar (1998)
$21 < Re < 324$; $0.32 < \varphi < 0.76$	$Sh = (0.53 - 0.58\varphi)Re^{0.53}Sc^{0.33}$		Costello *et al.* (1993)
$32 < Re < 1287$; $0.1 < \varphi < 0.7$	$Sh = (0.3045\varphi^2 - 0.3421\varphi$ $+ 0.0015)Re^{0.9}Sc^{0.33}$		Wu and Chen (2000)
$1 < Re < 25$; $\varphi = 0.7$	$Sh = 1.38Re^{0.34}Sc^{0.33}$	Cross flow	Yang and Cussler (1986)
$0.6 < Re < 49$; $\varphi = 0.003$	$Sh = 0.61Re^{0.363}Sc^{0.333}$		Cote *et al.* (1989)

22.5.1   *Mass transfer coefficient in lumen side of contactor*

In the case of liquid in the lumen side of the contactor, Kumar *et al.*, 2003 reported Equations (22.7)–(22.9) calculation of liquid side mass transfer coefficient (k_1):

$$\text{For } Gz < 10 \quad Sh = \frac{k_1 d_i}{D} = 3.67 \tag{22.7}$$

$$\text{For } Gz > 20 \quad Sh = 1.62(Gz)^{1/3} \tag{22.8}$$

$$\text{For entire range of } Gz \quad Sh = (3.67^3 + 1.62^3 Gz)^{1/3} \tag{22.9}$$

Gz is the Graetz number ($Gz = \frac{V_{liquid} d_i^2}{DL}$) and Sh is the Sherwood number ($\frac{k_1 d_i}{D}$) where d_i is the inner diameter of hollow fiber membrane and D is the bulk diffusivity (Kreulen *et al.*, 1993a).

22.5.2   *Mass transfer coefficient in shell side of contactor*

In the case of physical absorption, the mass transfer coefficient in terms of Sherwood number depends on the packing density, Reynolds (Re) and Schmidt (Sc) numbers. In addition, it depends on some other parameters such as the regularity of fibers in the module and the effect of the module wall (Mansourizadeh and Ismail, 2009). Some of the correlations reported for calculation of the shell side mass transfer coefficient are presented in Table 22.1.

22.5.3   *Membrane mass transfer coefficient*

In membrane contactor, the membrane makes an extra resistance in mass transfer between gas and liquid and its contribution to overall mass transfer resistance depends on different membrane characteristics among which the wettability of the pores is the main one. Three different cases for pore wetting were reported:

• Pores are completely filled by gas.
• Pores are filled by gas and liquid.
• Pores are completely filled by liquid.

The third one is the worst case. The membrane mass transfer coefficient can be calculated by Equation (22.1).

Table 22.2.   Results for ethylene absorption in silver salts solution (Nymeijer K. et al., 2004a).

Anion	T [K]	K_c [m^3 mol^{-1}][a]	Reference
NO_3^-	295–297	0.116 @ 298 K	Baker (1964)
CF_3COO^-	290.5–305.5	0.094 @ 298 K	Brandt (1959)
NO_3^-	303	0.076 @ 303 K	Clever (1970)
NO_3^-	293	0.119 @ 293 K	Crookes and Woolf (1973)
NO_3^-	293–353	0.086 @ 303 K	Temkin et al. (1964)
NO_3^-	298	0.098 @ 298 K	Trueblood and Lucas (1995)
NO_3^-	273–298	0.085 @ 298 K	Wasik and Tsang (1970)

22.6   THE TYPE OF ABSORBENT

There are two types of absorption processes in membrane contactors:

- Physical absorption where the solute gas(es) dissolves in absorbent.
- Chemical absorption where the solute gas(es) reacts with the absorbent after dissolution.

The latter one is usually preferred because of the higher absorption capacity and selectivity. In olefin/paraffin separation, silver salts solutions such as $AgCF_3COO$, $AgBF_4$ and $AgNO_3$ are usually used as absorbent because of their ability to make a complex with the unsaturated bond of olefins and its good solubility in water (Nymeijer K. et al., 2004a). Furthermore, these salt solutions have low vapor pressure and good thermal stability (Fallanza et al., 2011). The results for ethylene absorption using silver salts are presented in Table 22.2.

In addition, copper ion can be used for olefin/paraffin separation even though research in this field is rare. Chilukuri et al. (2007) studied the separation of propylene using two types of silver salts, $AgNO_3$ and $AgBF_4$, from technical and economical points of view and found that $AgBF_4$ dissociates completely and makes more concentration of Ag^+. In addition, it was found that even though increasing the concentration of Ag^+ will increase the absorption of propylene, it will be harder to break the olefin-Ag^+ complex at higher concentrations. Therefore, an optimum concentration of silver ion should be used. Nymeijer K. et al. (2004a) studied the formation of olefin-silver complex using 1 M and 3.5 M $AgNO_3$ solution. Fallanza et al. (2011) studied the mass transfer for separation of propylene in a contactor module with cross flow regime, using 0.25 M $BMIMBF_4$-Ag salt solution. In the case of liquid in the shell side, they found that overall mass transfer coefficient increases when liquid flow rate or salt concentration increases. Furthermore, they optimized the liquid flow rate based on the liquid side pressure drop and mass transfer flux and found that the mass transfer coefficient in the case of a cross flow regime is 17.6 times more than parallel flow regime and 17.4 times more than stirred tank.

Rajabzadeh et al. (2010) used $AgNO_3$ solution with different concentrations (0, 1000, 2000 and 4000 mol m^{-3}) and found that the propylene absorption flux increases as salt concentration or absorbent flow rate increases. Ortiz et al. (2010a; 2010b) studied the kinetics of reaction between $AgBF_4$ and propylene and found that the absorption rate is depending on both reactants and Ag^+ diffusivity will increase by increasing temperature and salt concentration. Furthermore, they studied the effect of solvent type on the absorption of propylene where water and ionic solvent ($BMIMBF_4$) were used as solvent and $AgBF_4$ was used as silver salt. In case of water, the propylene absorption flux is more at higher concentration of Ag^+ (e.g. 0.25 M) while for ionic solvent, the absorption flux is higher at lower concentration of Ag^+ (e.g. 0.1 M) as shown in Table 22.3.

One important characteristic of absorbent in contactor applications is the surface tension which reduces the ability of absorbent to wet the membrane; e.g. for CO_2 separation, some researches have been done on new absorbents such as potassium glycinate (Yan et al., 2007). Furthermore,

Table 22.3. The effect of solvent type on propylene absorption flux; silver salt: $AgBF_4$ (Ortiz *et al.*, 2010b).

	Propylene absorption flux $[\times 10^5 \, \mathrm{mol \, m^{-2} \, s^{-1}}]$				
	Solvent type				
	Water Silver salt concentration		$BMIMBF_4$ Silver salt concentration		
Feed gas composition [%v/v]	0.1 M	0.25 M	0 M	0.1 M	0.25 M
70% propane – 30% propylene	2.12	3.60	0.94	2.27	2.06
50% propane – 50% propylene	3.12	5.38	1.59	3.46	4.10
30% propane – 70% propylene	4.90	7.11	2.26	5.60	5.61

as silver salts are expensive, the regeneration of absorbent is critical in olefin/paraffin separation. Li and Chen (2005) studied the different factors in choosing the absorbent.

22.7 REACTIONS IN OLEFIN/PARAFFIN SEPARATION

22.7.1 *Complex formation between Ag^+ and olefin*

Olefins are unsaturated hydrocarbons which can make a complex with the metal ions such as Ag^+ (Nymeijer, 2003; Nymeijer D.C. *et al.*, 2004b; Nymeijer K. *et al.*, 2004a). These complexes can be broken via changing the operating conditions such as temperature and pressure, which makes these types of ions suitable in membrane contactors for olefin separation. A number of researches have been done on the kinetics and thermodynamics of reaction between Ag^+ and olefin (Nymeijer, 2003; Trueblood and Lucas, 1995) to reveal that the reactions proceed according to Equations (22.10)–(22.12):

$$Ag^+ + olefin \underset{K_1}{\longleftrightarrow} [Ag(olefin)]^+ \qquad (22.10)$$

$$[Ag(olefin)]^+ + Ag^+ \underset{K_2}{\longleftrightarrow} [Ag_2(olefin)]^{2+} \qquad (22.11)$$

$$[Ag(olefin)]^+ + olefin \underset{K_3}{\longleftrightarrow} [Ag(olefin)_2]^+ \qquad (22.12)$$

Generally, $[Ag(olefin)_2]^+$ and $[Ag_2(olefin)]^{2+}$ form at any concentrations of Ag^+ (Trueblood and Lucas, 1995) even though their concentration is low (Sunderrajan *et al.*, 1999).

22.7.2 *Kinetics and equilibrium constants*

If the silver salt dissociates completely in solvent, the absorption capacity will increase but because of the interaction between the anion and Ag^+, the dissociation of silver salt is not complete (Nymeijer K. *et al.*, 2004a). The degree of salt dissociation is presented in two terms, dissociation constant (K_{diss}) and dissociation degree (α), which are given by Equations (22.13) and (22.14):

$$K_{diss} = \frac{[Ag^+]_{aq}[NO_3^-]_{aq}}{[AgNO_3]_{aq}} \qquad (22.13)$$

$$\alpha = \frac{[Ag^+]_{aq}}{[AgNO_3]_{aq} + [Ag^+]_{aq}} \qquad (22.14)$$

Table 22.4.   Dissociation constant (K_{diss}) and degree of dissociation (α) at different temperatures for 1.0 and 3.5 M solutions of AgNO$_3$ (Nymeijer K. et al., 2004a).

| Temperature [K] | Salt concentration | | | |
| | 1 M AgNO$_3$ (aq) | | 3.5 M AgNO$_3$ (aq) | |
	K_{diss} [mol m^{-3}]	α [%]	K_{diss} [mol m^{-3}]	α [%]
275.5	213	37	177	20
278	274	40	228	22
283	449	48	373	28
288	722	56	600	34
293	1144	64	950	40

The degree of dissociation depends strongly on temperature and salt concentration and increases with reduction in salt concentration and increase in temperature as is presented in Table 22.4; that is the results reported by Nymeijer K. et al. (2004a).

Furthermore, they found in the reaction between Ag$^+$ and ethylene, that the equilibrium constant increases with reduction in temperature.

22.7.3   Gas absorption rate in olefin/paraffin separation

In olefin/paraffin separation by membrane contactor, the driving force for mass transfer between gas and liquid is the concentration gradient along the membrane. The chemical reaction between silver ion and olefin influences the mass transfer rate in the liquid phase. Equation (22.15) was reported for calculation of gas absorption flux when the gas side mass transfer resistance can be neglected (Nymeijer K. et al., 2004a):

$$R_{abs} = -K_1 E_a ([A]_l^i - [A]_l) a V_l \tag{22.15}$$

where a is the interfacial area per unit volume, K_1 is the liquid side mass transfer coefficient, $[A]_l^i$ is the concentration of ethylene at liquid interface, $[A]_l$ is the concentration of ethylene in the bulk of liquid, V_l is the volume of the liquid phase and E_a is the enhancement factor which is defined as the ratio of the absorption rate in the case of chemical absorption to the rate of absorption in the case of physical absorption at the same concentration differences of the solute gas(es) (Hogendoorn et al., 1997). The enhancement factor depends on the Hatta number (Ha) and infinite enhancement factor ($E_{a,\infty}$) as shown in Equation (22.16) (Ortiz, 2010):

$$E_a = \frac{-Ha^2}{2E_{a,\infty}} + \sqrt{\frac{Ha^4}{4(E_{a,\infty} - 1)^2} + \frac{E_{a,\infty} Ha^2}{E_{a,\infty} - 1} + 1} \tag{22.16}$$

A number of researches have been done to calculate the enhancement factor through analytical and numerical methods, in which the analytical ones are based on the mass transfer theories such as film theory, Higbie's theory and surface renewal theory but because of using many assumptions, analytical methods cannot give an exact solution even though they are easier to be used. On the other hand, in membrane contactor the pore size of membrane is small and there is no velocity gradient. Therefore, the mass transfer theories are not applicable and numerical methods should be used for calculation of the enhancement factor.

Researchers attempted to study the chemical reaction in the absorption process using the enhancement factor and Hatta number. Versteeg et al. (1989) reported the dependency between

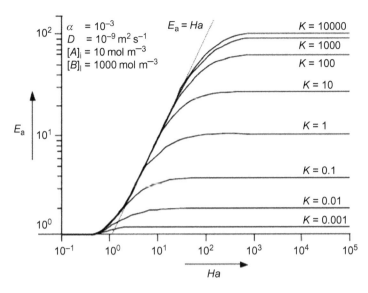

Figure 22.1.   Variation of enhancement factor versus Hatta number at different reaction equilibrium constants (Nymeijer D.C. *et al.*, 2004a).

the enhancement factor and Hatta number at different reaction equilibrium constants as was shown in Figure 22.1.

Nymeijer K. *et al.*, 2004a found that the actual enhancement factor is approximately equal to the infinite enhancement factor, indicating that the complex formation reaction is almost instantaneous. Hatta number and infinite enhancement factor can be calculated through Equations (22.17) and (22.18) (Fallanza *et al.*, 2011):

$$Ha^2 = \frac{\frac{2}{n+1}k_{1,1}(C_{C_3H_6}^{int,IL})^{n-1}C_{Ag^+}^{m}D_{C_3H_6,l}}{k_l^2} \tag{22.17}$$

$$E_{a,\infty} = \left(1 + \frac{D_{Ag^+}C_{Ag^+}}{D_{C_3H_6,l}C_{C_3H_6}^{int,IL}}\right)\left(\frac{D_{C_3H_6,l}}{D_{Ag^+}}\right)^{0.5} \tag{22.18}$$

where *Ha* is the Hatta number, *C* is the concentration, *D* is the diffusivity, $E_{a,\infty}$ is the infinite enhancement factor, $k_{1,1}$ is the kinetic constant of reaction and k_L is the liquid phase mass transfer coefficient.

22.8   OLEFIN/PARAFFIN SEPARATION USING MEMBRANE CONTACTOR

Separation of the olefin/paraffin mixture in membrane contactor is done by the selective absorption of olefins in silver salt solution. One of the most important drawbacks in contactor systems is the membrane wetting which decreases the mass transfer rate (Kreulen *et al.*, 1993b) as was reported that wetting of just 2% of membrane pores will increase the membrane mass transfer resistance up to 60% of the total mass resistance (Rangwala, 1996). The pore wetting depends on membrane structural parameters such as membrane pore size and pore size distribution, and the operating conditions such as transmembrane pressure difference, the liquid surface tension etc. One method to overcome this problem is the application of hydrophobic membranes such as PTFE or PP. These polymers are expensive or need complex process for membrane fabrication.

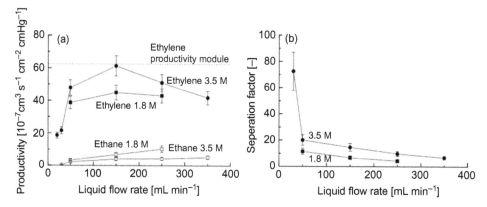

Figure 22.2.   The variation of (a): productivity and (b): ethylene over ethane separation factor versus liquid absorbent flow rate. Operating conditions for experiments: temperature $= 25°C$; silver salt $=$ AgNO$_3$; salt concentration $= 1.8$ M, 3.5 M; total feed pressure $= 0.3$ MPa (Nymeijer D.C. *et al.*, 2004a).

On the other hand, it is possible to reduce the membrane wettability by reduction in pore size as was presented in the Young-Laplace relation:

$$\Delta P = \frac{2\sigma_L \cos \theta}{r} \tag{22.19}$$

where σ_L is surface tension of liquid, θ is contact angle between liquid and membrane surface and r is the pore radius.

Bakeri *et al.* (2010) used polyetherimide (a relatively hydrophilic polymer) for the fabrication of membranes to be used in contactor applications and by manipulating the membrane characteristics, such as membrane pore size, porosity and tortuosity, could achieve higher performance than commercial hydrophobic membranes. Since PEI is much cheaper than PTFE, their method provided a more economical way of membrane manufacturing. Another method to overcome the membrane wettability is the coating of pores on the liquid side of membrane, using highly gas permeable elastomeric materials. Nymeijer D.C. *et al.* (2004a) applied 8 μm ethylene propylene diene monomer (EPDM) coating on the surface of a porous membrane and used the composite membrane in a contactor to separate 20:80 ethane/ethylene mixture. They found that the productivity of ethane kept increasing with liquid absorbent flow rate while for ethylene showed a maximum in productivity. Their results are presented in Figure 22.2.

Furthermore, Nymeijer K. *et al.* (2004b) applied a layer of sulfonated poly(ether ether ketone) (SPEEK) on the polypropylene hollow fiber membrane by the dip coating method and used the coated membrane in contactor for separation of 20:80 ethane/ethylene mixture, using 3.5 M silver nitrate solution as absorbent. The water in the absorbent makes the coating swell and prevents drying of the coating. Their results (Figs. 22.3 and 22.4) show that SPEEK coating provides higher selectivity compared to EPDM coating but the ethylene productivity is lower.

Even though SPEEK coating swells more than EPDM coating in the absorption process, it has lower ethylene permeability but as the SPEEK coating is saturated with silver ions, it is highly selective to ethylene. The effect of feed gas pressure on the ethylene productivity at various liquid flow rates of absorbent for the membranes with SPEEK and EPDM coatings was shown in Figure 22.5.

Furthermore, block copolymer of poly(ethylene oxide) (PEO) and poly(butylene terephthalate) (PBT) coating was applied on the polypropylene hollow fiber membrane (Nymeijer K. *et al.*, 2004c), and the result showed much higher productivity than SPEEK coating and much higher selectivity than EPDM coating. In other words, this coating showed higher productivity and selectivity than EPDM and SPEEK coatings and is most suitable to be used in olefin/paraffin

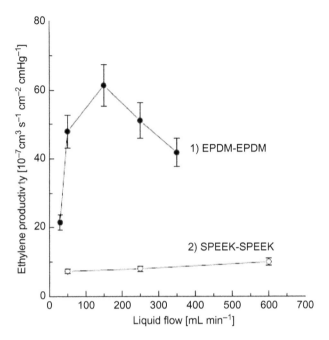

Figure 22.3.   Ethylene productivity as a function of the absorbent liquid flow rate for SPEEK- and EPDM-coated membranes (Nymeijer K. *et al.*, 2004b).

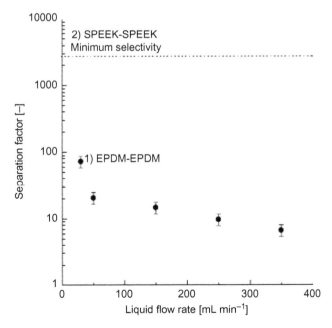

Figure 22.4.   Ethylene over ethane separation factor as a function of the absorbent liquid flow rate for SPEEK- and EPDM-coated membranes (Nymeijer K. *et al.*, 2004b).

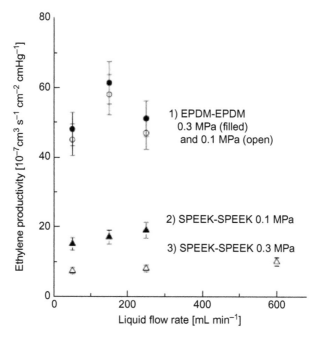

Figure 22.5.   Ethylene productivity as a function of the liquid absorbent flow rate and the feed gas pressure for SPEEK- or EPDM-coated hollow fiber membranes (Nymeijer K. *et al.*, 2004b).

Table 22.5.   The permeability and swelling properties of different coatings used for olefin separation.

Coating type	Permeability [barrer]			Swelling [wt%]	
	C_2H_6	C_2H_4	H_2O^a	H_2O	3.5 M AgNO$_3$
PEO/PBT	10.4	16.5	5.4×10^5	53.7	84.7
EPDM	40.2	46.4	1.5×10^3	0.3	0.1
SPEEK	NA^b	NA	1.2×10^5	88.0	51.0

^{a}3.5 M AgNO$_3$ (aq); ^{b}Not applicable
barrer: $10^{-10} \frac{cm^3 \cdot cm}{cm^2 \cdot s \cdot cmHg}$

separation by membrane contactor. The results for permeability and selectivity of the films of different coatings are presented in Table 22.5.

The high permeability of water through the coatings causes the permeate stream to be humid. Even though membrane coating can overcome the membrane wettability, adding a layer on the membrane surface increases membrane mass transfer resistance. Ceramic membrane was also proposed to reduce membrane wetting in olefin/paraffin separation (Faiz et al., 2012).

22.9   RECOMMENDATIONS

Even though many researches have been done in the field of membrane contactor and considering the important advantages of membrane contactor (such as higher absorption flux and selectivity) compared to traditional separation equipment and other membrane based separation processes such as gas separation membrane or facilitated transport membrane, it still needs more studies to be commercialized especially in the field of olefin/paraffin separation. Most of the researches for olefin/paraffin separation have been done at normal temperatures however it is still needed to study the separation process at higher temperatures as the reaction rate between silver ion and olefin increases with temperature even though the thermal stability of membrane should be considered. Although, silver salts are mostly used as absorbent, their high price limits the application in commercial scale. So, studies on other types of absorbent, such as copper ion, would be advantageous.

Application of other types of polymer for membrane fabrication can reduce the cost of process, e.g. PEI was used for membrane fabrication and presented higher performance than hydrophobic membranes (Cote *et al.*, 1989). Furthermore, surface modification using surface modifying macromolecules (Bakeri *et al.*, 2012a; 2012b) or hydrophobic nanoparticles can be applied to enhance the performance of membrane in contactor applications and reduce the process costs.

It is also essential to investigate the long term stability of membrane in contactor application for olefin/paraffin separation process and at different operating conditions such as temperature and pressure.

REFERENCES

Baker, B.B. (1964) The effect of metal fluoroborates on the absorption of ethylene by silver ion. *Inorganic Chemistry* 3 (2), 200–202.

Baker, R.W. (2012) *Membrane technology and applications.* 3rd edition. John Wiley & Sons, Chichester, UK.

Bakeri, Gh., Matsuura, T. & Ismail, A.F. (2011) The effect of phase inversion promoters on the structure and performance of polyetherimide hollow fiber membrane using in gas-liquid contacting process. *Journal of Membrane Science*, 383, 159–169.

Bakeri, Gh., Ismail, A.F. & Matsuura, T. (2010) Effect of polymer concentration on the structure and performance of polyetherimide hollow fiber membranes. *Journal of Membrane Science*, 363, 103–111.

Bakeri, Gh., Matsuura, T., Ismail, A.F. & Rana, D. (2012a). A novel surface modified polyetherimide hollow fiber membrane for gas-liquid contacting processes. *Separation & Purification Technology*, 89, 160–170.

Bakeri, Gh., Ismail, A.F., Rana, D. & Matsuura, T. (2012b). Development of high performance surface modified polyetherimide hollow fiber membrane for gas-liquid contacting processes. *Chemical Engineering Journal*, 198–199, 327–337.

Bessarabov, D.G. (1998) Phenomenological analysis of ethylene transport in a membrane contactor containing solutions of silver nitrate. *Desalination*, 115, 265–277.

Bessarabov, D.G., Sanderson, R.D., Jacobs, E.P. & Beckman, I.E. (1995) High-efficiency separation of an ethylene/ethane mixture by a large-scale liquid-membrane contactor containing flat-sheet non-porous polymeric gas-separation membranes and a selective flowing-liquid absorbent. *Industrial & Engineering Chemistry Research*, 34, 1769–1778.

Brandt, P. (1959) Addition compounds of olefins with metal salts. II. The reaction of ethylene with silver ions in aqueous solutions. *Acta Chemica Scandinavica*, 13, 1639–1659.

Burns, R.L. & Koros, W.J. (2003) Defining the challenges for C_3H_6/C_3H_8 separation using polymeric membranes. *Journal of Membrane Science*, 211, 299–309.

Ceresana (2011) Market study, propylene (UC-1705), January 2011. Konstanz, Germany.

Chang, J.W., Marrero, T.R. & Yasuda, H.K. (2002) Continuous process for propylene/propane separation by use of silver nitrate carrier and zirconia porous membrane. *Journal of Membrane Science*, 205, 91–102.

Chilukuri, P., Rademakers, K., Nymeijer, K., van der Ham, L. & van der Berg, H. (2007) Propylene/propane separation with a gas/liquid membrane contactor using a silver salt solution. *Industrial & Engineering Chemistry Research*, 46, 8701–8709.

Clever, H.L., Baker, E.R. & Hale, W.R. (1970) Solubility of ethylene in aqueous silver nitrate and potassium nitrate solutions – silver ion-ethylene association constant. *Journal of Chemical Engineering Data*, 15 (3), 411–413.

Costello, M.J., Fane, A.G., Hogan, P.A. & Schofield, R.W. (1993) The effect of shell-side hydrodynamics on the performance of axial flow hollow fiber modules. *Journal of Membrane Science*, 80, 1–11.

Cote, P., Bersillon, J.L. & Huyard, A. (1989) Bubble-free aeration using membranes: mass transfer analysis. *Journal of Membrane Science*, 47, 91–106.

Crookes, J.V. & Woolf, A.A. (1973) Competitive interactions in the complexing of ethylene with silver (I) salt solutions. *Journal of the Chemical Society, Dalton Transactions*, 12, 1241–1247.

Drioli, E., Criscuoli, A. & Curcio, E. (2006) *Membrane contactors: fundamental, applications and potentialities*. Elsevier, Amsterdam, The Netherlands.

Faiz, R. & Li, K. (2012) Olefin/paraffin separation using membrane based facilitated transport/chemical absorption techniques. *Chemical Engineering Science*, 73, 261–284.

Faiz, R., Fallanza, M., Ortiz, I. & Li, K. (2012) Olefin/paraffin separation using ceramic hollow fiber membrane contactors, Euromembrane 2012: 23–27 September 2012, London, United Kingdom. *Procedia Engineering*, 44, 662–666.

Fallanza, M., Ortiz, A., Gorri, D. & Ortiz, I. (2011) Improving the mass transfer rate in G-L membrane contactors with ionic liquids as absorption medium: recovery of propylene. *Journal of Membrane Science*, 385–386, 217–225.

Fallanza, M., Ortiz, A., Gorri, D. & Ortiz, I. (2012) Experimental study of the separation of propane/propylene mixtures by supported ionic liquid membranes containing Ag^+-RTILs as carrier. *Separation & Purification Technology*, 97, 83–89.

Feron, P.H.M. & Jansen, A.E. (2002) CO_2 separation with polyolefin membrane contactors and dedicated absorption liquids: performances and prospects. *Separation & Purification Technology*, 27, 231–242.

Fouad, E.A. & Bart, H.J. (2007) Separation of zinc by a non-dispersion solvent extraction process in a hollow fiber contactor. *Solvent Extraction and Ion Exchange*, 25, 857–877.

Gabelman, A. & Hwang, S.T. (1999) Hollow fiber membrane contactors. *Journal of Membrane Science*, 159, 61–106.

Ghosh, T.K., Lin, H.D. & Hines, A.L. (1993) Hybrid adsorption-distillation process for separating propane and propylene. *Industrial & Engineering Chemistry Research*, 32 (10), 2390–2399.

Hilyard, J. (2008) *International petroleum encyclopedia*. PennWell Corporation, Tulsa, OK.

Hirschfelder, J.H., Curtiss, C.F. & Bird, R.B. (1964) *Molecular theory of gases and liquids*. John Wiley and Sons, New York, NY.

Hogendoorn, J.A., Vas Bhat, R.D., Kuipers, J.A.M., van Swaaij, W.P.M. & Versteeg, G.F. (1997) Approximation for the enhancement factor applicable to reversible reactions of finite rate in chemically loaded solution. *Chemical Engineering Science*, 52 (24), 4547–4559.

Jiao, B., Cassano, A. & Drioli, E. (2004) Recent advances on membrane processes for the concentration of fruit juices: a review. *Journal of Food Engineering*, 63, 303–324.

Klaassen, R., Feron, P.H.M. & Jansen, A.E. (2005) Membrane contactors in industrial applications. *Chemical Engineering Research & Design*, 83, 234–246.

Kreulen, H., Smolders, C.A., Versteeg, G.F. & van Swaaij, W.P.M. (1993a) Microporous hollow fiber membrane modules as gas-liquid contactors. Part 1: Physical mass transfer processes. *Journal of Membrane Science*, 78, 197–216.

Kreulen, H., Smolders, C.A., Versteeg, G.F. & van Swaaij, W.P.M. (1993b) Determination of mass transfer rates in wetted and non-wetted microporous membranes. *Chemical Engineering Science*, 48, 2093–2102.

Kumar, P.S., Hogendoorn, J.A., Feron, P.H.M. & Versteeg, G.F. (2003) Approximate solution to predict the enhancement factor for the reactive absorption of a gas in a liquid flowing through a microporous membrane hollow fiber. *Journal of Membrane Science*, 213, 231–245.

Li, J.L. & Chen, B.H. (2005) Review of CO_2 absorption using chemical solvents in hollow fiber membrane contactors. *Separation & Purification Technology*, 41, 109–122.

Li, K., Kong, J.F., Wang, D. & Teo, W.K. (1999) Tailor-made asymmetric PVDF hollow fibers for soluble gas removal. *AIChE Journal*, 45 (6), 1211–1219.

Loeb, S. & Sourirajan, S. (1963) *Seawater demineralization by means of an osmotic membrane*. American Chemical Society, Los Angeles, CA.

Lonsdale, H.K. (1982) The growth of membrane technology. *Journal of Membrane Science*, 10, 81–181.

Lurgi & NPC-RT (2007) Operating Manual, M.T.P. Demo Plant, Introduction Section. Lurgi GmbH, Frankfurt, Germany.

Mansourizadeh, A. & Ismail, A.F. (2009) Hollow fiber gas-liquid membrane contactors for acid gas capture: a review. *Journal of Hazardous Materials*, 171, 38–53.

Nymeijer, D.C. (2003) *Gas-liquid membrane contactors for olefin/paraffin separation*. PhD Thesis, Twente University, Twente, The Netherlands.

Nymeijer, D.C., Visser, T., Assen, R. & Wessling, M. (2004a) Composite hollow fiber gas-liquid membrane contactors for olefin/paraffin separation. *Separation & Purification Technology* 37, 209–220.

Nymeijer, D.C., Folkers, B., Breebaart, I., Mulder, M.H.V. & Wessling, M. (2004b) Selection of top layer materials for gas-liquid membrane contactors. *Journal of Applied Polymer Science*, 92, 323–334.

Nymeijer, K., Visser, T., Brilman, W. & Wessling, W. (2004a) Analysis of the complexation reaction between Ag^+ and ethylene. *Industrial & Engineering Chemistry Research*, 43, 2627–2635.

Nymeijer, K., Visser, T., Assen, R. & Wessling, M. (2004b) Superselective membranes in gas-liquid membrane contactors for olefin/paraffin separation. *Journal of Membrane Science*, 232, 107–114.

Nymeijer, K., Visser, T., Assen, R. & Wessling, M. (2004c) Olefin-selective membranes in gas-liquid membrane contactors for olefin/paraffin separation. *Industrial & Engineering Chemistry Research*, 43, 720–727.

Ortiz, A. (2010) *Process intensification in the separation of olefin/paraffin mixtures*. VDM Verlag Dr. Müller, Saarbrücken, Germany.

Ortiz, A., Galán, L.M., Gorri, D., de Haan, A.B. & Ortiz, I. (2010a) Kinetics of reactive absorption of propylene in $RTIL-Ag^+$ media. *Separation & Purification Technology*, 73, 106–113.

Ortiz, A., Gorri, D., Irabien, A. & Ortiz, I. (2010b) Separation of propylene/propane mixtures using Ag^+-RTIL solutions – evaluation and comparison of the performance of gas-liquid contactors. *Journal of Membrane Science*, 360, 130–141.

Ortiz, A., Galan Sanchez, L.M., Gorri, D., De Haan, A.B. & Ortiz, I. (2010c) Reactive ionic liquid media for the separation of propylene/propane gaseous mixtures. *Industrial & Engineering Chemistry Research*, 49 (16), 7227–7233.

Perry, R.H. & Green, D. (1984) *Perry's chemical engineers handbook*. 6th edition. McGraw Hill, New York, NY.

Porter, M.C. (1990) *Handbook of industrial membrane technology*. William Andrew Publishing/Noyes, Park Ridge, NJ.

Prasad, R. & Sirkar, K.K. (1998) Dispersion-free solvent extraction with microporous hollow-fiber modules. *AIChE Journal*, 34, 177–188.

Qi, Z. & Cussler, E.L. (1985a) Microporous hollow fibers for gas absorption. Part 1: Mass transfer in the liquid. *Journal of Membrane Science*, 23, 321–332.

Qi, Z. & Cussler, E.L. (1985b) Microporous hollow fibers for gas absorption. Part 2: Mass transfer across the membrane. *Journal of Membrane Science*, 23, 333–345.

Rajabzadeh, S., Teramoto, M., Al-Marzouqi, H., Kamio, E., Ohmukai, Y., Maruyama T. & Matsuyama, H. (2010) Experimental and theoretical study on propylene absorption by using PVDF hollow fiber membrane contactors with various membrane structures. *Journal of Membrane Science*, 346, 86–97.

Rangwala, H.A. (1996) Absorption of carbon dioxide into aqueous solutions using hollow fiber membrane contactors. *Journal of Membrane Science*, 112, 229–240.

Safarik, D.J. & Eldridge, R.B. (1998) Olefin/paraffin separations by reactive absorption: a review. *Industrial & Engineering Chemistry Research*, 37, 2571–2581.

Scholz, W. & Lucas, M. (2003) Techno-economic evaluation of membrane filtration for the recovery and re-use of tanning chemicals. *Water Research*, 37, 1859–1867.

Schöner, P., Plucinski, P., Nitsch, W. & Daiminger, U. (1998) Mass transfer in the shell side of cross flow hollow fiber modules. *Chemical Engineering Science*, 53, 2319–2326.

Simioni, M., Kentish, S.E. & Stevens, G.W. (2011) Membrane stripping: desorption of carbon dioxide from alkali solvents. *Journal of Membrane Science*, 378, 18–27.

Sunderrajan, S., Freeman, B.D. & Hall, C.K. (1999) Fourier transform infrared spectroscopic characterization of olefin complexation by silver salts in solution. *Industrial & Engineering Chemistry Research*, 38, 4051–4059.

Temkin, O.N., Ginzburg, A.G. & Flid, R.M. (1964) Soluble complexes of unsaturated hydrocarbons with metal salts and their effect on catalytic reactions. *Kinetics and Catalysis*, 5, 195–200.

Trueblood, K.N. & Lucas, H.J. (1995) Coordination of silver ion with unsaturated compounds. V. Ethylene and propene. *The Journal of the American Chemical Society*, 74, 1338–1339.

Versteeg, G.F., Kuipers, J.A.M., van Beckum, F.P.H. & van Swaaij, W.P.M. (1989) Mass transfer with complex reversible chemical reactions. 1. Single reversible chemical reaction. *Chemical Engineering Science*, 44 (10), 2295–2310.

Wasik, S.P. & Tsang, W. (1970) Gas chromatographic determination of partition coefficients of some unsaturated hydrocarbons and their deuterated isomers in aqueous silver nitrate solutions. *Journal of Physical Chemistry*, 74 (15), 2970–2976.

Wu, J. & Chen, V. (2000) Shell-side mass transfer performance of randomly packed hollow fiber modules. *Journal of Membrane Science*, 172, 59–74.

Yan, S.P., Fang, M.X., Zhang, W.F., Wang, S.Y., Zu, Z.L., Luo, Z.Y. & Cen, K.F. (2007) Experimental study on the separation of CO_2 from flue gas using hollow fiber membrane contactors without wetting. *Fuel Processing Technology*, 88, 501–511.

Yang, M.C. & Cussler, E.L. (1986) Designing hollow-fiber contactors. *AIChE Journal* 32, 1910–1916.

Zheng, J.M., Dai, Z.W., Wong, F.S. & Xu, Z.K. (2005) Shell side mass transfer in a transverse flow hollow fiber membrane contactor. *Journal of Membrane Science*, 261, 114–120.

CHAPTER 23

The reliability of the conventional gas permeation testing method for characterizing the porous asymmetric membranes: a conceptual review

Seyed A. Hashemifard, Ahmad Fauzi Ismail & Takeshi Matsuura

23.1 INTRODUCTION

Attempts have been made to employ extensively asymmetric membranes with porous skin layer for ultrafiltration (UF), microfiltration (MF), membrane distillation (MD) and, more recently, membrane contactor (MC) as separation processes in various industries. Selection of an adequate membrane for a specific application requires a deep understanding of its permeation mechanism and also information on surface porosity, bubble point, mean pore size, and pore-size distribution. The pore size and its distribution have been measured using various methods, best exemplified by gas permeation testing, bubble point method, liquid displacement method, solute probe techniques, and atomic force microscopy (Lee *et al.*, 2002). Each method has its own unique distinction by which it is suitable for specific types of membranes, hence, poses some limitations for the other types of membranes. The important aspect is the accuracy of the method in measuring or estimating the pore size, which in turn arises from a strong theoretical background. One of the easiest and the most frequently used approaches is the well-known conventional gas permeation testing (GPT).

Gas permeation testing was introduced for the first time by Carman in 1956 for characterization of porous media (Carman, 1956). Later Yasuda and Tsai (1974) employed a constant volume permeation testing setup in order to characterize a polymeric membrane made from polysulfone. A porous membrane was placed in a Millipore high-pressure cell, and the assembled cell was connected to a stainless steel tank (16.4×10^{-3} m^3 capacity) which had a pressure transducer. After the tank was pressurized to a predetermined pressure, the ball valve between the tank and the permeability cell was opened and the pressure decay was recorded. It was empirically found that the initial stage of pressure decay was not representative of the steady flux. In order to avoid the initial transient stage of pressure decay, the tank was pressurized to a pressure of approximately 50% higher than the pressure needed for the measurement. The gas was then allowed to permeate through the test membrane before recording the pressure. The slopes of the pressure decay curve were measured and recorded at various pressures on a linear strip-chart recorder and were used to plot against the mean pressure. Wang *et al.* (1995) adopted the technique in an easier way, i.e. constant pressure gas permeation testing. Two kinds of pure non-sorbable gases (H$_2$ and N$_2$) were used to determine the permeability and the separation factor of the asymmetric membrane similar to those employed in Monsanto's prism system. A mathematical model was developed to relate the gas permeation characteristics to the membrane structural parameters (the mean pore size and the surface porosity) of the asymmetric membrane. The models were applied to characterize the asymmetric polysulfone flat and hollow fiber membranes prepared in their laboratory.

Many researchers (Altena *et al.*, 1983; Bird *et al.*, 2007; Schofield *et al.*, 1990; Uchytil *et al.*, 1992; 1995; Wang *et al.*, 1990) applied the gas permeation testing for integrated porous skin layer membrane characterization for different purposes, e.g. UF, MF, MD and MC. Khayet and Matsuura (2001) fabricated polyvinylidene fluoride (PVDF) flat-sheet membranes for membrane distillation (MD) with different pore sizes and porosities from casting solutions containing 15 wt% of PVDF, dimethylacetamide, and different concentrations of pure water. They used the conventional GPT method to estimate the membranes pore size. Mansurizadeh and Ismail (2011) used this technique to measure the pore size of a PVDF membrane for the purpose of CO_2 removal by a MC. Bakeri *et al.* (2010) applied the technique to measure the pore size and effective surface porosity of MC membranes prepared from different concentrations of polyetherimide (dissolved in *N*-methyl-2-pyrrolidone). The concentration of polymer in the spinning dope was varied from 10 to 15 wt% and water was used as the internal and external coagulant. Increasing PEI concentration in the spinning dope decreased the mean pore size and the effective surface porosity at the skin layer and the bulk void fraction. Tailor-made asymmetric PVDF hollow-fiber membranes and their membrane modules were fabricated by Li *et al.* (1999) to remove soluble gas H_2S. Hollow fibers with different morphological structures were prepared using the phase-inversion process, and were characterized by the gas permeation technique. Results revealed that the skin location is largely dependent on the coagulation medium. The addition of a substantial amount of ethanol to the coagulation medium would greatly affect the formation of the skin.

Even though all the above mentioned works are very important since they are providing the information on the porous structure at the membrane surface, the methods they employed are prone to some inaccuracy caused by inadequate experimental and theoretical techniques. In this work we intend to conduct a thorough analysis of the conventional GPT, especially in its theoretical aspect. As well, we intend to present an improved approach.

23.2   THEORETICAL BACKGROUND

The diffusion and flow of gases through porous media such as integrated porous skin layer membranes can occur by different mechanisms. For solids with relatively narrow pore size distributions, several distinct regimes of behavior are observed, depending upon the ratio of the mean free path of gas molecules to the mean pore diameter, which is known as Knudsen number:

$$Kn = \frac{\lambda}{2r_p} \qquad (23.1)$$

where, Kn is the Knudsen number, r_p is the pore radius [m]. λ is gas mean free path [m] given by kinetic theory of gases (Bird *et al.*, 2007):

$$\lambda = \frac{1}{\sqrt{2}} \frac{k_B T}{\pi \sigma^2 P} \qquad (23.2)$$

where k_B is the Boltzmann constant (equal to 1.38×10^{-23} J K^{-1}), T is the absolute temperature [K], σ is the collision diameter [m] and P is system mean pressure [Pa]. When an isothermal condition is assumed, the mass transfer in the pores of the porous medium can take place by Knudsen diffusion, molecular diffusion or viscous flow. Different driving forces are responsible for each diffusion mechanism. Gradient of total pressure (dp/dx) leads to a viscous flow and/or Knudsen diffusion, whereas, a concentration gradient (dX_i/dx) leads to molecular diffusion and/or Knudsen diffusion (Kast and Hohenthanner, 2000).

There are several issues associated with gas permeability through the capillary porous membrane namely: (i) characterization of flow conditions applying the Knudsen number, (ii) describing the mechanism involved and the governing theory, (iii) distinguishing the contribution of every individual mechanism in the overall gas flow, and (iv) validity of the proposed modeling approach (Civan, 2010).

Three different regimes in integrated porous skin layer membranes can be considered, depending on the Knudsen number, Kn, which is used as a criteria to distinguish the flow regimes. It should however be noted that, according to different researchers, there is an order of magnitude difference in the Knudsen number upper or lower limits that is applicable to each flow regime (Bird *et al.*, 2007; Knudsen, 1909). Therefore, we adopt the criteria proposed by Roy *et al.* (2003)

Table 23.1. Categorizing the flow regimes in capillary channels according to the Knudsen number criteria (Civan, 2010).

Characteristics	Flow regime	Knudsen number limit
Molecule-wall collision	Free molecular	$Kn \geq 10$
Transition from molecule-wall collision to molecule-molecule collision	Transition	$0.1 \leq Kn < 10$
Molecule-molecule collision	Slip phenomena (slip on the wall)	$0.001 < Kn < 0.1$
	Continuum (friction on the wall)	$Kn \leq 0.001$

as shown in Table 23.1. These limits were determined experimentally by Schaff and Chambre (1961) for capillary pipes possessing different flow regimes.

If the pore size or mean pressure is so small that $Kn \geq 10$ then the flow regime is in the free molecular region. In this region, the mean free path is large compared with the pore diameter (Bird et al., 2007; Civan, 2010), whilst the pore size or mean pressure is so large that $Kn \leq 0.001$ the flow regime is in continuum region. In the continuum region the mean free path of the gas is small compared with the pore diameter. The flow mechanism is considered as transition regime when the mean free path is comparable with the pore diameter or $0.1 \leq Kn \leq 10$.

In the free molecular region, i.e. at high values of Kn, molecule-wall collision predominates the molecule-molecule collision and the flow is governed by Knudsen diffusion as a result of either concentration gradient or total pressure gradient (Rangarajan et al., 1984). The free molecular region is modeled by the Knudsen diffusion equation:

$$N_K = \frac{2r_pA_p}{3RT} \left(\frac{8RT}{\pi M}\right)^{0.5} \frac{\Delta P}{l_p} \tag{23.3}$$

where N_K is the Knudsen flow rate [mol s^{-1}], r_p is the pore radius [m], A_p is pore cross sectional area [m^2], M is the molecular weight, ΔP is the transmembrane pressure difference [Pa], R is the universal gas constant (equal to 8.314 Pa m^3 mol^{-1} K^{-1}) and l_p is the pore length [m].

In contrast, in the continuum region, the molecule-molecule collision predominates over the molecule-wall collision. In this case, molecular diffusion is responsible for the mass transfer in a gas mixture. In a pure gas system there is no molecular diffusion since $dX_i/dx = 0$ (Kast and Hohenthanner, 2000; Wakao et al., 1965), thus, a gradient of total pressure results in viscous flow, which is described by Hagen-Poiseuille equation:

$$N_V = \frac{r_p^2 A_p P}{8\mu RT} \frac{\Delta P}{l_p} \tag{23.4}$$

where N_V is the viscous flow rate [mol/s] and μ is the gas viscosity [Pa s].

Special care should be taken in the transition region where $0.1 \leq Kn \leq 10$. In this regime, according to the dusty gas theorem, a combination of various transport mechanisms: Knudsen diffusion, molecular diffusion and viscous flow are considered (Wakao et al., 1965). As mentioned above, in the pure gas systems, molecular diffusion plays no role in the mass transfer. Consequently, a combination of the Knudsen and viscous flow governs the system flow. In this region, gas molecules collide with each other and the pore-walls while travelling through the pore medium.

When the pore radius approaches the mean free path of gas molecules and also at low pressure ($0.001 \leq Kn \leq 0.1$), the frequency of collision between gas molecules and solid walls increases, since gas molecules are often so far apart, they slip onto the channel's wall almost without interaction (no friction loss) (Klinkenberg, 1941; Renksizbulut et al., 2006). This additional flux due to the gas flow at the wall surface, which is called "slip flow", becomes effective to enhance the total flow rate. This phenomenon is also known as Klinkenberg effect (Tanikawa et al., 2006; Ursin and Zolotukhin, 1997). What is important and has been widely ignored in asymmetric porous membrane is that, to what extent each flow mechanism is contributing to total gas flow rate?

The three mechanisms (Knudsen, slip and viscous) considered all involve diffusion through the gas phase. If there is significant adsorption on the pore wall, (in the case of condensable gases, e.g. carbon dioxide, ammonia, sulfur dioxide, etc.) there is the possibility of an additional gas permeation due to diffusion through the adsorbed phase or "surface diffusion". Surface diffusion is an activated process, but the diffusional activation energy is generally smaller than the heat of adsorption. As a result, surface diffusion increases with decreasing temperature. Therefore, surface diffusion is generally insignificant at temperatures which are high relative to the normal boiling point of the adsorbate gas such as nitrogen, methane, hydrogen and helium (Karger and Ruthven, 1992).

When UF, MF, and MC membranes are characterized via the GPT method, the pressures are usually in the range, where the transition regime is applicable, which means Knudsen flow, viscous flow as well as slip flow are contributing to the total flow.

Most attempts at describing the Knudsen/viscous transition region involve a combination of Equations (23.3) and (23.4). For example, Weber (1954) proposed the combination of Knudsen and viscous flow permeabilities, with the Knudsen permeability being multiplied by $(\pi/4+\lambda/2r_p)/(1+\lambda/2r_p)$, to describe the transition region for the flow along capillary tubes. Creutz (1974) adopted a similar approach, including an adjustable parameter to fit the experimental data. Various authors have integrated these (and other) capillary equations over the range of pore sizes in a porous medium with reasonable success (for example Schneider, 1975). Others adopted a similar approach including adjustable parameters to fit experimental data, e.g. Shofield et al. (1990) stated that, for a range of gases and membranes, flux can be expressed by a simple equation, $J = aP^b\Delta P$, where a is the membrane mass transfer constant and b is a measure of the extent of viscous flow. As can be seen in all of these attempts, the effect of the slip flow was not taken into consideration.

The authors believe that the contribution of the slip flow at the wall should be included when the transport of membrane pores is discussed, particularly for the ranges of pressures and pore sizes we are dealing with. Scott and Dullien (1962) by using modern kinetic theory concepts described the slip flow mechanism by:

$$N_S = \frac{r_p A_p}{RT}\left(\frac{\pi RT}{8M}\right)^{0.5}\frac{\Delta P}{l_p} \tag{23.5}$$

Wakao et al. (1965), after a rigorous algebraic formulation showed that for a pure gas system, total gas flow N, is given by:

$$N = \frac{cA_p}{RT}\frac{\Delta P}{l_p} \tag{23.6}$$

where:

$$C = \left(\frac{1}{1+\frac{1}{Kn}}\right)\frac{2}{3}r_p\left(\frac{8RT}{\pi M}\right)^{0.5}+\left(\frac{1}{1+Kn}\right)\left[r_p\left(\frac{\pi RT}{8M}\right)^{0.5}+\frac{r_p^2 P}{8\mu}\right] \tag{23.7}$$

where on the right hand side, the first term is the Knudsen flow, the first term in the square bracket is the slip flow and the second term is the viscous flow. The path of a molecule in a tube is interrupted either by collision with the tube wall or with another molecule. Wakao et al. (1965) also proposed the probability of collisions with the wall be φ, that is:

$$\varphi = \frac{\text{Wall collision frequency}}{\text{Total collision frequency}} \tag{23.8}$$

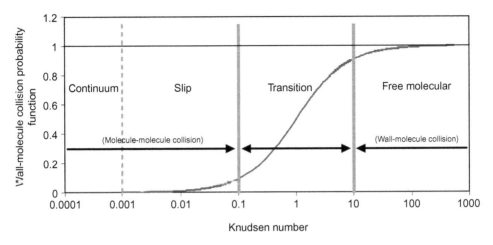

Figure 23.1.   Wall-molecule probability function (φ) versus Knudsen number (Kn), combined with the domain of the individual mechanism.

With this concept the total flow rate through the tube is:

$$N = \text{(probability of wall-molecule collisions)} N_K$$
$$+ \text{(probability of molecule-molecule collisions)} (N_S + N_V) \qquad (23.9)$$

and

$$N = \varphi N_K + (1 - \varphi)(N_S + N_V) \qquad (23.10)$$

where, N_K is the Knudsen flow rate, N_S is the slip flow rate, and N_V is the viscous flow rate. By comparing Equations (23.3)–(23.7) with Equation (23.10) they concluded that:

$$\varphi = \frac{1}{1 + \frac{1}{Kn}} \qquad (23.11)$$

φ was also called wall-molecule collision probability function. Figure 23.1 illustrates φ versus Kn. The domain of the individual mechanism associated with the corresponding range of Knudsen number is shown in Figure 23.1 as well.

Scott and Dullien (1962) derived an approximate expression similar to Equation (23.7) by applying the kinetic theory. Their equations were derived by supposing the molecules to be hard spheres. With this assumption, the results could be expressed in terms of a mean free path.

Their results are the same as that of Wakao *et al.* (1965) except for a different probability function defined by:

$$\varphi = \exp\left[-\sinh\left(\frac{1}{Kn}\right)\right] \qquad (23.12)$$

The two expressions for φ, i.e. Equations (23.11) and (23.12), state that φ is a function of the Knudsen number. Eventually the final form of the total flow rate over the entire range of the Knudsen number including the transition regime is given by:

$$N = \left\{ \varphi \frac{2r_p}{3RT}\left(\frac{8RT}{\pi M}\right)^{0.5} + (1 - \varphi)\left[\frac{r_p}{RT}\left(\frac{\pi RT}{8M}\right)^{0.5} + \frac{r_p^2 P}{8\mu RT}\right] \right\} \frac{A_P \Delta P}{l_p} \qquad (23.13)$$

Equation (23.13) has the ability to predict the gas flow for different flow regimes corresponding to various ranges of the Knudsen number and also match the end limits. When φ approaches to

Table 23.2.   Collision diameter and the coefficient in the φ equation for several gases (Reid and Sherwood, 1966).

Gas species	σ [Å]	Coefficient [m Pa]$^{-1}$
CH$_4$	3.82	315.30
CO$_2$	3.941	335.59
N$_2$	3.798	311.68
Ar	3.542	271.08
O$_2$	3.467	259.72
He	2.55	140.50
H$_2$	2.93	185.50

$T = 298$ K.

unity, N becomes nearly equal to N_K, whereas N becomes nearly equal to $(N_S + N_V)$ when φ approaches to zero, Furthermore by increasing the system pressure, slip flow will be negligible as compared to the viscous flow and eventually N will be equal to N_V.

23.3   RESULTS AND DISCUSSION

23.3.1   Analyzing the Wakao et al. (1965) model for GPT

By substituting Equation (23.1) and (23.2) for the Knudsen number of Equation (23.11):

$$\varphi = \frac{1}{1 + \left(\frac{2\sqrt{2}\pi\sigma^2}{k_B T}\right) r_p P} \tag{23.14}$$

As can be seen from Equation (23.14), φ is a function of pore radius, system operating parameters, i.e. temperature and system mean pressure, along with the property of the testing gas, i.e. collision diameter of gas:

$$\Phi = f(T, P, r_p, \text{ gas properties}) \tag{23.15}$$

Thus for a particular gas and at an isothermal condition, φ is only a function of system mean pressure and pore size:

$$\Phi = f(P, r_p) \tag{23.16}$$

For example, for nitrogen gas and at room temperature ($T = 298$ K) Equation (23.14) is reduced to:

$$\varphi = \frac{1}{1 + 311.68 \, r_p P} \tag{23.17}$$

Table 23.2 summarizes the coefficient preceding $r_p P$ in Equation (23.17) for some other gases.

According to Equation (23.17) when r_p or P becomes large φ approaches zero, which indicates a pure viscous flow, i.e. Equation (23.4). In contrast, if r_p or P approaches zero φ approaches unity, which corresponds to pure Knudsen flow, i.e. Equation (23.3). Figure 23.2 illustrates φ as a function of pore size and mean pressure for nitrogen gas. From Figure 23.2 and Equation (23.17), φ is always between zero and unity limits; $0.0 < \varphi < 1.0$. As can be seen, by increasing pore size at any pressure φ decreases, which means the contribution of viscous flow is enhanced. The same trend is observed for increasing the mean pressure at every pore size. At high pressures and large pore sizes, φ approaches zero. Also by decreasing pore size and mean pressure φ tends to approach unity. From Equation (23.13) and Figure 23.2, it is concluded that during the gas permeation testing by changing the upstream pressure, the fraction of viscous and Knudsen flow

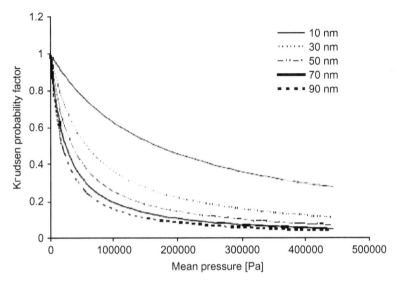

Figure 23.2.   Wall-molecule probability function (φ) of nitrogen versus system mean pressure for various
pore radius at $T = 298$ K.

is continuously changing. It is noted that when r_p and the pressure are kept constant, φ is only
a function of the gas species which is utilized as the testing gas. As a result, even at the same
conditions, different gases may experience various flow regimes, which is due to the difference
in their mean free path.

23.3.2   Predictions of the model for integrated porous skin layer membranes

The permeance, J [mol^{-1} m^{-2} Pa^{-1}] through a membrane possessing porous skin layer is as
follows:

$$ J = \frac{1}{RT} \left\{ \varphi \frac{2r_p}{3} \left(\frac{8RT}{\pi M} \right)^{0.5} + (1 - \varphi) \left[r_p \left(\frac{\pi RT}{8M} \right)^{0.5} + \frac{r_p^2 P}{8\mu} \right] \right\} \frac{\varepsilon}{l_p} \tag{23.18} $$

where ε is the membrane surface porosity.

Wakao *et al.* (1965) showed that their model exhibited good agreement with their experimental
results and also with the experimental results reported by Knudsen for small capillaries. To the
best of our knowledge, this concept has not been documented to model and characterize the
performance of asymmetric membranes to estimate the membrane structural parameters such as
mean pore size and effective surface porosity. Therefore, their model will be used as a principal
model throughout the present study. To estimate the asymmetric membrane with porous skin layer,
the mass transfer resistance of the support layer is roughly considered to be negligible.

Figure 23.3, shows the model prediction for porous skin layer membranes with different mean
pore size ranging from 10 to 90 nm. According to the figure there is a minimum in permeance,
particularly when the pore size is large. This trend was indeed observed experimentally. For
example, Figure 23.4 shows experimental J versus P plot for a flat sheet asymmetric membrane
prepared from a dope containing 15% polyetherimide in N-methyl-2-pyrrolidone. Referring to
Figures 23.3 and 23.4, the diagram can be divided into three major regions namely: a decreasing
trend, an almost plateau region with a minimum, and a nearly constant increasing trend (the linear
part). Most of our experiments exhibited a similar trend.

Figure 23.5 depicts the contribution of each flow regime to the total permeance at different
pressures. At low pressure (high Kn) the system is controlled by the Knudsen diffusion, and at high

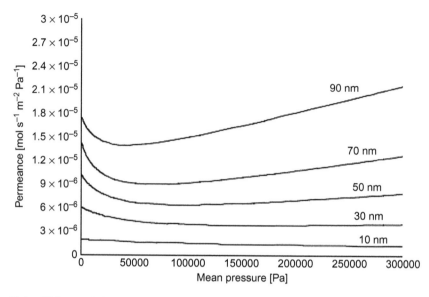

Figure 23.3.   Wakao model: permeance (J) of nitrogen versus system mean pressure (P) for various pore radii, at $T = 298$ K.

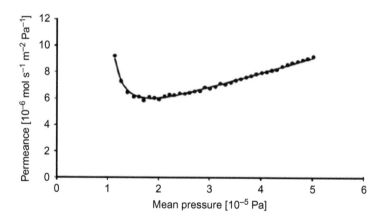

Figure 23.4.   Permeance (J) versus system mean pressure (P) of nitrogen at $T = 298$ K, for a system of 15% polyetherimide concentration (dissolved in N-methyl-2-pyrrolidone) asymmetric flat sheet membrane.

pressure (low Kn) the system is governed by the viscous flow mechanism, whereas at intermediate pressures the Knudsen, slip and viscous flow are contributing almost equally. Experimental data as well as theoretical work have demonstrated a nonlinear dependence of mass flow rate on the pressure drop for the slip flow regime. In fact, the minimum point in the J versus P indicates the area of transition from Knudsen flow toward viscous flow which the slip flow mechanism on the pore walls also interfere the total flow mechanism. This nonlinear dependence of permeance on pressure was also reported by other researchers (Altena et al., 1983; Beskok and Karniadakis, 1999; Beskok et al., 1995; Dullien, 1979).

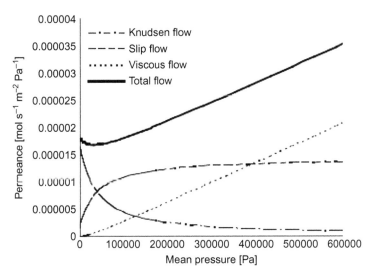

Figure 23.5.   Role of different mechanisms by varying system mean pressure for $r_p = 90$ nm, $\varepsilon l = 1600$ m^{-1}; nitrogen was used as testing gas at $T = 298$ K.

23.3.3   Drawbacks of the conventional GPT method

The conventional GPT method to determine the mean pore size and effective porosity is based on a model in which the Knudsen and viscous flow contribute equally to the total permeance, while neglecting the slip flow contribution. The applicable equation is therefore (Altena *et al.*, 1983):

$$J = J_K + J_V = \frac{1}{RT} \left[\frac{2r_p}{3} \left(\frac{8RT}{\pi M} \right)^{0.5} + \left(\frac{r_p^2 P}{8\mu} \right) \right] \frac{\varepsilon}{l_p} \tag{23.19}$$

Although this method has been used frequently, the following problems will arise when the experimental data are fitted to the model.

First of all, referring to Equation (23.18), the conventional GPT method, is not attainable by simplifying the Wakao *et al.* (1965) model. In the conventional GPT method it is assumed that the Knudsen and viscous mechanism have equal contribution in the overall permeance. If we consider that φ is equal to 0.5 which states the Knudsen and viscous mechanism have the same degree of contribution in the overall permeance, then the gas permeation equation reduces to:

$$J = \frac{1}{RT} \left\{ 0.5 \frac{2r_p}{3} \left(\frac{8RT}{\pi M} \right)^{0.5} + 0.5 \left[r_p \left(\frac{\pi RT}{8M} \right)^{0.5} + \frac{r_p^2 P}{8\mu} \right] \right\} \frac{\varepsilon}{l_p} \tag{23.20}$$

Even though, the slip term is neglected in Equation (23.20):

$$J = \frac{1}{RT} \left[0.5 \frac{2r_p}{3} \left(\frac{8RT}{\pi M} \right)^{0.5} + 0.5 \left(\frac{r_p^2 P}{8\mu} \right) \right] \frac{\varepsilon}{l_p} \tag{23.21}$$

Comparing the two last equations, keeping the mean pore size and effective surface porosity constant, obviously, the predictions of the conventional gas permeation method in relation to gas permeance is twice the predictions by the Wakao *et al.* (1965) model. This is depicted schematically in Figure 23.6.

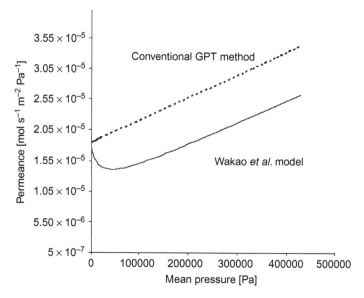

Figure 23.6.   Conventional GPT method against Wakao *et al.* (1965) model for nitrogen at $T = 298$ K, for $r_p = 90$ nm, $\varepsilon/l_p = 1600$ m^{-1}.

Based on the Equation (23.19) the pore radius r_p and the effective porosity, ε/l_p of the membrane are obtained as follows: Plot J versus P. This is supposed to be a linear line according to Equation (23.19). From the intercept with the y axis (I) and slope of the straight line (S):

$$r_{p(conv.GPT)} = \left(\frac{16}{3}\right)\mu\left(\frac{S}{I}\right)\left(\frac{8RT}{\pi M}\right)^{0.5} \qquad (23.22)$$

and

$$\varepsilon/l_{p(conv.GPT)} = \frac{8\mu RTS}{r_{p,conv.GPT}^2} \qquad (23.23)$$

The J versus P plot often shows a nonlinear relationship as mentioned earlier, Figure 23.6. Then, the method is either inapplicable in the region where J decreases with P (negative slopes), or leads to erroneous r_p and ε/l_p, values (nonlinear positive slopes).

At low P, φ is still large enough and the Knudsen flow has its considerable influence. As a result, the trend is not linear and the slope is not constant relative to the change in pressure. At slopes close to zero, by a small change in the slope of J versus P, the predictions can be varied dramatically, and hence, the results of the conventional GPT method can be very erroneous. Suppose the experimental data in the intermediate P region of 1.6 to 3.5 × 10^5 Pa are as shown in Figure 23.7a (replot of Fig. 23.4 in this narrower region) r_p and ε/l_p strongly depend on the range of P adopted for the interpolation. This is quite obvious from the linear plots made in the lower, middle and upper range of P. Table 23.3 shows even more clearly how different r_p and ε/l_p values may become, depending on which portion of the J versus P plot is used. Even when you use a very high pressure range of P (above 3.5 × 10^5 Pa), where J versus P plot may is linear, incorrect r_p and ε/l_p values are obtained by the above approach.

Table 23.3 indicates that the results of the conventional GPT method can be very different depending on the slope and the interception of the J versus P diagram. As a matter of fact, since the GPT experiment is not conducted by a standard procedure, the conventional GPT method is unfortunately an arbitrary approach to determine the membrane structural parameters. That is why even when the membrane is fabricated under the identical conditions using the same dope

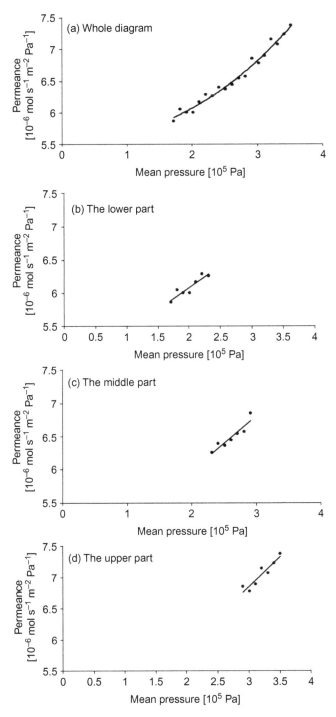

Figure 23.7.   Permeance (J) versus system mean pressure (P) of nitrogen at $T = 298$ K, for a system of 15% polyetherimide (dissolved in NMP) flat sheet asymmetric membrane, (a) the whole positive sloped part, (b) the lower part, (c) the middle part, and (d) the upper part.

Table 23.3. r_p and (ε/l_p) as a function of the slope and intercept of J versus P for different part of the diagram.

J versus P	Slope (S)	Intercept (I)	r_p [nm]	ε/l_p [m^{-1}]
Lower part	7.594×10^{-12}	4.538×10^{-6}	76.7	463.2
Middle part	8.088×10^{-12}	4.368×10^{-6}	84.9	403.1
Upper part	1.163×10^{-11}	3.299×10^{-6}	161.4	160.0

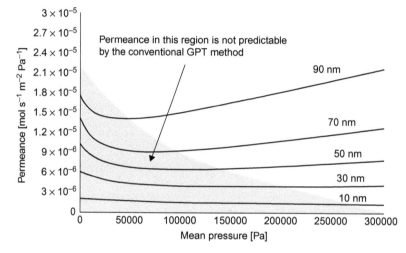

Figure 23.8. Incapability of the conventional GPT method to predict the membrane. Parameters at negative and zero slopes; The system is nitrogen at $T = 298$ K.

formulation, different values are reported as a result of conventional GPT approach. It is important to remember that all of the preceding discussion is valid only if we are in the linear region of the J versus P, i.e. when $Kn \ll 1$. Otherwise, in the transition region, which produces negative or zero slopes the two Equations (23.19) and (23.21) are not capable to make any prediction (see Fig. 23.8). This has been also pointed out by the other researchers (Bird *et al.*, 2007; Uchytil *et al.*, 1995). In such a case, the effect of φ on the slope of the J versus P should be taken into consideration. This issue will be discussed in the future work more in detail.

According to Equation (23.18), which seems more complete than Equation (23.19) since experimental J versus P plots are often not linear but nonlinear, φ approaches zero at high pressure and J becomes:

$$J = J_S + J_V = \frac{1}{RT}\frac{\varepsilon}{l_p}\left[r_p\left(\frac{\pi RT}{8M}\right)^{0.5} + \frac{r_p^2 P}{8\mu}\right]$$
(23.24)

Then, r_p and ε/l_p values which are obtained by the I and S of the straight line will become:

$$r_{p(\text{Modif.GPT})} = 8\mu\left(\frac{S}{I}\right)\left(\frac{\pi RT}{8M}\right)^{0.5}$$
(23.25)

$$\varepsilon/l_{p(\text{Modif.GPT})} = \frac{8\mu RTS}{r_{p,\text{Modif.GPT}}^2}$$
(23.26)

$$\frac{r_{p(\text{Modif.GPT})}}{r_{p(\text{Conv.GPT})}} = \frac{3\pi}{16} \approx 0.58$$
(23.27)

$$\frac{(\varepsilon/l_p)_{\text{Modif.GPT}}}{(\varepsilon/l_p)_{\text{Conv.GPT}}} = \left(\frac{16}{3\pi}\right)^2 \approx 2.88$$
(23.28)

Therefore, much smaller r_p and much larger ε/l_p are obtained by using Equation (23.18). The best approach seems to use the entire range of P that is covered by experiments and fit the data to Equation (23.18). Implementation of Equation (23.18) and comparing the predicted values for structural properties of asymmetric porous membrane with the experimental points is the topic of our future study.

23.4   CONCLUSION

A theoretical analysis along with some experimental data was presented in order to study the capability of the conventional GPT method with reference to the membrane structural parameters. Although the conventional GPT method is relatively easy and fast to be implemented, it suffers from several conceptual drawbacks. First of all, it is based on a rather weak theoretical background. That is the reason why the conventional GPT does not necessarily describe the experimental data very well. It was shown that simple addition of the Knudsen and viscous terms may lead to erroneous conclusions. By comparing the conventional GPT method with the Wakao *et al.* (1965) model, it was concluded that the modified values of r_p and (ε/l_p) were different from the conventional ones by the factor of $3\pi/16$, and $(16/3\pi)^2$, respectively. In addition, the slope of J versus P plot depends on the adopted region of the mean pressure. In the other words, depending on which region of the J versus P plot is selected, different r_p and (ε/l_p) will be achieved. Finally, at the regions which the slope is negative or zero, the conventional GPT method is simply not applicable. Since the GPT experiment was not conducted by a standard procedure, the conventional GPT method was unfortunately an arbitrary approach to determine the structural parameters. That is why, sometimes with the same membrane fabrication conditions and the same dope formulation, the structural parameters obtained by the conventional GPT method do not agree with each other. The best approach seems to use the entire range of P that is covered by experiments and fit the data to Equation (23.18). Implementation of Equation (23.18) and comparing the predicted values for structural properties of asymmetric porous membrane with the experimental points is the topic of our future study.

ACKNOWLEDGEMENT

The authors are thankful to Advance Membrane Technology Centre (AMTEC), UTM for support in research management activities for this chapter.

REFERENCES

Altena, F.W., Knoef, H.A.M., Heskamp, H., Bargeman, D. & Smolders, C.A. (1983) Some comments on the applicability of gas permeation methods to characterize porous membranes based on improved experimental accuracy and data handling. *Journal of Membrane Science*, 12, 313–322.
Bakeria, Gh., Ismail, A.F., Shariaty-Niassar, M. & Matsuura, T. (2010) Effect of polymer concentration on the structure and performance of polyetherimide hollow fiber membranes. *Journal of Membrane Science*, 363, 103–111.
Beskok, A. & Karniadakis, G.E. (1999) Report: a model for flows in channels, pipes, and ducts at micro and nano scales. *Thermal Engineering*, 3, 43–77.
Beskok, A., Trimmer, W. & Karniadakis, G.E. (1995) Rarefaction and compressibility and thermal creep effects in gas microflows. *IMECE 95, Proceedings ASME Dynamic Systems and Control Division*, 57, 877–892.
Bird, R.B., Stewart, W.E. & Lightfoot, E.N. (2007) *Transport phenomena*. Revised 2nd edition. John Wiley & Sons, New York, NY.
Carman, P.C. (1956) *Flow of gases through porous media*. Butterworth Publications, London, UK.
Civan, F. (2010) A review of approaches for describing gas transfer through extremely tight porous media, porous media and its applications in science, engineering, and industry. In: Vafai, K. (ed.) *Proceedings*

of 3rd International Conference, 20–25 June 2010, Montecatini, Italy. American Institute of Physics, Melville, NY.

Creutz, E. (1974) The permeability minimum and the viscosity of gases at low pressure. *Nuclear Science and Engineering*, 53, 107–109.

Dullien, F.A.L. (1979) *Porous media, fluid transport and pore structure.* Academic Press, New York, NY.

Karger, J. & Ruthven, D.M. (1992). *Diffusion in zeolites and other microporous materials.* Wiley, New York, NY.

Kast, W. & Hohenthanner, C.R. (2000) Mass transfer within the gas-phase of porous media. *International Journal of Heat and Mass Transfer*, 43, 807–823.

Khayet, M. & Matsuura, T. (2001) Preparation and characterization of polyvinylidene fluoride membranes for membrane distillation. *Industrial & Engineering Chemistry Research*, 40, 5710–5718.

Klinkenberg, L.J. (1941) The permeability of porous media to liquids and gases. *American Petroleum Institute, Drilling and Productions Practices*, 200–213.

Knudsen, M. (1909) Die Gesetze der Molekularströmung und der Innerenreibungsströmung der Gase durch Röhren. *Annalen der Physik*, 28, 75–130.

Lee, S., Park, G., Amy, G., Hong, S.-K., Moon, S.-H., Lee, D.-H. & Cho, J. (2002) Determination of membrane pore size distribution using the fractional rejection of nonionic and charged macromolecules. *Journal of Membrane Science*, 201, 191–201.

Li, K., Kong, J.F., Wang, D. & Teo, W.K. (1999) Tailor-made asymmetric PVDF hollow fibers for soluble gas removal. *AIChE Journal*, 45, 1211–1219.

Mansourizadeh, A. & Ismail, A.F. (2011) A developed asymmetric PVDF hollow fiber membrane structure for CO_2 absorption. *International Journal of Greenhouse Gas Control*, 5, 374–380.

Rangarajan, R.M., Mazid, M.A., Matsuura, T. & Sourirajan, S. (1984) Permeation of pure gases under pressure through asymmetric porous membranes: membrane characterization and prediction of performance. *Industrial and Engineering Chemistry Process Design and Development*, 23, 79–87.

Reid, R.C. & Sherwood, T.K. (1966) *The properties of gases and liquids.* McGraw-Hill, New York, NY.

Renksizbulut, M., Niazmand, H. & Tercan, G. (2006) Slip-flow and heat transfer in rectangular microchannels with constant wall temperature. *International Journal of Thermal Sciences*, 45, 870–881.

Roy, S., Raju, R., Chuang, H.F., Cruden, B.A. & Meyyappan, M. (2003) Modeling gas flow through icrochannels and nanopores. *Journal of Applied Physics*, 93, 4870–4879.

Schaaf S.A. & Chambre, P.L, (1961) *Flow of rarefied gases.* Princeton University Press, Princeton, NJ.

Schneider, P. (1975) Permeation of simple gases through a porous medium. *Collection of Czechoslovak Chemical Communications*, 40, 3114–3122.

Schofield, R.W., Fane, A.G. & Fell, C.J.D. (1990) Gas and vapour transport through microporous membranes. I. Knudsen-viscous transition. *Journal of Membrane Science*, 53, 159–171.

Scott, D.S. & Dullien, F.A.L. (1962) The flow of rare field gas. *AIChE Journal*, 8, 113–117.

Tanikawa, W., Shimamoto, T. & Klinkenberg, L.J. (2006) Effect for gas permeability and its comparison to water permeability for porous sedimentary rocks. *Hydrology and Earth System Sciences Discussions*, 3, 1315–1338.

Uchytil, P., Nguyen, X.Q. & Broz, Z. (1992) Characterization of membrane skin defects by gas permeation method. *Journal of Membrane Science*, 73, 47–53.

Uchytil, P., Wagner, Z., Roček, J. & Brož, Z. (1995) Possibility of pore size determination in separation layer of ceramic membrane using permeation method. *Journal of Membrane Science*, 103, 151–157.

Ursin, J.R. & Zolotukhin, A.B. (1997) *Fundamentals of petroleum reservoir engineering.* Høyskoleforlaget Norwegian Academic Press, Oslo. Norway.

Wakao, N., Otani, S. & Smith, J.M. (1965) Significance of pressure gradients in porous materials. Part I. Diffusion and flow in fine capillaries. *AIChE Journal*, 11, 435–439.

Wang, D.L., Xu, R.X., Jiang, G.L. & Zhu, B.L. (1990) Determination of surface dense layer structure parameters of the asymmetric membrane by gas permeation method. *Journal of Membrane Science*, 52, 97–108.

Wang, D., Li, K. & Teo, W.K. (1995) Effects of temperature and pressure on gas permselection properties in asymmetric membranes. *Journal of Membrane Science*, 105, 89–101.

Weber, S. (1954) *Danske Matematisk Fysiske Meddelelser*, 28: 1.

Yasuda, H. & Tsai, J.T. (1974) Pore size of microporous polymer membrane. *Journal of Applied Polymer Science*, 18, 805–819.

CHAPTER 24

Modeling of concurrent and counter-current flow hollow fiber membrane modules for multi-component systems

Serene Sow Mun Lock, Kok Keong Lau & Azmi Mohd Shariff

24.1 INTRODUCTION

The combined United States market for membranes employed in gas and liquid separating applications is reported to be worth approximately US$ 2.2 billion in 2013 (BCC Research, 2014). Based on forecasts of analysts, the gas separation and pervaporation market, which is solely based in the United States, is expected to boost from US$ 247 million in 2014 to US$ 380 million in 2019, achieving a compound annual growth rate of 7.9% during the five-year period (BCC Research, 2014). The rapid growth of membrane separation techniques is due to the fact that they exhibit many advantages in comparison to the conventional separation technologies for gases, such as occupying a relatively smaller footprint, chemical free operation, cost effectiveness, high process flexibility and high energy efficiency (Baker and Lokhandwala, 2008). The promising expansion of membrane gas separation is also attributed to improved membrane performance due to development of new membrane material with enhanced permeability and selectivity over recent years (Marriott and Sorensen, 2003). Among all the commercially available membrane types, hollow fiber membrane modules are exceptionally favorable in industrial applications because a large membrane area can be packed within a small volume, which allows for more effective separation (Baker, 2002).

In order to assist in the optimal design of the hollow fiber membrane module in industrial application, such as the required module specification and operating conditions, an accurate mathematical model in characterizing membrane permeation process is vital (Baker and Lokhandwala, 2008; Makaruk and Harasek, 2009). It can be adapted to minimize the technical risks that are inherent in the development of the newer process. The availability of the mathematical model also eliminates the need for the time consuming and costly pilot plant and experimental studies to evaluate the required conditions in order to optimize the membrane process behavior. In addition, the mathematical model can also be applied in the process of membrane material development by determining the favorable characteristics of various operating conditions and the scale-up of the fabricated materials in industry scale.

The currently available model characterizing separation within hollow fiber membrane module is analyzed to determine the pros and cons associated to each method, as well as outlining the gap presented. The findings are summarized in Table 24.1.

Based on a review of the published literature, it is concluded that the favorable mathematical model should be easy to be implemented to ensure stability and convergence of the solution procedure. In addition, over-complexity associated to the mathematical modeling often requires excessive level of detail or algorithms that constraint the applicability of the mathematical model. However, the methodology should not be over-simplified, such as that of the averaging method (Pettersen and Lien, 1999), which reduces the model prediction accuracy. In this context, the stage-wise coupled with mass conservation approach is found to be appealing since it involves merely reducing the membrane module into a number of entities, in which the mass transfers are assumed to be constant and independent of one another (Thundyil and Koros, 1997). Non-ideal effects, such as pressure drop, can also be incorporated easily in conjunction with the mass

Table 24.1.   Summary of major developments in methodology adapted to characterize the separation performance of hollow fiber membrane module.

Published literature	Study domain	Research gap
Pan (1986)	– Mathematical modeling of multi-component high-flux, asymmetric hollow fiber membranes by formulating numerical integration of governing differential equations over relevant boundary conditions. – Proposed that all gas separation performances can be approximated by the cross-flow mechanism regardless of the bulk flow configuration.	– Instability when solving the differential equations (Makaruk and Harasek, 2009). – Shortcoming in predicting higher stage cuts performance due to the assumption of negligible bulk flow configuration (Makaruk and Harasek, 2009).
Thundyil and Koros (1997)	– Succession of states method to characterize the mass balances of radial cross-flow, counter-current and concurrent flow configuration within the hollow fiber membrane module by dividing the module into many finite elements, while performing mass balance independently on each element.	– The equations presented are merely applicable to the binary system separation.
Coker *et al.* (1998)	– Mathematical modeling of the multi-component hollow fiber membrane contactors adapting the stage-wise approach to transform the conservation differential mass balance equations to a first order finite difference solution procedure.	– Required the conservation equations to be fitted into the form of a tridiagonal matrix and to be solved simultaneously using the complicated Thomas algorithm. – Experimental validation of the method and sensitivity of the technique to initial estimates were not provided (Chowdhury *et al.*, 2005).
Pettersen and Lien (1999)	– Averaging method by using the logarithmic mean pressure driving force to describe the overall symmetric hollow fiber module performance.	– Accuracy of the method, which was developed based on the assumption of constant driving force along the length of the membrane permeator, was only in good agreement with the exact solution at lower stage cuts.
Marriott and Sorensen (2003)	– Mathematical modeling of the multi-component separation within the hollow fiber and spiral wound membrane module by developing rigorous mass, momentum and energy balances to characterize the system, which were later solved simultaneously using the orthogonal collocation numerical method.	– Required the knowledge of diffusion and dispersion coefficients in the fluid phase, which was not available at the initial stage of design process.

conservation equations as compared to mathematical modeling adapting differential equations (Thundyil and Koros, 1997).

However, there are still limited mathematical models developed based on the stage-wise with mass balance approach that are applicable to multi-component system currently since the methodology is evolved from binary system separation. Although the binary component system is often assumed for model development purpose to provide the fundamental knowledge in understanding the realistic modeling of separation processes, usage of the models is limited. It is highly desirable to extend the approach to incorporate the multi-component system to encompass more practical functions and higher prediction capability since real membrane separation processes usually involve more than two components.

Therefore, in this work, an experimentally validated multi-component hollow fiber membrane mathematical model is developed to characterize the counter-current and concurrent flow configuration based on a "multi-component progressive cell balance" methodology. The chapter also demonstrates the case study of CO_2 removal from natural gas through adaptation of hollow fiber membrane module to compare the performance of the different flow configurations based on their separation efficiency.

24.2 METHODOLOGY

24.2.1 *Multi-component progressive cell balance methodology*

The "multi-component progressive cell balance methodology" involves reducing the problem into a number of finite entities. The outlet condition of the entity is computed based on the specified inlet condition provided and will be the inlet condition of the subsequent entity. Therefore, from one entity, the computation proceeds to the next and is completed over the entire membrane module. Coker *et al.* (1998), Marriot and Sorensen (2003), Soni *et al.* (2009) and Thundyil and Koros (1997) highlighted that the counter-current and concurrent flow configuration can be sufficiently described using the one dimension approach along the axial direction of the membrane module since the characteristic of the fibers in the radial direction can be assumed to be the same.

The membrane bundle specifications are provided by the end user to determine the active membrane area of each finite entity based on the predetermined number of entities, m. It is done by dividing the membrane module along the longitudinal direction. Among the input parameters required to determine the active membrane area of each entity are as followed:

- Active length of the fiber bundle, L
- Radius of the fiber bundle, R
- Inner and outer diameter of the hollow fibers, d_i and d_o
- Packing density, $Ø$, or porosity, ε, or number of fibers, n_f, of the membrane module, which are related to one another, as depicted as (Li *et al.*, 2004):

$$Ø = 1 - \varepsilon = n_f \frac{A_f}{A_{module}} \tag{24.1}$$

The area calculation is adapted from Rautenbach as well as Thundyil and Koros published literature, in which the active area for permeation of each entity, A_m, is determined from the volume (Rautenbach, 1990; Thundyil and Koros, 1997), whereby the geometry is defined as a cylinder with radius, R, and axial thickness, $\Delta z = L/m$:

$$A_m = \frac{4\pi R^2 (\Delta z)(1 - \varepsilon)}{d_o} \tag{24.2}$$

In addition, several assumptions have been outlined when developing the "multi-component progressive cell balance" mathematical model to simplify the modeling approach. These assumptions are provided as following:

- The model assumes no mixing in the tube and shell sides and can be sufficiently described by plug flow (Pan and Habgood, 1978; Pettersen and Lien, 1994; Stern and Walawender, 1969; Thundyil and Koros, 1997; Weller and Steiner, 1950). Therefore, the mass balance of each plug can be solved independently provided the previous plug state is known.
- The shell side pressure variations are negligible while the permeate side pressure drop can be adequately described using the Hagen-Poiseuille equation (Coker *et al.*, 1998; Davis, 2002; Pan, 1986; Thundyil and Koros, 1997).
- Operation within the hollow fiber membrane module is isothermal (Davis, 2002; Pan, 1986; Thundyil and Koros, 1997).

(a)

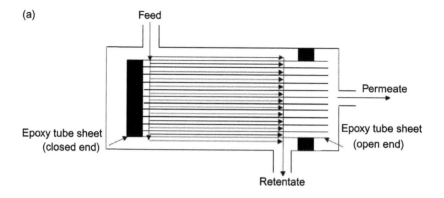

Feed

Permeate

Epoxy tube sheet
(closed end)

Epoxy tube sheet
(open end)

Retentate

(b)

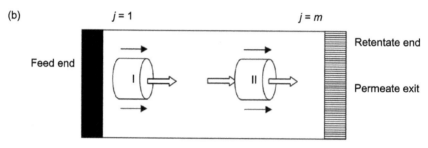

$j = 1$ $j = m$

Feed end

Retentate end

Permeate exit

Figure 24.1.   Schematic representation of (a) concurrent hollow fiber membrane module and (b) entities within the multi-component progressive cell balance methodology for mathematical modeling.

24.2.2   Multi-component progressive cell balance in concurrent flow mechanism

Figure 24.1a shows the schematic representation of the concurrent hollow fiber membrane module with the feed introduced through the shell side. The feed gas flows in the axial direction parallel to the hollow fibers until the retentate end, while the permeate stream diffuses into the fibers in the same direction with the feed. The end of the fiber bundle near the feed gas is sealed with epoxy glue while the other end allows the fibers to protrude through the seal for permeate gas to be separated from the retentate. Figure 24.1b depicts the schematic representation of the multi-component progressive cell balance approach to characterize the mass balance of the concurrent hollow fiber membrane module. From Figure 24.1b, there are two types of entity, which are Type I with feed on only the shell side, and Type II with feed on both the shell and tube sides.

24.2.3   Multi-component progressive cell balance in concurrent flow mechanism: Type I entity with feed on the shell side

The solution algorithm of the concurrent flow pattern is relatively simple because the computation proceeds from the feed end using the feed known composition to the retentate end. For a binary system separation, such as that of Thundyil and Koros's (1997) work, the permeate composition at the membrane module sealed end-sheet, y_1, can be easily calculated using Equation (24.3) (Brubaker and Kammermeyer, 1954; Pan and Habgood, 1978; Thundyil and Koros, 1997):

$$y_1 = \frac{(\alpha - 1)(\beta x_1 + 1) + \beta - \left[((\alpha - 1)(\beta x_1 + 1) + \beta)^2 - 4\alpha\beta x_1(\alpha - 1)\right]^{0.5}}{2(\alpha - 1)} \qquad (24.3)$$

Based on computation of the permeate composition near the sealed end, the calculation proceeds throughout the entire hollow fiber membrane module. However, Equation (24.3) is only confined to the binary separation.

For the multi-component system, a simple yet versatile iterative technique is required to ensure quick convergence of the solution. The following solution procedure has been implemented, which is characterized based on solution-diffusion model. According to the solution-diffusion model, separation is achieved among different components because some gases are more soluble in the membrane material, and pass more readily through the membrane than other components in the gas. The governing flux equation of the solution-diffusion model that characterizes the transport mechanism of component n for a Type I entity within the concurrent flow hollow fiber membrane module can be sufficiently described using Fick's law of diffusion (Baker, 2012), as provided in expression (24.4). The driving force for separation of the longitudinal flow is the partial pressures of the shell side retentate to the tube side permeate leaving the entity, as described by Barrer *et al.* (1962), Coker *et al* (1998), and Graham (1866).

$$y_n(1)\theta^* Q_f = P_n A_m \left\{ \frac{p_h}{1-\theta^*} \left[x_{f,n} - \theta^* y_n(1) \right] - p_l y_n(1) \right\} \tag{24.4}$$

Coupled with the Newton bisection methodology in Visual Basic subroutine, the guessed permeate composition, $y_n(1)$, is initiated from both the low and the high sides to ensure quick convergence. The solution algorithm is as follows:

- The value of the permeate composition of the first component, $y_1(1)$, is assumed.
- The local stage cut, θ^*, is estimated from Equation (24.4).
- $Q_T(1)$ is computed based on the estimated θ^*.
- The sum of all the compositions in the permeate stream is calculated and compared to unity. This procedure is iterated until the summation meets unity within pre-defined tolerance, ϵ, of 0.1%, as shown in expression (24.5).

$$\delta = abs \left(\sum_{n=1}^{N} y_n(1) - 1 \right) \ll \epsilon \tag{24.5}$$

- The local retentate composition is determined from Equation (24.6), which followed from the Type I entity balance:

$$x_n(1) = \frac{\left[x_{f,n} - \theta^* y_n(1) \right]}{(1 - \theta^*)} \tag{24.6}$$

24.2.4 *Multi-component progressive cell balance in concurrent flow mechanism: Type II entity with feed on the shell and tube sides*

For the Type II entity, the characterization of the transport across the concurrent flow hollow fiber membrane module with shell and tube sides feed using Fick's law of diffusion is described by (24.7)

$$y_n(j)\theta^* Q_S(j-1) - y_n(j-1)Q_T(j-1)$$

$$= P_n A_m \left\{ \frac{p_h[x_n(j-1) - y_n(j)\theta^* + y_n(j-1)Q_T(j-1)/Q_S(j-1)]}{1 + \dfrac{Q_T(j-1)}{Q_S(j-1)} - \theta^*} - p_l y_n(j) \right\} \tag{24.7}$$

The same solution procedure as outlined in the previous section has been applied to determine the permeate composition followed by the local retentate composition of each Type II entity using (24.8).

$$x_n(j) = \frac{x_n(j-1) - y_n(j)\theta^* + \dfrac{y_n(j-1)Q_T(j-1)}{Q_S(j-1)}}{1 + \dfrac{Q_T(j-1)}{Q_S(j-1)} - \theta^*} \tag{24.8}$$

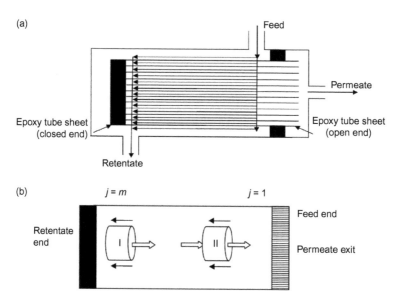

Figure 24.2.   Schematic representation of: (a) counter-current hollow fiber membrane module and (b) entities within the multi-component progressive cell balance methodology for mathematical modeling.

After determining the local permeate and retentate composition of each entity, the flow rates in the shell and tube sides of the hollow fiber membrane module are determined adapting the following equations. Only with the exception of the Type I concurrent flow entity in direct contact with the feed, the indices $(j-1)$ are replaced with the feed condition, e.g. Q_f and $x_{f,n}$:

$$\Delta Q = \sum_{n=1}^{N} P_n A_m (p_h x_n(j) - p_l y_n(j)) \tag{24.9}$$

$$\Delta Q_n = P_n A_m (p_h x_n(j) - p_l y_n(j)), \quad n = 1, 2, \dots, N \tag{24.10}$$

$$Q_s(j) = Q_s(j-1) - \Delta Q \tag{24.11}$$

$$Q_T(j) = Q_T(j-1) + \Delta Q \tag{24.12}$$

As mentioned previously, the tube side pressure drop is characterized by the Hagen-Poiseuille equation, as presented in Equation (24.13), while incorporation of the Hagen-Poiseuille equation within the multi-component progressive cell balance approach is depicted by (24.14) (Thundyil and Koros, 1997):

$$\frac{dp_l^2}{dz} = \frac{25.6 RT\omega}{\pi d_i^4} \left(\frac{Q_T}{n_f} \right) \tag{24.13}$$

$$p_l(j) = \sqrt{[p_l(j-1)^2 - 25.6 \frac{RT\omega Q_T(j)}{\pi d_i^4 n_f} \Delta z]} \tag{24.14}$$

24.2.5   Multi-component progressive cell balance in counter-current flow mechanism

Figure 24.2a shows the schematic representation of the counter-current hollow fiber membrane module with the feed introduced through the shell side. The feed flows axially along the hollow fibers until the retentate end to be collected at the retentate outlet. The permeate stream into the fibers flows in the opposite direction with respect to the feed. The end of the fiber bundle near

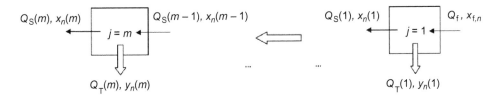

Figure 24.3.   Schematic representation of cross flow module for m stages as initial guesses for counter-current flow simulation.

the feed gas is the tube sheet to enable the permeate stream to exit the membrane module, while the end of the fiber bundle near the retentate end is sealed.

The complexity associated with the mathematical modeling of the counter-current hollow fiber membrane module arises because the permeation at the stage where the feed gas enters the module is dependent upon the composition of the gas that has already permeated. This information is initially not known. Therefore, the composition and flow rates of each entity have to be obtained using cross flow results as initial guesses, which is proposed in Figure 24.3. A similar approach has been employed by Coker *et al.* (1998).

Starting from the feed end, the solution procedure outlined previously has been adopted to compute the initial guesses of the entities. Using the cross-flow solution as an initial guess, improved estimates of $x_n(j)$ and $y_n(j)$ are obtained from Equations (24.15) and (24.16) respectively, which followed from the mass balance of each entity:

$$x_{n,\text{new}}(j) = \frac{x_{n,\text{old}}(j-1)Q_{\text{S,old}}(j-1) + A_m P_n p_l y_{n,\text{old}}(j)}{Q_{\text{S,old}}(j) + A_m P_n p_h} \tag{24.15}$$

$$y_{n,\text{new}}(j) = \frac{y_{n,\text{old}}(j+1)Q_{\text{T,old}}(j+1) + A_m P_n p_h x_{n,\text{old}}(j)}{Q_{\text{T,old}}(j) + A_m P_n p_l} \tag{24.16}$$

The improved estimates of $x_n(j)$ and $y_n(j)$ are employed to calculate the shell and tube sides flow rate, starting from the feed end to a series of entities until the retentate end, such as that demonstrated in Figure 24.2b adapting the following equations:

$$\Delta Q_{\text{new}} = \sum_{n=1}^{N} P_n A_m (p_h x_{n,\text{new}}(j) - p_l y_{n,\text{new}}(j)) \tag{24.17}$$

$$\Delta Q_{n,\text{new}} = P_n A_m (p_h x_{n,\text{new}}(j) - p_l y_{n,\text{new}}(j)), \quad n = 1, 2, \ldots, N \tag{24.18}$$

$$Q_{\text{S,new}}(j) = Q_{\text{S,new}}(j-1) - \Delta Q_{\text{new}} \tag{24.19}$$

$$Q_{\text{T,new}}(j) = Q_{1,\text{new}}(j+1) + \Delta Q_{\text{new}} \tag{24.20}$$

The updated tube side flow rate, $Q_T(j)$, obtained through iteration has been adapted to calculate the tube side pressure profile, as depicted in Equation (24.14). Similarly, for the entity in direct contact with the feed side, the indices $(j-1)$ are replaced with the feed condition, e.g. Q_f and $x_{f,n}$. The above procedures are iterated until the changes in the component permeate and retentate flow rates are within a defined tolerance limit, ϵ, of 0.1%.

24.3   RESULTS AND DISCUSSION

24.3.1   *Mathematical model validation*

To demonstrate the applicability of the simulation model, it has been validated using published experimental data by Tranchino *et al.* (1989) for concurrent separation and Sada *et al.* (1992)

Table 24.2.   Input parameters adapted for model validation (Sada *et al.*, 1992; Tranchino *et al.*, 1989).

Published literature	Tranchino *et al.* (1989)	Sada *et al.* (1992)
Flow configuration	Concurrent, shell side feed	Counter-current, shell side feed
Feed pressure [MPa]	0.4053	1.570
Permeate pressure [MPa]	0.1013	0.1013
Feed composition	60.0% CO_2, 40.0 CH_4	50.0% CO_2, 10.5% O_2, 39.5% N_2
Permeance [GPU]	$P_{CO_2} = 9.44$, $P_{CH_4} = 2.63$	$P_{CO_2} = 61.01$, $P_{O_2} = 17.99$, $P_{N_2} = 3.91$
Fiber active length [cm]	15	26
Number of fibers	6	270
Inner and outer diameter of fiber [μm]	735/389	156/63

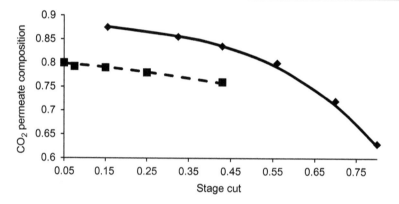

Figure 24.4.   Model validation with published literature using Tranchino *et al.* (1989) experimental data for concurrent flow (■), and Sada *et al.* (1992) experimental data for counter-current flow (◆). (——— Counter-current simulation model; – – – – Concurrent simulation model).

for counter-current separation. The simulation conditions adapted for the model validation are summarized in Table 24.2, while the model validation is provided in Figure 24.4.

As illustrated in Figure 24.4, the suggested simulation model gives good approximation to the published experiment data for both the counter-current flow and concurrent flow, which demonstrates the accuracy of the mathematical model.

24.3.2   *Comparison between concurrent and counter-current flow*

To compare the performance of the counter-current and concurrent flow, the hollow fiber membrane module is simulated as a simple single stage membrane. The simulations are run under 1,000,000 cm³(STP) s⁻¹ feed gas flow rate with 5.0 MPa pressure. Feed gas containing 10% CO_2, 70% methane, 10% ethane and 10% nitrogen, corresponding to the sweetening of natural gas with low quantities of acid gas is simulated. The ratio of the feed pressure to the permeate pressure, β, is maintained at 50, analogous to atmospheric pressure for the permeate stream. The fiber bundle is assumed to be 50% packed with hollow fibers of inner diameter 100 μm and outer diameter 250 μm (Thundyil and Koros, 1997). The material of fabrication of the membrane is assumed to be the commonly used cellulose triacetate for acid gas separation with permeance values as followed: $P_{CH_4} = 2.86$ GPU, $P_{C_2H_6} = 2$ GPU and $P_{N_2} = 3.57$ GPU (Coker *et al.*, 1998; Kundu *et al.*, 2012). High selectivity and low selectivity membranes are simulated using α_{CO_2/CH_4}. 40 and 20 respectively. The active length of the hollow fibers, L, is altered within the range of 100–500 cm while the radius of the fiber bundle, R, is changed within the range of 10–25 cm to vary the membrane area. The performance of the hollow fiber membrane module for the 10% CO_2 case is presented in Figure 24.5a–c.

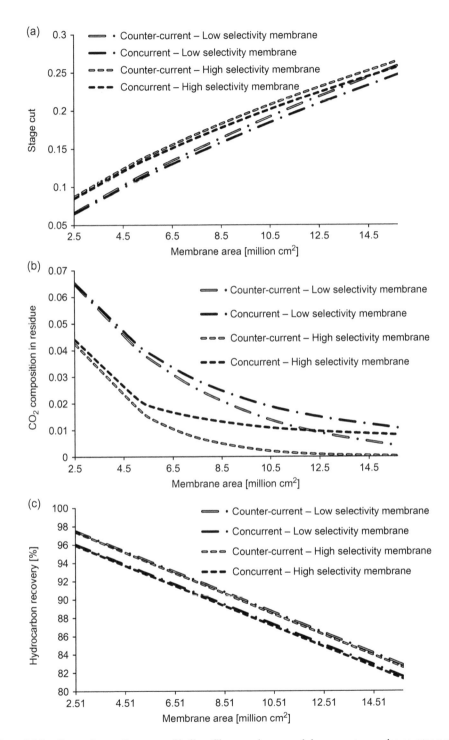

Figure 24.5.   Separation performance of hollow fiber membrane module permeators under counter-current and concurrent flow configuration as a function of membrane area for low and high selectivity membrane: (a) stage cut, (b) CO_2 composition in the residue and (c) fraction of hydrocarbon recovered (CO_2 feed composition $= 10\%$).

As depicted in Figure 24.5a–c, in order to analyze the performance and separation efficiency of the hollow fiber permeators under different flow configuration, the membrane area is plotted against the stage cut, the CO_2 composition in the residue stream and the fraction of hydrocarbon recovered.

The effect of the availability of membrane area to the stage cut is depicted in Figure 24.5a. The stage cut is defined as the flow rate of permeate over the flow rate of the feed. When more membrane area is available for permeation, the readily permeable CO_2 permeates more through the hollow fibers. Therefore, the stage cut increases when the permeate flow increases. It is also observed from Figure 24.5a that the stage cut of the higher selectivity membrane is greater than that of the lower selectivity membrane. Similarly, it is attributed to more permeation of CO_2 since the membrane has a higher CO_2 permeability characteristic, which further increases the permeate flux. The superiority of the higher selectivity membrane to the lower selectivity membrane in terms of stage cut becomes less substantial when the membrane area is further increased since the availability of the permeable CO_2 in the shell side decreases.

Figure 24.5b demonstrates the effect of the membrane area to the CO_2 residue composition. It is observed that the CO_2 residue composition decreases with the membrane area attributed to the permeation of more CO_2 to the permeate stream when more membrane area is available. The higher selectivity membrane exhibits more pronounced impact by allowing more CO_2 to transport through the hollow fibers when provided with the same membrane area as compared to the lower selectivity membrane. The less significant impact of the CO_2 change at higher membrane area as described previously is demonstrated through the slow decrement of the CO_2 residue composition, which is especially substantial for the high selectivity membrane.

The effect of the membrane area to the hydrocarbon recovery is depicted in Figure 24.5c. Hydrocarbon recovery is defined as the percentage of the hydrocarbon recovered in the residue stream as compared to the total amount of hydrocarbon contained in the feed. It is depicted that the hydrocarbon recovery decreases with increment in the membrane area. This is due to availability of more membrane area that also enables permeation of more hydrocarbons to the permeate stream. It is also depicted from Figure 24.5c that the hydrocarbon recovery of the high and low selectivity membrane is similar since the membrane under investigation exhibits similar characteristics to permeation of methane. However, the percentage hydrocarbon recovery of the higher selectivity membrane decreases slightly in comparison to the lower selectivity membrane at high membrane area. This is attributed to the excessive membrane area that further leads to increased hydrocarbons lost in the permeate stream when there is reducing availability of the more permeable CO_2 in the shell side.

Viewing from the aspect of the flow configuration, the counter-current flow is demonstrated to provide higher separative performance by displaying outputs, such as increased stage cut, decreased CO_2 concentration in the residue stream, and improved hydrocarbon recovery when provided with the same membrane area for permeation as compared to the concurrent flow. The superiority of the counter-current flow is attributed to the partial pressure driving force distribution. The opposite flow of the tube and shell sides contributes to the slowly declining pressure gradient that enables the pressure driving force to be maximized along the module length (Ho and Sirkar, 1992). Therefore higher permeation of the CO_2 is achieved for the counter-current flow while retaining the hydrocarbon in the residue stream in comparison to the concurrent pattern (Ho and Sirkar, 1992; Makaruk and Harasek, 2009; Thundyil and Koros, 1997).

Parameter sensitivity of the feed composition is also conducted by altering the feed composition to 60% CO_2, 20% methane, 10% ethane and 10% nitrogen. This simulation condition is analogous to the natural gas processing in enriched oil recovery with high impurities concentration. The membrane selectivity, α_{CO_2/CH_4}, is maintained at 20 in this case study to determine the effect of harsh feed condition (e.g. high impurities composition) on lower performance membranes (Lock *et al.*, 2015). The performance of the hollow fiber membrane module for the 60% CO_2 case as compared to the 10% case is presented in Figure 24.6a–b.

It is depicted from Figure 24.6a–b that the performances at the higher feed concentrations is merely an amplification of the results obtained at the lower feed concentrations, such as increased

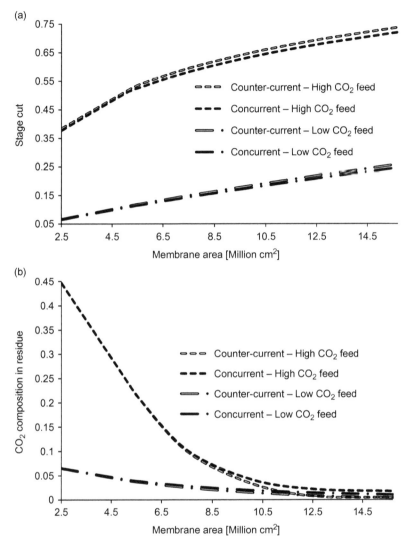

Figure 24.6.   Separation performance of hollow fiber membrane module permeators under counter-current and concurrent flow configuration as a function of membrane area for 10% and 60% CO_2 feed composition: (a) stage cut and (b) CO_2 composition in the residue ($\alpha_{CO_2/CH_4} = 20$).

stage cut and higher concentration of the CO_2 in the residue stream. The simulation results are consistent with the observation by Thundyil and Koros (1997). The increased stage cut is ascribed to the higher permeate flux with more permeable CO_2, while the higher CO_2 concentration in the retentate is due to the availability of the higher CO_2 in the feed to be removed.

Similarly, the counter-current flow is demonstrated to be more effective in removing the CO_2 in comparison to the concurrent flow attributed to the partial pressure driving force as discussed previously. However, it is depicted from Figure 24.6a–b that the superiority of the counter-current to the concurrent flow is more pronounced when the CO_2 feed concentration is higher. This is attributed to the more substantial impact of the CO_2 flux permeation with more driving force when the CO_2 feed composition is increased, which further increases the efficiency of the counter-current in comparison to the concurrent configuration.

24.4   CONCLUSION

A solution algorithm coupled with Newton Bisection methodology has been established for the "multi-component progressive cell balance" approach to characterize the multi-component separation performance of the hollow fiber membrane module. The solution algorithm has been suited for the longitudinal flow configuration, which encompasses the counter-current and concurrent, by characterizing the membrane module into many pre-defined number of entities in the axial direction. The advantages of the solution algorithm are it is simple, ensures stable convergence and allows the non-linearity of permeate pressure drop to be incorporated easily in conjunction with the mass conservation equations. The accuracy of the simulated model has been validated with published experimental results, in which the simulated data exhibit good agreement with the laboratory observation. The performance of the counter-current and concurrent membrane module fashion has been evaluated based on the stage cut, CO_2 residue concentration and the percentage hydrocarbon recovery with respect to the membrane area for separation. The simulation sensitivity has been investigated by altering the membrane selectivity properties and feed CO_2 composition. It is observed that the performance of the counter-current flow is better than that of the concurrent flow due to the slowly declining pressure gradient that maximizes the driving force for permeation.

ACKNOWLEDGEMENT

This work was done with the financial and technical support from CO_2 management (MOR) research group, Universiti Teknologi PETRONAS.

NOMENCLATURE

A_f	cross section area of fiber [cm^2]
A_m	area of permeation for one entity [cm^2]
A_{module}	cross section area of membrane module [cm^2]
d_i	inner diameter of the hollow fiber [cm]
d_o	outer diameter of the hollow fiber [cm]
L	active length of the fiber bundle [cm]
n_f	number of fibers
p_h	pressure on the shell side [cmHg]
p_l	pressure on the tube side [cmHg]
P_n	mix gas permeance of each component in the feed [GPU, 1 GPU $= 10^{-6}$ cm^3 (STP)/(cm^2 cmHg s)]
Q_f	feed flow rate [cm^3 (STP) s^{-1}]
$Q_S(j)$	shell side flow rate of each entity [cm^3 (STP) s^{-1}]
$Q_T(j)$	tube side flow rate of each entity [cm^3 (STP) s^{-1}]
R	radius of the fiber bundle [cm]
x_1	mole fraction of component on shell side at the sealed end-sheet for binary gas system
$x_{f,n}$	feed composition of each component
$x_n(j)$	shell side composition of each entity
y_1	mole fraction of component on tube side at the sealed end-sheet for binary gas syem
$y_n(j)$	tube side composition of each entity
ΔQ	flow rate that permeates through the membrane [cm^3 (STP) s^{-1}]
Δz	thickness of entity in the axial direction [cm]
$\in$	convergence criteria
$\emptyset$	packing factor of fibers within the membrane bundle

GREEK SYMBOLS

α selectivity of the higher permeance component to the lower permeance component
β ratio of the pressure on the shell side to the pressure on the tube side
δ difference between the check conditions with the datum
ε porosity
θ^* local stage cut of each entity
ω viscosity $[cm^2\,s^{-1}]$

REFERENCES

Baker, R.W. (2002) Future directions of membrane gas separation technology. *Industrial & Engineering Chemistry Research*, 41, 1393–1411.

Baker, R.W. (2012) *Membrane technology and applications*. 3rd edition. John Wiley and Sons Ltd., Chichester, UK.

Baker, R.W. & Lokhandwala, K. (2008) Natural gas processing with membranes: an overview. *Industrial & Engineering Chemistry Research*, 47, 2109–2112.

Barrer, R.M., Barrie, J.A. & Raman, N.K. (1962) Solution and diffusion in silicone rubber. I.A. Comparison with natural rubber. *Polymer*, 3, 595–603.

BCC Research (2014) Membrane technology for liquid and gas separations. BCC Research, New York, NY.

Brubaker, D.W. & Kammermeyer, K. (1954) Separation of gases by plastic membranes – permeation rates and extent of separation. *Industrial & Engineering Chemistry*, 46, 733–739.

Chowdhury, M.H.M., Feng, X., Douglas, P. & Croiset, E. (2005) A new numerical approach for a detailed multicomponent gas separation membrane model and AspenPlus simulation. *Chemical Engineering & Technology*, 28, 773–782.

Coker, D.T., Freeman, B.D. & Fleming, G.K. (1998) Modeling multicomponent gas separation using hollow-fiber membrane contactors. *AICHE Journal*, 44 (6), 1289–1302.

Davis, R.A. (2002) Simple gas permeation and pervaporation membrane unit operation models for process simulators. *Chemical Engineering & Technology*, 25, 717–722.

Graham, T. (1866) On the absoprtion and dialytic separation of gases by colloid septa: action of a septum of caoutchouc. *Philosophical Magazine*, 32, 399–439.

Ho, W.S.W. & Sirkar, K.K. (1992) *Membrane handbook*, Volume I. Springer Science and Business Media, New York, NY.

Kundu, P.K., Chakma, A. & Feng, X. (2012) Simulation of binary gas separation with asymmetric hollow fibre membranes and case studies of air separation. *The Canadian Journal of Chemical Engineering*, 90, 1253–1268.

Li, D.R. Wang & Chung, T.S. (2004) Fabrication of lab-scale hollow fiber membrane modules with high packing density. *Separation and Purification Technology*, 40, 15–30.

Lock, S.S.M., Lau, K.K. & Shariff, A.M. (2015) Effect of recycle ratio on the cost of natural gas processing in counter-current hollow fiber membrane system. *Journal of Industrial and Engineering Chemistry*, 21, 542–551.

Makaruk, A. & Harasek, M. (2009) Numerical algorithm for modeling multicomponent multipermeator systems. *Journal of Membrane Science*, 344, 258–265.

Marriott, J. & Sorensen, E. (2003) A general approach to modeling membrane modules. *Chemical Engineering Science*, 58, 4975–4990.

Pan, C.Y. (1986) Gas separation by high-flux, asymmetric hollow-fiber membrane. *AICHE Journal*, 32, 2020–2026.

Pan, C.Y. & Habgood, H.W. (1978) Gas separation by permeation. Part I. Calculation methods and parametric analysis. *The Canadian Journal of Chemical Engineering*, 56, 197–209.

Pettersen, T. & Lien, K.M. (1994) A new robust design model for gas separating membrane modules, based on analogy with counter-current heat exchangers. *Computers & Chemical Engineering*, 18, 427–439.

Rautenbach, R. (1990) *Handbook of industrial membrane technology*. Noyes Publication, Park Ridge, NJ. pp. 349–400.

Sada, E., Kumuzawam, H., Wang, J.S. & Koizumi, M. (1992) Separation of carbon dioxide by asymmetric hollow fiber membrane of cellulose triacetate. *Journal of Applied Polymer Science*, 45, 2181–2186.

Soni, V., Abildskov, J., Jonsson, G. & Gani, R. (2009) A general model for membrane-based separation processes. *Computers & Chemical Engineering*, 33, 644–659.

Stern, S. & Walawender, W. (1969) Analysis of membrane separation parameters. *Journal of Separation Science*, 4, 129–159.

Thundyil, M.J. & Koros, W.J. (1997) Mathematical modeling of gas separation permeators – for radial crossflow, countercurrent, and cocurrent hollow fiber membrane modules. *Journal of Membrane Science*, 125, 275–291.

Tranchino, L., Santarossa, R., Carta, F., Fabiani, C. & Bimbi, L. (1989) Gas separation in a membrane unit: experimental results and theoretical predictions. *Separation Science and Technology*, 24, 1207–1226.

Weller, S. & Steiner, W. (1950) Separation of gases by fractional permeation through membranes. *Journal of Applied Physics*, 21, 279–285.

Part VI
Membranes for other applications

CHAPTER 25

Preparation of sulfonated polyether ether ketone (SPEEK) and optimization of degree of sulfonation for using in microbial fuel cells

Mostafa Ghasemi, Wan Ramli Wan Daud, Ahmad Fauzi Ismail & Takeshi Matsuura

25.1 INTRODUCTION

Recently by depletion of fossil fuels and environmental pollution, fuel cells have attracted attention due to low emissions of pollutants and high efficiency. Microbial fuel cells (MFCs) as a novel technology, convert the chemical energy in organic compounds into electricity and biohydrogen, by the aid of microorganisms as biocatalyst (Ghasemi et al., 2011). Generally, MFC is a device that simultaneously produces energy and treats the wastewater (Ghasemi et al., 2013a). Also MFC can be applied for bioremediation of some chemicals. MFC, similar to all other fuel cells and batteries, consists of two parts (anode, cathode) which are separated by a proton exchange membrane (PEM). The performance of a MFC depends on several factors including, type of electrodes, cathode catalyst, PEM, distance of the electrodes, microorganisms etc., but among them membrane has one of the most significant roles as it separates anode and cathode and must support the transfer of protons from anode to cathode while, at the same time, should inhibit transferring of media from anode to cathode (Ghasemi et al., 2013b). One of the big obstacles of commercialization of MFC is high capital cost of the PEMs which cover about 40% of the expense of MFC (Ghasemi et al., 2013c; 2013d). There are a lot of researches about different types of PEM (nano-composite, ultrafiltration etc.) in MFCs, but due to complication of the MFC system and existence of several factors, the proper membrane has not been found yet (Lim et al., 2012).

Recent researches into developing suitable and cost efficient PEMs in fuel cells have focused on aromatic polymers such as polyimide (PI), polybenzimidazole (PBI) and poly ether ether ketone (PEEK) composite membranes (Ghasemi et al., 2013e). Among the various polymeric membranes, with diverse mechanical, thermal and electrical properties PEEK was shown to be one of considerable promise due to long term stability, cost efficiency and performance. The studies found that PEEK has good thermal stability, high proton conductivity and mechanical stability. All of these properties depend on degree of sulfonation of the polymer (Leong et al., 2013).

As we mentioned, there are a lot of works about various types of membranes for PEM in MFC but there is not much study about the application of sulfonated poly ether ether ketone (SPEEK) in MFCs, although it was used widely in other types of fuel cell (Ilbeygi et al., 2015). Our group has studied some types of membranes in MFCs. Ghasemi et al. (2012) developed and applied activated carbon nano-fiber/Nafion as PEM in MFC. It produced the power of 1.5 times more than Nafion 117, a PEM traditionally used in MFC. Also in another study Rahimnejad et al. (2012) fabricated Fe_3O_4/PES nano-composite membranes and compared its performance with Nafion 117. The Fe_3O_4/PES PEM also produced the power of 1.3 times more than Nafion 117. However, the big obstacle for using this type of membrane in MFC is their porous nature which allows the passage of media from anode to cathode and also oxygen from cathode to anode, disturbing the long term performance of MFC.

The objective of this study is to maximize the performance of MFC by using sulfonated polyether ether ketones (SPEEKs) of different degrees of sulfonation as PEM.

25.2   MATERIALS AND METHODS

25.2.1   *Synthesis of SPEEK*

For the preparation of SPEEK, 20 g of PEEK powder (Goodfellow Cambridge Ltd., UK) was added slowly to 500 mL of 95–98% concentrated sulfuric acid (R & M Chemicals, Essex, UK) under vigorous stirring so that the entire PEEK was dissolved completely. Stirring was continued at an elevated temperature of 80°C for 1, 2, 3 and 4 hours to obtain SPEEKs with different degrees of sulfonation. The SPEEKs so synthesized were named as SPEEK1, SPEEK2, SPEEK3 and SPEEK4, respectively, depending on the reaction periods. The SPEEK solution was then poured into a large excess of ice water to precipitate the SPEEK polymer, followed by filtration by a Whatman filter paper and washing with deionized water. Finally, the SPEEK was dried at 70°C to remove any remaining water before use.

25.2.2   *Determination of the degree of sulfonation*

The degree of sulfonation (DS) was measured by 1H nuclear magnetic resonance (FT-NMR ADVANCE 111 600 MHz with cryoprobe) spectroscopic analysis (Bruker, Karlsruhe, Germany). Before measurement, the SPEEK was dissolved in dimethyl sulfoxide (DMSO-d_6). The DS can be calculated by the following equation:

$$\frac{DS}{S - 2DS} = \frac{A_1}{A_2} \quad (0 < DS < 1) \tag{25.1}$$

where S is the total number of hydrogen atoms in the repeat unit of the polymer before sulfonation, which is 12 for PEEK, $A_1(H_{13})$ is the peak area of the distinct signal and A_2 is the integrated peak area of the signals corresponding to all other aromatic hydrogen. To calculate DS in percent (DS%), the answer for DS has to be multiplied by 100.

25.2.3   *Fabrication of membranes*

Ten grams of SPEEK was dissolved in 90 grams of NMP and the mixture was stirred for 24 h for complete dissolution of the polymer. The solution so prepared was cast on a glass plate by using a casting knife to a thickness of 80–120 μm. The cast film was then dried in a vacuum oven at 60°C for 6 h. The membrane was separated from the glass plate by immersing in to water and kept in deionized water until it was used.

25.2.4   *MFC configuration*

Two cubic shaped chambers were constructed from Plexiglas, with a height of 10 cm, a width of 6 cm, and length of a 10 cm (giving a working volume of 420 mL). They were separated by a proton exchange membrane (PEM). Oxygen was continuously fed to the cathode side by an air pump at a rate of 80 mL min^{-1}. Both the cathode and the anode had a surface area of 12 cm^2. The cathode was made of carbon paper, coated with 0.5 mg cm^{-2} Pt and the anode (as described above) was plain carbon paper.

25.2.5   *Enrichment*

Palm oil mill effluent (POME, Indah Water Consortium) anaerobic sludge was used for the inoculation of MFCs. The media contained 5 g of glucose, 0.07 g of yeast extract, 0.2 g of KCl, 1 g of NaH$_2$PO$_4$ · 4H$_2$O, 2 g of NH$_4$Cl, 3.5 g of NaHCO$_3$ (all from the Merck company), 10 mL of Wolfe's mineral solution and 10 mL of Wolfe's vitamin solution (added per liter). All experiments were conducted in an incubator at 30°C. Furthermore, the cathode chamber contained a phosphate buffer solution which consisted of 2.76 g L^{-1} of NaH$_2$PO$_4$, 4.26 g L^{-1} of Na$_2$HPO$_4$, 0.31 g L^{-1} of NH$_4$Cl, and 0.13 g L^{-1} of KCl (all from the Merck company).

25.2.6 Analysis and calculation

FTIR spectroscopy was performed by Nicolet 5700 FTIR (Thermo Electron, USA) used to identify the functional group of the membranes before pre-treatment, after pre-treatment, and after fouling. Scanning Electron Microscopy (SEM, Supra 55vp-Zeiss, Germany) was implemented to observe the attachment of microorganisms onto the surface of the anode electrode. Moisture should be removed from the biological samples (POME mix culture sludge) by critical drying. They were then coated with a conductive material (such as gold or carbon), with a thickness of approximately 20–50 nm, in order to make them conductive for the SEM analysis.

To measure the chemical oxygen demand (COD), a sample was first diluted 10 times and 2 mL of the diluted sample was mixed with a digestion solution of a high-range COD reagent. Then, the sample was heated at 150°C for 2 h in a thermoreactor (DR B200), before reading the absorbance with a spectrophotometer (DR 2800). The voltage was measured using a multimeter (Fluke 8846A), and the power density curve was obtained by applying different loads to the system and calculating the power at different loads.

The current was obtained by:

$$I = \frac{V}{R} \tag{25.2}$$

where, I is the current [A], V is the voltage [V], and R is the applied external resistance [Ω].

The power density was calculated using:

$$P = R \times I^2 \tag{25.3}$$

The Coulombic efficiency (CE) was calculated as the current over time; until the maximum theoretical current was achieved, using:

$$CE = \frac{M \int_0^t I \, dt}{F b V_{an} \Delta COD} \tag{25.4}$$

where, M is the molecular weight of oxygen (32), F is Faraday's constant, $b = 4$ indicates the number of electrons exchanged per mole of oxygen, V_{an} is the volume of liquid in the anode compartment, and ΔCOD is the change in chemical oxygen demand (COD) over time, 't'.

25.2.7 Pre-treatment of PEMs

The pre-treatment of Nafion 117 was conducted by boiling in distilled water, 3% hydrogen peroxide, or 0.5 M sulfuric acid for 1 h consecutively and then stored in water until being used in the system. SPEEK is also kept in water after fabrication until use (Shahgaldi *et al.*, 2014).

25.3 RESULTS AND DISCUSSION

25.3.1 Attachment of microorganisms on anode electrode

Figure 25.1 shows the attachment of microorganisms on the anode electrode in MFC. As can be seen in the figure, different types of microorganisms are attached on the electrode surface. It is one of the benefits of using mixed culture microorganisms as the media than single culture in MFCs. The attached microorganisms play the role of biocatalyst to transfer the electrons from media to the electrode.

25.3.2 Degree of sulfonation of SPEEK

Figure 25.2 shows the NMR spectra of the SPEEK2. From the figure, the degree of sulfonation for this SPEEK is 41% applying Equation (25.1). Table 25.1 summarizes the *DS* of all the SPEEKs. It should be mentioned that more than 4 h of sulfonation, make the SPEEK soluble in water.

Figure 25.1.   Attachment of microorganisms on anode electrode surface.

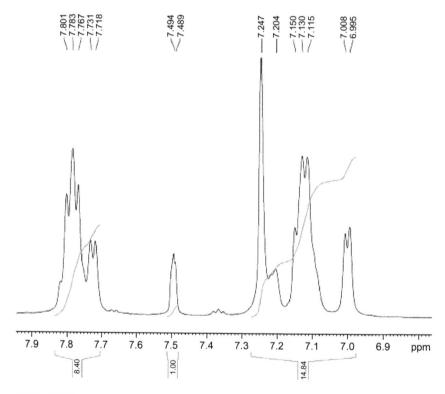

Figure 25.2.   NMR spectra for determination of degree of sulfonation (*DS*) of SPEEK2 polymer.

25.3.3   *FTIR analysis*

Figure 25.3 shows FTIR spectra of PEEK and SPEEK. The peak observed at $1224\,cm^{-1}$ in PEEK and SPEEK represents the presence of C–O–C with aromatic ring and the peak at $1490\,cm^{-1}$ for PEEK corresponds to the presence of C–C aromatic ring. The figure also shows a large band from

Table 25.1.   Degree of sulfonation (*DS*) of SPEEKs synthesized in this work.

Membrane	Time of sulfonation [h]	DS
SPEEK1	1	20.8
SPEEK2	2	41
SPEEK3	3	63.6
SPEEK4	4	76

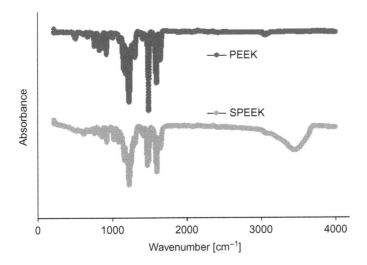

Figure 25.3.   FT-IR spectra for PEEK and SPEEK polymer.

3200–3500 cm^{-1} for SPEEK, which represents the O–S–O stretching vibration of sulfonic acid groups.

25.3.4   *Power density and polarization curve*

Figure 25.4 shows the power density of the MFCs constructed using the membranes under study. As the figure shows the MFC working with SPEEK3 ($DS = 63.6\%$) produced the highest power density (68.6408 mW m^{-2}) among all the tested membranes, followed by SPEEK4 ($DS = 76\%$, 54.83 mW m^{-2}), SPEEK2 ($DS = 41\%$, 42.56 mW m^{-2}) and SPEEK1 ($DS = 20.8\%$, 29.48 mW m^{-2}). The results can be due to the fact that the sulfonation enhances proton transport cation exchange capacity. However, sulfonation also enhances the transport of metal cations such as Na$^+$, K$^+$ etc. which are present in the media. That is the reason why the power density of SPEEK4 membrane is not the highest among all the SPEEK membranes.

Figure 25.5 shows the polarization curve of different MFCs. The internal resistance of the systems can be calculated from this figure as the slope of the line of current density vs. voltage. The results showed that the smallest internal resistance belongs to the system with SPEEK ($DS = 63\%$) but the internal resistance of the other three SPEEKs are also nearly the same as that of SPEEK3.

25.3.5   *COD removal and CE*

Figure 25.6 shows the *CE* and COD removal of the studied systems. As the figure shows the highest *CE* (26%) belongs to the system working with SPEEK3 as PEM, followed by SPEEK4

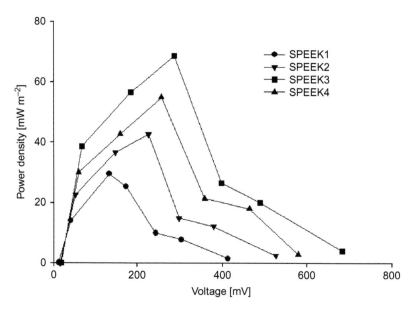

Figure 25.4.   Power density of MFC systems with different *DS* of SPEEK membranes.

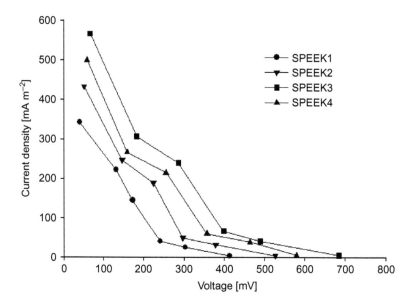

Figure 25.5.   Polarization curve for MFC systems with different *DS* of SPEEK membranes.

(18.5%) SPEEK2 (18%) and SPEEK1 (14%). Thus the order of SPEEK appearance in the *CE* level is the same as the order in the power density. This is because MFC's performance depends on PEM's ability to conduct proton while preventing the crossover of oxygen, which enables to keep the anaerobic (Ghoreishi *et al.*, 2014).

The figure also shows that all the systems have high COD removal (>80%), indicating the high potential of MFC as a new method for wastewater treatment.

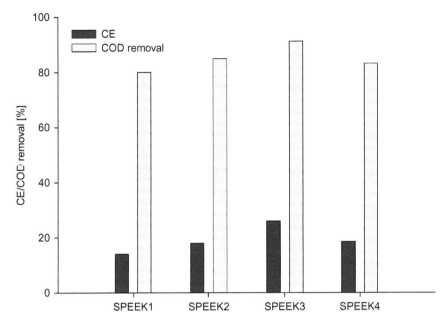

Figure 25.6.   COD removal and Coulombic efficiency of MFCs system with different *DS* of SPEEK membranes.

25.4   CONCLUSION

MFCs were constructed using PEMs made of SPEEKs with different degrees of sulfonation. The MFCs were then tested for their performance of them in power generation and COD removal. A maximum power density of 68.6408 mW m^{-2} was obtained by SPEEK3 whose *DS* was 63.6%. Thus, the SPEEK membrane can be a proper alternative to replace Nafion® 117 membrane in MFC system applications. Also it is a novel PEM for commercialization of MFC.

REFERENCES

Ghasemi, M., Shahgaldi, S., Ismail, M., Kim, B.H., Yaakob, Z. & Daud, W.R.W. (2011) Activated carbon nanofibers as an alternative cathode catalyst to platinum in a two-chamber microbial fuel cell. *International Journal of Hydrogen Energy*, 36, 13,746–13,752.

Ghasemi, M., Shahgaldi, S., Ismail, M., Yaakob, Z. & Daud, W.R.W. (2012) New generation of carbon nanocomposite proton exchange membranes in microbial fuel cell systems. *Chemical Engineering Journal*, 184, 82–89.

Ghasemi, M., Daud, W.R.W., Hassan, H.S.A., Oh, S.-E., Ismail, M., Rahimnejad, M. & Jahim, J.M. (2013a) Nano-structured carbon as electrode material in microbial fuel cells: a comprehensive review. *Journal of Alloys and Compounds*, 580, 245–255.

Ghasemi, M., Daud, W.R.W., Rahimnejad, M., Rezayi, M., Fatemi, A., Jafari, Y., Somalu, M.R. & Manzour, A. (2013b) Copper-phthalocyanine and nickel nanoparticles as novel cathode catalysts in microbial fuel cells. *International Journal of Hydrogen Energy*, 38, 9533–9540.

Ghasemi, M., Daud, W.R.W., Mokhtarian, N., Mayahi, A., Ismail, M., Anisi, F., Sedighi, M. & Alam, J. (2013c) The effect of nitric acid, ethylenediamine, and diethanolamine modified polyaniline nanoparticles anode electrode in a microbial fuel cell. *International Journal of Hydrogen Energy*, 38, 9525–9532.

Ghasemi, M., Daud, W.R.W., Ismail, M., Rahimnejad, M., Ismail, A.F., Leong, J.X., Miskan, M. & Ben Liew, K. (2013d) Effect of pre-treatment and biofouling of proton exchange membrane on microbial fuel cell performance. *International Journal of Hydrogen Energy*, 38, 5480–5484.

Ghasemi, M., Ismail, M., Kamarudin, S.K., Saeedfar, K., Daud, W.R.W., Hassan, S.H.A., Heng, L.Y., Alam, J. & Oh, S.-E. (2013e). Carbon nanotube as an alternative cathode support and catalyst for microbial fuel cells. *Applied Energy*, 102, 1050–1056.

Ghoreishi, K.B., Ghasemi, M., Rahimnejad, M., Yarmo, M.A., Daud, W.R.W., Asim, N. & Ismail, M. (2014) Development and application of vanadium oxide/polyaniline composite as a novel cathode catalyst in microbial fuel cell. *International Journal of Energy Research*, 38, 70–77.

Ilbeygi, H., Ghasemi, M., Emadzadeh, D., Ismail, A.F., Zaidi, S.M.J., Aljlil, A., Jaafar, J., Martin, D. & Keshani, S. (2015) Power generation and wastewater treatment using a novel SPEEK nanocomposite membrane in a dual chamber microbial fuel cell. *International Journal of Hydrogen Energy*, 40, 477–487.

Leong, J.X., Daud, W.R.W., Ghasemi, M., Liew, K.B. & Ismail, M. (2013) Ion exchange membranes as separators in microbial fuel cells for bioenergy conversion: a comprehensive review. *Renewable and Sustainable Energy Reviews*, 28, 575–587.

Lim, S.S., Daud, W.R.W., Md Jahim, J., Ghasemi, M., Chong, P.S. & Ismail, M. (2012) Sulfonated poly(ether ether ketone)/poly(ether sulfone) composite membranes as an alternative proton exchange membrane in microbial fuel cells. *International Journal of Hydrogen Energy*, 37, 11,409–11,424.

Rahimnejad, M., Ghasemi, M., Najafpour, G.D., Ismail, M., Mohammad, A.W., Ghoreyshi, A.A. & Hassan, S.H.A. (2012) Synthesis, characterization and application studies of self-made Fe_3O_4/PES nanocomposite membranes in microbial fuel cell. *Electrochimica Acta*, 85, 700–706.

Shahgaldi S., Ghasemi, M., Daud, W.R.W., Yaakob, Z., Sedighi, M., Alam, J. & Ismail, A.F. (2014) Performance enhancement of microbial fuel cell by PVDF/Nafion nanofibre composite proton exchange membrane. *Fuel Processing Technology*, 124, 290–295.

CHAPTER 26

Performance of polyphenylsulfone solvent resistant nanofiltration membrane: effects of polymer concentration, membrane pretreatment and operating pressure

Nur Aimie Abdullah Sani, Woei Jye Lau & Ahmad Fauzi Ismail

26.1 INTRODUCTION

The presence of trace amount of organic acids such as naphthenic acid in crude oil and distilled fractions could cause corrosion problems damaging pipelines and separation equipment in the oil and gas industry (Yépez, 2007). Therefore, it is desirable to reduce the acidity of crude oil by removing the amount of naphthenic acid. Generally, naphthenic acid is removed by extraction with methanol and then the methanol is recovered using distillation. However, these conventional separation processes are associated with many significant drawbacks such as high energy consumption and need for large amounts of solvents (Subramanian et al., 2004).

Membrane technology especially nanofiltration (NF), experiences increasing attention compared to these separation techniques as it offers many unique advantages. For instance, it consumes less energy and requires no additive during the treatment process, leading to relatively low operating cost (Dobrak et al., 2010). NF of organic solutions also known as solvent resistant nanofiltration (SRNF) is applied to separate compounds dissolved in solvents with molecular weight (MW) in the range from 200 to 1400 Da with simultaneous passing of the organic solvent through the membrane.

It must be pointed out that the applications of NF membrane in organic solution are not very successful compared to their uses in aqueous solution. The use of polymeric membranes in SRNF has been employed by a growing number of researchers (Machado et al., 1999; Soroko et al., 2011; Tsarkov et al., 2012; Van der Bruggen et al., 2002a; 2002b), however, these membranes show severe performance loss due to their chemical instability in organic solvents. Among the problems, infinite flux caused by either the membrane swelling or dissolving (Raman et al., 1996), zero flux due to membrane collapse (Raman et al., 1996), poor selectivity or rejection (Subramanian et al., 1998) and membrane performance deterioration as a function of filtration time are always reported (Bridge et al., 2002).

Several solutions have been proposed in the literature to overcome the recurrent problems. Aerts et al. (2006) have prepared plasma modified polydimethylsiloxane (PDMS) membrane subjected for the separation of dyes in alcohol solvents such as methanol, ethanol and isopropanol. Bitter et al. (1988) have used halogen-substituted silicon rubber for the separation of solvents from hydrocarbons dissolved in the toluene and methyl ethyl ketone. However, both approaches were not practical for industrial applications due to the significant flux reduction after a short period of operation. In view of this, the aim of this study is to evaluate the third member of the polysulfone family – polyphenylsulfone (PPSF) for organic solvent NF application. In this work, the properties of PPSF NF membranes were investigated by varying the polymer concentration in dope solution, the duration of membrane pretreatment using methanol solvent and the operating pressure during organic solvent application. The influences of these three variables on the performance of the membranes were characterized with respect to methanol flux and dye rejection (Reactive Black 5, RB5). Performance comparison was also made to compare the performances of in-house made PPSF membranes with a commercially available SRNF membrane, i.e. NF030306F (SolSep BV, CA).

Table 26.1.   Condition for PPSF membranes pretreatment prior to testing.

Designation	Solvent used in pretreatment	Duration
Case 1	Methanol	5 min
Case 2		1 day

26.2   EXPERIMENTAL

26.2.1   *Materials*

PPSF polymer pellets with MW of 11,044 gmol^{-1} purchased from Sigma-Aldrich, Malaysia was used as the main membrane forming material for SRNF preparation. N-methyl-2-pyrrolidinone (NMP) obtained from Merck, Malaysia was used as solvent to dissolve the polymer in pellet form. Methanol (analytical grade >99%) used in this study was supplied by Merck, Malaysia and is the common solvent used in deacidification of crude oil. Reactive Black 5 (RB5) with MW of 991 g mol^{-1} from Sigma-Aldrich, Malaysia was selected to represent naphthenic acid which has MW in the range of 180–2400 g mol^{-1}.

26.2.2   *Preparation of membranes*

PPSF membrane was prepared by dissolving a pre-weighed quantity of the polymer pellets in NMP at room temperature to form 17, 21 and 25 wt% polymer dope solution. Then, the solution was stirred overnight to ensure complete polymer dissolution. The solution was left at least 24 h to remove air bubbles before it was used for membrane casting. The polymer solution was cast on a glass plate without any non-woven support using a casting knife at room temperature. It was subsequently immersed in a non-solvent bath of tap water and kept for 24 h. The membrane was subjected to air drying process at room temperature for at least 24 h prior to use. The morphology of prepared membrane was observed by scanning electron microscope (SEM) (TM3000, Hitachi, Japan). Samples of SEM analysis were prepared by fracturing the membrane in liquid nitrogen. The commercial membrane used for the comparison purpose was NF030306F manufactured by SolSep BV, CA and was supplied in dry form. This membrane was made of silicone polymer.

26.2.3   *Experimental procedure*

The separation performances of the membranes were carried out using dead-end stirred cell (Sterlitech HP4750, Sterlitech Corporation, USA) with maximum capacity of 300 mL. The active membrane area was approximately 14.6 cm^2. A nitrogen cylinder equipped with a two-stage pressure regulator was connected to the top of the stirred cells to supply the desired pressure for filtration experiments. In order to minimize concentration polarization during the experiments, a Teflon-coated magnetic stirring bar was used and was controlled at 700 rpm on top of the active side of membrane.

To study the effect of membrane pretreatment process on membranes performance, the membranes were first subjected to two different pretreatment conditions as shown in Table 26.1 followed by compaction at pressure of 3.0 MPa for about 1 h. The experiment was then performed at 0.5 and 2.5 MPa using pure methanol.

To study the effect of operating pressure on membrane flux and rejection, the pressure was varied between 0.5 and 2.5 MPa and before the experiment, the membrane was compacted at pressure of 2.6 MPa with the tested solvent for about 1 h. The permeate was collected after 30 min of experiment when flux had achieved steady-state and measured every 10 min for up to 2 h. The flux (J) [L h^{-1} m^{-2}] was calculated by the following equation where V, A and t are permeate

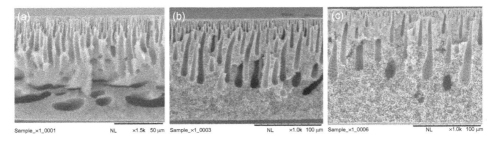

Figure 26.1. SEM images of membranes prepared from different polymer concentrations, (a) 17 wt%, (b) 21 wt% and (c) 25 wt%.

volume, membrane area and time, respectively. Three flux measurements were made and the average value was reported.

$$J = \frac{V}{At} \tag{26.1}$$

With respect to dye rejection determination, the experiment was carried out by filtering methanol containing RB5 at initial concentration of $10 \, \text{mg} \, \text{L}^{-1}$. The rejection rate, R [%] of the dyes by the membranes was calculated using:

$$R\,[\%] = \left(1 - \frac{C_p}{C_f}\right) \times 100 \tag{26.2}$$

where c_p is the dye concentration $[\text{mg} \, \text{L}^{-1}]$ of permeates and c_f is the initial concentration $[\text{mg} \, \text{L}^{-1}]$. The dye concentrations in the permeate and feed stream were measured using UV-vis spectrophotometer (DR5000, Hach Company, USA). A blank wavelength scan with pure methanol was first performed prior to permeate sample analysis. The wavelength of maximum absorbance (λ_{max}) for RB5 is 592 nm.

26.3 RESULTS AND DISCUSSION

26.3.1 *Effect of polymer concentration on membrane morphology and methanol flux*

Figure 26.1 shows the SEM images of the cross sectional structure of the PPSF membrane prepared at a different polymer concentration i.e. 17, 21 and 25 wt%. Generally, an integral asymmetric structure with finger-like macrovoids supported by a spongy structure is often observed in flat-sheet type polymeric membranes fabricated from the phase inversion method. It can be seen that with increasing the PPSF concentration from 17 to 25 wt%, the morphology of the membrane became less porous with less finger-like pores and macrovoids. It is found that the sponge-like structure was dominant in 25 wt% PPSF membrane compared to the significant amount of finger-like structure observed in membranes prepared from 17 and 21 wt% PPSF. Typically, viscosity of polymer solution increases exponentially with polymer concentration, which will delay the exchange rate between the solvent and water during the polymer precipitation process, leading to a higher polymer concentration at the interface between the polymer solution and the non-solvent (water). This as a result, develops a membrane with less porous morphology coupled with less finger-like pores and microvoids, as evidenced in this study (Darvishmanesh *et al.*, 2011a; 2011b).

Figure 26.2 shows the influence of the polymer concentration on the methanol flux at operating pressure of 0.5 MPa. As can be seen, the flux decreased as the polymer concentration in the dope solution increased. The PPSF membrane prepared with 17 wt% polymer concentration has the highest flux ($16.77 \, \text{L} \, \text{m}^{-2} \, \text{h}^{-1}$) followed by 21 wt% ($12.38 \, \text{L} \, \text{m}^{-2} \, \text{h}^{-1}$) and 25 wt% ($0.80 \, \text{L} \, \text{m}^{-2} \, \text{h}^{-1}$). This phenomenon can be explained by the increase in the entire membrane resistance, which mainly resulted from reduced surface pore size and suppression of finger-like

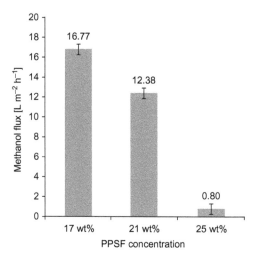

Figure 26.2.   Effect of polymer concentration on methanol flux.

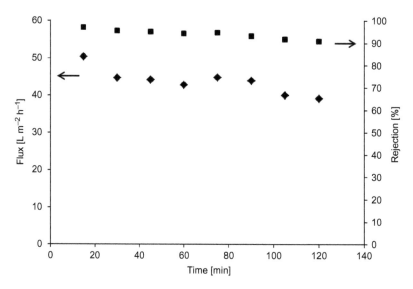

Figure 26.3.   Methanol flux and RB5 rejection of 17 wt% PPSF membrane as a function of time (conditions: initial feed concentration $= 10\,mg\,L^{-1}$ in methanol, pressure: 0.5 MPa).

microvoids. Compared to the 17 and 21 wt% PPSF membrane, in which the cross sectional structure was dominated by finger-like microvoids, the development of sponge-like morphology as shown in the 25 wt% PPSF membrane has played a role in increasing solvent transport resistance and reducing methanol productivity.

26.3.2   *Performance of PPSF membrane in RB5 removal*

To assess the performance of the as-prepared membranes in SRNF, 17 wt% of PPSF membranes was tested using methanol solution containing $10\,mg\,L^{-1}$ of RB5 at the pressure of 0.5 MPa and the results are shown in Figure 26.3. As observed, both flux and rejection tended to decrease with filtration time. The decrease of methanol flux from initial 50.3 to $39.1\,L\,m^{-2}\,h^{-1}$ at the end

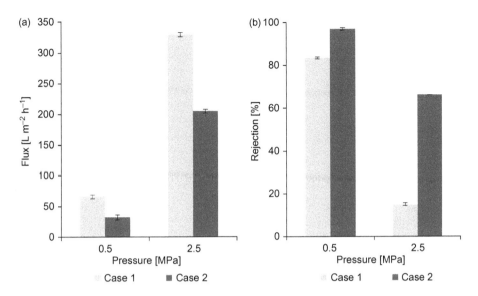

Figure 26.4.   Effect of pretreatment process on PPSF membrane performances: (a) flux of methanol and (b) rejection of RB5.

of the experiment was caused by the pressure compaction on the membrane. This phenomenon can be related to the finger-like macrovoids structure of the membrane as shown in Figure 26.1a. As suggested by Persson *et al.* (1995), the finger-like macrovoids structure is more affected by compaction than a sponge-like structure. The compaction preferentially occurs in the bulk layer where most of the pore volume, i.e. large pores and macrovoids are situated (Jonsson, 1977). Another possible explanation of the reduction of flux is the pore blockage by solvent molecules, and similar observation has been reported elsewhere (Whu *et al.*, 2000; Yang *et al.*, 2001). In comparison to membrane flux, it was found that the rejection ability of PPSF membrane was not significantly affected throughout the entire experimental period. Its rejection remained in the range of 90.8–96.9%.

26.3.3   *Effect of pretreatment process duration on membrane performance*

Membrane pretreatment process plays an important role in membrane performance for non-aqueous systems. Pretreating the membrane before testing becomes critical because the solvent-membrane interactions significantly affect the performance of the membrane. The purpose of pretreating the membrane with organic solvent is to make the membrane in stable condition prior to any experiment. It is because the sudden exposure of membrane to the solvent of filtration may result in inconsistent flux and sudden swell of the membrane. Figure 26.4 presents the effect of pretreatment conditions on the solvent flux and dye rejection of 17 wt% PPSF membrane at two different operating pressures. As shown, fluxes of methanol significantly changed when the membranes were pretreated with different treatment conditions. A longer period of immersion causes the membranes to be more resistant towards methanol flux probably due to the swelling of the membrane structure. The rejection of dye was measured to evaluate the effect of the pretreatment process on membrane pore size. The significant increase in the dye rejection from Case 1 to 2 indicates decrease in the pretreated membrane pore size. Ebert and Cuperus (1999) reported that when a porous membrane swells, the pores become narrower and therefore increases the membrane rejection. With respect to the effect of pressure on the performance of the pretreated membrane, it was found that the methanol flux and the dye rejection were increased as the pressure

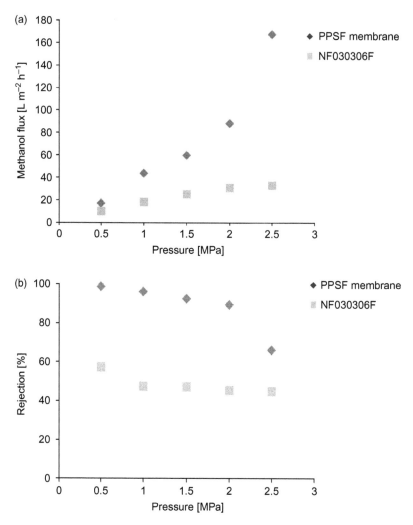

Figure 26.5. Performance comparison between the in-house made PPSF membrane and the commercial
SRNF membrane as a function of operating pressure, (a) permeate flux and (b) RB5 rejection.

increased from 0.5 to 2.5 MPa. The highest methanol flux and dye rejection were obtained at
2.5 MPa by case 2 membrane (328.8 L m^{-2} h^{-1}) and case 1 membrane (96.9%), respectively.

26.3.4 *Effect of operating pressure on membrane performance: comparison of PPSF with commercial membranes*

The effect of operating pressure on membrane performance was examined in the range of 0.5 to
2.5 MPa at room temperature. Figure 26.5 compares the performances of 17 wt% PPSF membrane
with the commercially available SRNF membrane – NF030306F. The experimental results showed
that the methanol flux of both membranes increased by increasing the operating pressure. For
PPSF membrane, the flux increased from 16.8 to 167.2 L m^{-2} h^{-1} when increasing pressure from
0.5 to 2.5 MPa. This flux increment is 1.7–5.1 times greater than the one reported for commercial NF030306F membrane. The greater flux of PPSF membrane could be due to its greater
hydrophilicity (contact angle: 63.9 ± 1.7°) compared to NF030306F membrane (86 ± 1.2°).

It was previously reported that the lower solvent flux of a hydrophobic membrane is due to the decrease in surface energy when a polar solvent like methanol is used (Machado *et al.*, 1999).

With respect to the membrane separation performance at different operating pressure, it was found that for both membranes, the rejection of dye was decreased with the increase in operating pressure. Obviously, the rejection of PPSF membrane was much greater than that of NF030306F membrane, irrespective of the pressure applied. The higher separation ability of PPSF membrane can be explained by the fact that the membrane possesses much smaller molecular weight cut off (MWCO) in comparison to the NF030306F membrane. It must be also noted that the excellent separation rate of PPSF membrane was negatively affected when it was tested at high operating pressure and this decreasing trend is likely caused by the permeation of dye molecules through the membrane matrix and/or enlargement of pore structure at high filtration pressure.

26.4   CONCLUSION

In the present study, PPSF membranes of different properties were successfully prepared using the phase inversion method. The effect of polymer concentration, membrane pretreatment and operating pressure on membrane performance were investigated. An increase in polymer concentration from 17 to 25 wt% caused the membrane flux to decrease, owing to the increase in the entire membrane resistance, which mainly resulted from reduced surface pore size and suppression of finger-like macrovoids. With respect to pretreatment conditions, it is reported that different immersion periods in methanol did affect the properties of PPSF membrane during filtration process. The longer the immersion period the more resistant becomes the membrane towards methanol flux probably due to the swelling of the membrane structure. Increasing operating pressure from 0.5 to 2.5 MPa was also reported to negatively affect the separation performance of PPSF membrane against RB5, although solvent flux was increased. This decreasing trend of rejection could be due to the permeation of dye molecules through the membrane matrix to the permeate side and/or enlargement of pore structure when membrane was operated at high filtration pressure.

ACKNOWLEDGEMENT

The authors gratefully acknowledge the financial support of this work by the Ministry of Education (MOE) under Long-term Research Grant Scheme with grant no. of R.J130000-7837.4L803. N.A.A. Sani alsothanks the MOE for the MyBrain15 (MyPhD) sponsorship received during her PhD studies.

REFERENCES

Aerts, S., Vanhulsel, A., Buekenhoudt, A., Weyten, H., Kuypers, S., Chen, H., Bryjak, M., Gevers, L.E.M., Vankelecom, I.F.J. & Jacobs, P.A. (2006) Plasma-treated PDMS-membranes in solvent resistant nanofiltration: characterization and study of transport mechanism. *Journal of Membrane Science*, 275, 212–219.

Bitter, J.G., Haan, J.P. & Rijkens, H.C. (1988) Process for the separation of solvents from hydrocarbons dissolved in the solvents. US Patent, 4,748,288.

Bridge, M.J., Broadhead, K.W., Hlady, V. & Tresco, P.A. (2002) Ethanol treatment alters the ultrastructure and permeability of PAN-PVC hollow fiber cell encapsulation membranes. *Journal of Membrane Science*, 195, 51–64.

Darvishmanesh, S., Tasselli, F., Jansen, J.C., Tocci, E., Bazzarelli, F., Bernardo, P., Luis, P., Degrève, J., Drioli, E. & Van der Bruggen, B. (2011a) Preparation of solvent stable polyphenylsulfone hollow fiber nanofiltration membranes. *Journal of Membrane Science*, 384, 89–96.

Darvishmanesh, S., Jansen, J.C., Tasselli, F., Tocci, E., Luis, P., Degrève, J., Drioli, E. & Van der Bruggen, B. (2011b) Novel polyphenylsulfone membrane for potential use in solvent nanofiltration. *Journal of Membrane Science*, 379, 60–68.

Dobrak, A., Verrecht, B., Van den Dungen, H., Buekenhoudt, A., Vankelecom, I.F.J. & Van der Bruggen, B. (2010) Solvent flux behavior and rejection characteristics of hydrophilic and hydrophobic mesoporous and microporous TiO$_2$ and ZrO$_2$ membranes. *Journal of Membrane Science*, 346, 344–352.

Ebert, K. & Petrus Cuperus, F. (1999) Solvent resistant nano-filtration membranes in edible oil processing. *Membrane Technology*, 1999, 5–8.

Jonsson, G. (1977) Methods for determining the selectivity of reverse osmosis membranes. *Desalination*, 24, 19–37.

Machado, D.R., Hasson, D. & Semiat, R. (1999) Effect of solvent properties on permeate flow through nanofiltration membranes. Part I. Investigation of parameters affecting solvent flux. *Journal of Membrane Science*, 163, 93–102.

Persson, K.M., Gekas, V. & Trägårdh, G. (1995) Study of membrane compaction and its influence on ultrafiltration water permeability. *Journal of Membrane Science*, 100, 155–162.

Raman, L., Cheryan, M. & Rajagopalan, N. (1996) Deacidification of soybean oil by membrane technology. *Journal of the American Oil Chemists' Society*, 73, 219–224.

Soroko, I., Lopes, M.P. & Livingston, A. (2011) The effect of membrane formation parameters on performance of polyimide membranes for organic solvent nanofiltration (OSN). Part A. Effect of polymer/solvent/non-solvent system choice. *Journal of Membrane Science*, 381, 152–162.

Subramanian, R., Nakajima, M. & Kawakatsu, T. (1998) Processing of vegetable oils using polymeric composite membranes. *Journal of Food Engineering*, 38, 41–56.

Subramanian, R., Nakajima, M., Raghavarao, K.S.M.S. & Kimura, T. (2004) Processing vegetable oils using nonporous denser polymeric composite membranes. *Journal of the American Oil Chemists' Society*, 81, 313–322.

Tsarkov, S., Khotimskiy, V., Budd, P.M., Volkov, V., Kukushkina, J. & Volkov, A. (2012) Solvent nanofiltration through high permeability glassy polymers: effect of polymer and solute nature. *Journal of Membrane Science*, 423, 65–72.

Van der Bruggen, B., Geens, J. & Vandecasteele, C. (2002a) Fluxes and rejections for nanofiltration with solvent stable polymeric membranes in water, ethanol and *n*-hexane. *Chemical Engineering Science*, 57, 2511–2518.

Van der Bruggen, B., Geens, J. & Vandecasteele, C. (2002b) Influence of organic solvents on the performance of polymeric nanofiltration membranes. *Separation Science and Technology*, 37, 783–797.

Whu, J.A., Baltzis, B.C. & Sirkar, K.K. (2000) Nanofiltration studies of larger organic microsolutes in methanol solutions. *Journal of Membrane Science*, 170, 159–172.

Yang, X.J., Livingston, A.G. & Freitas dos Santos, L. (2001) Experimental observations of nanofiltration with organic solvents. *Journal of Membrane Science*, 190, 45–55.

Yépez, O. (2007) On the chemical reaction between carboxylic acids and iron, including the special case of naphthenic acid. *Fuel*, 86, 1162–1168.

CHAPTER 27

Application of membrane technology in degumming and deacidification of vegetable oils refining

Noor Hidayu Othman, Ahmadilfitri Md Noor, Mohd Suria Affandi Yusoff,
Ahmad Fauzi Ismail, Pei Sean Goh & Woei Jye Lau

27.1 INTRODUCTION

World consumption of oils and fats has grown steadily during last 25 years. The 'Oil World' statistics indicate that oils and fats consumption has increased from 152 million metric tons in 2007 to 183 million tons in 2012 (Basiron, 2013). Vegetable oils and fats are not only used for human consumption, but also in animal feed, medical purposes and certain technical applications. Production of the major oils, derived from palm, soybean, rapeseed and sunflower seed, grew by over 70% and accounted for 88% of the increase in world output of all oils and fats (R.E.A., 2013). According to the 2012 revision of the official United Nations population estimates and projections, the world population of 7.2 billion in mid-2013 is projected to increase by almost one billion people within the next twelve years, reaching 8.1 billion in 2025, and to further increase to 9.6 billion in 2050 and 10.9 billion by 2100 (UN, 2013). Thus, there will be new challenges to the food producers and processors to develop new technologies, processing methods, and agricultural techniques to meet food demands of a growing population.

The vegetable oils industry involves the extraction and processing of oils and fats from a variety of fruits, seeds, and nuts. The objectives of refining are to remove the undesirable constituents (e.g.: FFA, PLs, metal ions, color pigments and others) from crude oils, to preserve valuable vitamins (e.g., tocopherols and tocotrienols – natural antioxidants), to protect the oils against degradation, and ensuring a good quality and stability of end product (Lin and Koseoglu, 2005). However, several drawbacks are identified from conventional refining processes such as high energy usage, losses of neutral oil, large amounts of water and chemical are used and heavily contaminated effluents are produced (Gibon *et al.*, 2007).

Membrane technology is a mature industry and has been successfully applied in various food industries for separation of undesirable fractions from valuable components. Commercial membrane separation processes are offered in the areas of nitrogen production and waste-water treatment application. The developing membrane application in vegetable oils processing includes solvent recovery, degumming, deacidification, pigment removal, wax removal and extraction of minor components (tocopherols and carotenoids) (Cheryan, 2005).

The application of membrane technology in vegetable oils refining has been receiving attention in the light of various inherent advantages associated with membrane processes. It seems to be a simple and promising tool to refine vegetable oils due to low energy consumption, ambient temperature operation, no addition of chemicals, and retention of nutrients and other desirable components that contributes to cost and energy effectiveness and eco-friendliness (de Morais Coutinho *et al.*, 1990; Koseoglu and Engelgau, 1990; Raman *et al.*, 1994a).

27.1.1 *Membrane technology and its basic principles*

Membranes can be defined as semipermeable barriers that separate two phases and restrict the transport of various substances in a specific way. The primordial function of a membrane is to act as a selective barrier, allowing the passage of certain components and the retention of others from a determined mixture, implying the concentration of one or more components both in the permeate and in the retentate (Ladhe and Kumar, 2010). Its selectivity is related to the dimensions of the molecule or particle of interest for separation and the pore size, as also the solute diffusivity in the matrix and the associated electric charges (Strathmann, 1990).

In the chemical process industry one often encounters the problem of separating a mixture in its components. Membranes can in principle carry out most of the separation processes and can complement or form an alternative for processes like distillation, extraction, fractionation, adsorption etc. The basic membrane systems commercially available include microfiltration (MF), ultrafiltration (UF), nanofiltration (NF), and reverse osmosis (RO), each of which depending on the nature of particle or on the molecular size of the solutes that are separated. Generally, MF addresses the largest macromolecules from 200,000 to 1 million molecular weight (MW) (500–2 million angstrom units); UF with middle-range molecules from 10,000–300,000 MW (40–2000 angstrom units); NF with small molecules of 15,000–150 MW (80–80 Angström units); and RO with ions and smallest molecules up to 600 MW (20 Angström units) (Lin and Koseoglu, 2005).

The success of using membranes is closely related to the intrinsic properties of the membrane. Interfacial interactions between membrane surface, surrounding environment and solutes govern membrane performance to a great extent (Cheryan, 1998). There are four essential features required for the efficient operation of a membrane process: (ii) selectivity which determines product purity, (ii) flux which determines the product throughput, (iii) rate of separation which determines the equipment size and (iv) durability/ease of cleaning which determines the operating cost. The use of membranes in industrial applications is growing at a rapid rate. Equipment suppliers and academicians have generated a great deal of information about the performance, capabilities, and applications of membrane systems (Cheremisinoff, 2002).

The factor that distinguishes between the common membrane separation processes of MF, UF, NF and RO is the application of hydraulic pressure as the driving force for mass transport. Nevertheless the nature of the membrane controls which components will permeate and which will be retained, since they are selectively separated according to their molar masses or particle size (Strathmann, 1990). In a conventional filtration system, the fluid flow, be it liquid or gaseous is perpendicular to the membrane surface, such that solutes deposit on it, requiring periodic interruption of the process to clean or substitute the filter. In tangential membrane filtration the fluid flow is parallel to the membrane surface, and the solutes that tend to accumulate on the surface are stripped away due to the high velocity, making the process more efficient (Muralidhara, 2010).

27.2   VEGETABLE OILS AND DEGUMMING

27.2.1   *Overview and principle of degumming process*

Crude oil obtained by screw pressing and solvent extraction of oilseeds will throw a deposit of so-called gums on storage. The chemical nature of these gums has been difficult to determine. These gums consist mainly of phosphatides but also contain entrained oil and mill particles. They are formed when the oil absorbs water that causes some of the phosphatides to become hydrated and thereby oil-insoluble. Accordingly, hydrating the gums and removing the hydrated gums from the oil before storing the oil can prevent the formation of a gum deposit (Paulson *et al.*, 1984). Five basic principles of degumming processes are described as water degumming, dry degumming, acid degumming, acid refining and soft degumming as shown in Figure 27.1.

Water degumming is the oldest degumming treatment and also forms the basis of the production of commercial lecithin. It is the simplest method for phosphatide reduction where the hot water is mixed into warm oil (70–90°C). Hydratable phospholipids (PLs) absorb water to agglomerate into a gum phase, which is, after a certain holding time, separated in separators or decanters. The procedure may be sufficient for some types of vegetable oils, e.g. 99% of PLs are removed by heating oil to 80–85°C with water and then filtering at 55–60°C (Deffense, 2011a). The classic

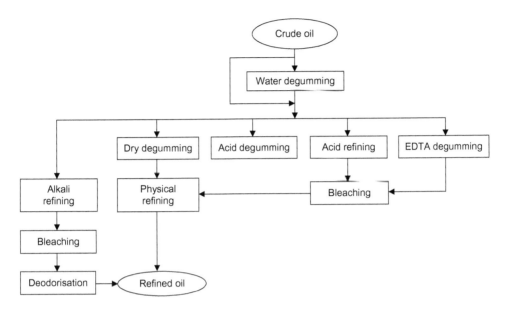

Figure 27.1. Overview of degumming routes (Deffense, 2011b).

water degumming process leads to a considerable loss of neutral oil, large amount of wastewater and energy consumption. Besides, non-hydratable salts of phosphatidic acids are not removed. In most cases, it is not sufficiently efficient for the preliminary purification of oils before physical refining (Vintilă, 2009).

Dry degumming process is mainly for oils and fats that are low in phosphatides (20 mg kg^{-1} of phosphorus) such as palm oil, palm kernel oil and olive oil. In physical refining of palm oil, the crude palm oil (CPO) is mixed with 0.04–0.1% phosphoric acid (concentration 85%) for about 10–20 min. Then 1–2% bleaching earth is added under vacuum at a temperature of 80–120°C. After a suitable contact time, the spent bleaching earth is removed by filtration. The function of phosphoric acid is to disrupt non-hydratable phosphatides by decomposing magnesium and calcium complexes. Then, it coagulates the phosphatides, sequestrates iron and copper, before being removed by adsorption on bleaching earth (Molík and Pokorný, 2000). Citric acid can also be used but due to economic reasons, phosphoric acid is mainly used by Malaysia refiners. The amount of acid introduced into the oil is quite critical. A low dosage of phosphoric acid may lead to the oil darkening during deodorization and off-flavor problems due to phosphatides breakdown caused by thermal instability (Deffense, 2011b). An excess of phosphoric acid also promotes darkening of refined oil due to the reverse reaction of phosphatidic acid (PA) into phosphatides (Patterson, 1992).

Acid degumming processes usually lead to lower residual phosphorus and iron contents, than those of water degumming. It is preferred for processing rapeseed and sunflower oils. In acid degumming; phosphoric acid was used as it forms no dissociable phosphates with calcium, iron, or magnesium ions. The use of 0.05–0.20% phosphoric acid was found satisfactory for the degumming of palm oil but not for most other vegetable oils (Thiagarajan and Tang, 1991). For degumming of soybean oil, the application of phosphoric acid and water was recommended, first with a concentrated acid and then with diluted acid (Faessler, 1998). Citric acid is an alternative to phosphoric acid. It is more expensive than phosphoric acid, but the removal of non-hydratable phospholipids is more efficient. Heating of crude rapeseed oil with 0.1% citric acid to 70°C with removal of the precipitate by centrifugation resulted in degummed oil containing only 2 mg phosphorus and 0.3 mg iron in 1 kg oil (Aly, 1992). Sodium citrate was reported as more efficient for removing calcium, magnesium, or iron from non-hydratable phosphatides (NHP) than free citric acid (Segers, 1994).

Acid refining is applicable for both water degummed oil and oils that have not yet been degummed (Molík and Pokorný, 2000). Similar to acid degumming, non-hydratable phospholipids are firstly decomposed by a sufficiently strong acid. Instead of diluting the degumming acid with water, the hydratability of phospholipids is achieved by the addition of a base, such as sodium hydroxide, caustic soda, or sodium silicate. Only partially degumming acid was neutralized to prevent soap formation (Vintilă, 2009). Various processes have been developed based on partial neutralization of PA by sodium hydroxide without touching the free fatty acids such as TOP degumming by Vandemoortele (original process) (Dijkstra and Opstal, 1989); Uni-degumming by Unilever as a post-treatment like in the TOP degumming has been added to the Super degumming (Sande van de and Segers, 1989); special degumming developed on crude oils by Alfa Laval (Johansson *et al.*, 1988); UF degumming using lower temperature than in the TOP and low agitation by Krupp (Rohdenburg *et al.*, 1993); and the Impac degumming using additionally a wetting agent by De Smet (2013). TOP is a Dutch acronym derived from "Totaal Ontslijmings Process" meaning total degumming process.

Soft degumming is another option to remove phosphatides from crude palm oil with high phosphatide and iron contents (Gibon and Tirtiaux, 1998). In this case, the oil is heated (75–85°C) and mixed with a water solution (2%) containing a complexing molecule [ethylenediaminetetracetic acid (EDTA)] and, eventually, a wetting compound [sodium dodecyl sulfate (SDS)]. EDTA is able to form strong chelates with magnesium, calcium and iron. As a consequence, non-hydratable phospholipids will be more easily removed. After 20 min of retention time followed by centrifugation, less than 1 mg kg^{-1} of phosphorus can be achieved. Iron is reduced to below 0.1 mg kg^{-1} as it forms an oil-insoluble complex with EDTA (Lin and Koseoglu, 2005). The efficiency of the process depends on the degree of dispersion and contact between the chelating agent (EDTA) and the NHP. A detergent, sodium lauryl sulfate (SLS) is used to facilitate the contact between the NHP in the oil phase and the water phase solution containing the chelating agent. However the separation of both phases was made more difficult due to the high stability of the emulsion (Jamil *et al.*, 2000).

27.2.2   *Membrane degumming of vegetable oils*

Membrane separation is primarily a size-exclusion based pressure-driven process. It separates different components according to the molecular weights or particle sizes and shapes of individual components. Sometimes it depends on their interactions with the membrane surfaces and other components of the mixture (Cheryan, 1986). The performance of membrane separation is affected by membrane composition, temperature, pressure, velocity of flow and interactions between components of the feedstock and with the membrane surface (Subramanian and Nakajima, 1997).

Theoretically, PLs and triacylglycerols (TG) have almost similar molecular weights (MW) of about 700 Da and 900 Da, respectively which makes them difficult to be separated by a membrane. However PLs are surfactants in nature, having both hydrophilic and hydrophobic groups. It tends to form reverse micelles with MW of 20 kDa or more in non-aqueous systems such as vegetable oils and oil/hexane micelle that enable their separation by using of appropriate membrane (Gupta, 1977; Lin *et al.*, 1997; Ochoa *et al.*, 2001; Pagliero *et al.*, 2001; Subramanian *et al.*, 2001).

According to the previous researches on the processing of vegetable oils using membrane technology, the degumming process can be conducted using 3 categories of membranes; polymeric membrane, nonporous polymeric composite membrane and ceramic membrane.

27.2.3   *Membrane degumming using polymeric membrane*

In the last decade, the use of membrane in the degumming of vegetable oils has been a subject of research. Some of the alternatives are based on removal of the PLs from the micelle (oil/hexane) and others from solvent-free oil. Sen Gupta in his pioneering work demonstrated that ultrafiltration (UF) membranes could be used for degumming hexane-oil miscella (Gupta, 1977). Since then, many researchers from various countries followed this technique using porous and non-porous membrane with different types of source oils.

Different types of polymeric materials were studied including polyvinylidenfluoride (PVDF), polyethersulfone (PES), polysulfone (PSF) and polyimide (PI) in crude soybean oil filtration. The UF cell resulted up 99% PLs removal by PVDF membrane. A sharp initial decrease in permeate flux especially by PSF membrane explained the membrane behavior in term of concentration polarization effects and internal fouling (Pagliero *et al.*, 2001). Similar behavior was observed in a study with PVDF and polyimide (PI) membranes in the same UF cell. PVDF membrane allowed larger pure hexane fluxes than PI at the same transmembrane pressure in the whole temperature range analyzed. Further study has been made to minimize the surface fouling between the membrane structure and membrane-solvent interaction effects. PVDF and PI membranes gave selectivity values and permeate color that did not differ significantly from each other. The PLs rejections more than 98% were obtained in all cases while PVDF membrane had higher productivity than other membranes (Ochoa *et al.*, 2001; Pagliero *et al.*, 2001; 2004).

However, under the influence of the solvent, micelle can greatly increase the effective particle size of the PLs and enable them to be wholly retained by membrane. Due to hydrophobicity of PI membrane, lower fluxes are indicated in the case of relatively polar solvents, such as water and alcohols. Higher fluxes are evident in the case of non-polar solvent, such as hexane and acetone. Various solvents (water, alcohols, acetone and hexane) fluxes were measured with PLs that were rejected above 90% (Kim *et al.*, 2002).

Besides soybean oil, cottonseed and rice bran oil, the removal of PLs was also studied in sunflower oil using a cross-flow UF membrane. Two tubular PES membranes with different MWCO (4000 and 9000 Da) were used for degumming of sunflower oil micelle. Both membranes presented approximately the same rejection of phospholipids (95–97%), although the 9000 Da membrane showed higher miscella permeate flux, lower oil rejection and higher FFA rejection (García *et al.*, 2006).

The flat sheet PES UF membrane was also investigated for their ability to selectively separate PLs miscella from the desolventized crude soybean oil and from the crude soybean oil/hexane (miscella). The maximum of 77% and 85% PLs were recorded from the disolventized crude soybean oil and oil/hexane micelle, respectively. The use of miscella presents an important advantage to the industrial level, where system productivity is measured mainly by permeate flux and solute (phospholipids) removal (de Moura *et al.*, 2005).

27.2.4 *Membrane degumming using nonporous polymeric composite membrane*

Nonporous membranes offer several advantages over UF membranes, namely higher rejection of PLs, carotenoids and chlorophylls. Recently, the use of nonporous polymeric composite membrane was actively investigated for processing of hexane/oil micelle.

Flat sheet polymeric composite membranes (NTGS-2200) with silicon as an active layer and PI as a support layer from Nitto Denko, Japan were used to degum crude rice bran oil and water-degummed soybean oils after dilution with hexane at required pressure of 3 MPa. During membrane processing, the reduction in PLs varied from 93.8 to 96.4% and from 63.9 to 77.3% in rice bran and soybean oils, respectively (Saravanan *et al.*, 2006).

Similar polymeric composite membrane (NTGS-2200) was used in processing of CPO and palm olein under both undiluted and hexane-diluted conditions. The membrane showed very high selectivity between TG and PLs even under hexane-diluted conditions and the PLs rejection was nearly complete (95–100%). However the oil flux was very low in both undiluted (0.12–0.14 kg m^{-2} h^{-1}) and hexane diluted (0.9–1.1 kg m^{-2} h^{-1}), making it industrially not applicable (Arora *et al.*, 2006).

The efficacy of non-porous membrane was further investigated using two composite membranes; NTGS-2200 and NTGS-2100 with polydimethylsiloxane (PDMS) as active layer and PI as support layer. Both membranes showed very high reduction (above 99%) of PL content in undiluted crude oils (Subramaniana *et al.*, 2004). However, due to very low oil flux, hexane-diluted condition was applied for processing of crude rice bran oil. The percentage rejection of PL varied from 95.1 to 98.8%, which indicated that the membrane posed a very high selectivity for PL even under hexane-diluted (Manjula and Subramanian, 2009; Subramaniana *et al.*, 2004).

Generally, the PL content is well above the critical micelle concentration (CMC) in crude solvent-extracted oils. In such systems, size exclusion may provide a synergistic effect in the rejection (Subramaniana *et al.*, 2004). The nonporous polymeric composite of hydrophobic membrane (NTGS-2200) was very effective in rejecting PLs (98–100%) in crude rice bran oil and groundnut oil. Even in hexane-diluted conditions, the PLs rejection by the membrane was nearly as effective as that of undiluted systems (Manjula *et al.*, 2011).

27.2.5   *Membrane degumming using ceramic membrane*

Ceramic membrane is an inorganic membrane. It shows great mechanical resistance and supports high pressures and can tolerate the entire pH range (0–14), high temperatures of over 400°C. It is also chemically inert (Cuperus and Nijhuis, 1993). A tubular ceramic membrane (MWCO: 15 kDa) made of ZrO_2 was used to treat a crude soybean oil-hexane micelle using UF module 92–93% of PLs were successfully rejected with a 25% crude soybean oil-hexane mixture (Marenchino *et al.*, 2006).

Direct membrane filtration to pretreat CPO was studied using ceramic MF membranes (alpha alumina: 0.45 μm and 0.2 μm) and UF membranes (Zirconia: 50 nm and 20 nm). The MF ceramic membranes and UF ceramic membranes were able to reject 14–56.8% PLs and 60–78.1% PLs respectively as compared to that of bleached oil with 34.4% PLs rejection. Although the PLs rejection is not higher, the membrane pre-treatment technique showed better prospect than conventional bleaching method (Iyike *et al.*, 2004).

Corn oil is distinct from other vegetable oils since it has a higher content of waxes that require removal. A multichannel ceramic membrane constructed from alumina (support and filter element) was used for degumming a crude corn oil and crude soybean oil in an oil/hexane micelle, respectively. The membrane was initially conditioned with a solvent of high polarity and continuing up to a nonpolar solvent using pure water, ethyl alcohol, hexane and a combination of binary solutions before filtration process. The ceramic membrane showed good performance in the filtration of both corn oil/hexane miscella and soybean oil/hexane micelle with the maximum rejection of 93.5% PLs and 99% PLs, respectively (de Souza *et al.*, 2008; Ribeiro *et al.*, 2008).

Another attempt for direct membrane filtration using ceramic UF membrane with relatively small in MWCO (6 kDa or smaller) was conducted to refine sunflower seed oil. The highest PLs retention of 97% was achieved. It can be stated that the higher the pressure, the greater was the flux, but PL rejection decreased as well. Direct filtration seems not to be efficient enough because of the membrane fouling due to the high solid and particle content of the pressed oil (Koris and Marki, 2006).

27.3   VEGETABLE OILS AND DEACIDIFICATION

27.3.1   *Conventional method and principle of deacidification process*

The removal of free fatty acids (deacidification) is the most critical step in vegetable oil refining. It represents the main economic impact on oil production which determines the quality of the final refined product. Any inefficiency in this process has a great impact on the subsequent process operations. Chemical, physical and micelle deacidification have been used industrially for deacidification (Bhosle and Subramanian, 2005).

Chemical deacidification of palm oil was introduced in the early 1970s in Malaysia (PORAM, 2013). The deacidification process accomplished by neutralizing the FFA with sodium hydroxide (NaOH) to remove the FFA in the form of soap-stock as well as any remaining PLs. Sometime an excess of NaOH is required to reduce color of the refined oil and to ensure the removal of trace elements (Lin and Koseoglu, 2005). Any residual soap will be removed by the addition of hot water and subsequent centrifugation. FFA content in the crude oil has direct impact in the loss of significant amounts of triglycerides due to saponification (Snape and Nakajima, 1996). However, for oils with high acidity, chemical refining causes high losses of neutral oil due to saponification and emulsification. In spite of having several disadvantages; chemical deacidification is still industrially used upon request by customers.

Physical deacidification emerged as a better alternative to chemical deacidification in the late 1970s. Since then it was mainly used by modern refineries in Malaysia. Physical deacidification uses steam distillation where the FFA and other volatile components are distilled off from the oil by using an effective stripping agent under suitable processing conditions (Ceriani and Meirelles, 2006). As a consequence, oil losses are reduced, the quality of FFA is improved, and the operation is simplified. Despite all the advantages of reducing the loss of neutral oil, this process consumed high energy due to high deodorization temperatures in the range of 240–260°C. Besides, incomplete removal of undesirable components during pre-treatment of the crude oil has to be compensated by the increased use of bleaching earth which will in turn increase the operational cost (Bhosle and Subramanian, 2005).

Miscella deacidification is another alternative to conventional chemical and physical refining. In this process, miscella (typically 40–60% oil in hexane) is mixed with sodium hydroxide solution for neutralization, reaction with phosphatides, and decolorization. The separated micelle and the soap stock are both evaporated to remove the solvent (Hodgson, 1996). This micelle refining process results in less oil loss and lighter oil color. However, micelle refining is not widely used in industry because of several disadvantages. All equipment must be totally enclosed and explosion-proof, which increases investment considerably. Besides, the refining must be carried out at the solvent mills to be effective and economical (Anderson, 1996).

27.3.2   Membrane deacidification of vegetable oils

FFA in principle is almost impossible to be removed by membranes itself due to the molecular size of FFA (<300 Da) is much smaller than that of TG (~900 Da) (Cheryan, 2005). Theoretically, the ideal process would be to use a membrane with pores so precise that they could effectively separate the FFAs from the TG (Gibon *et al.*, 2007). Most of the previous researchers had reported the use of solvents in membrane deacidification of vegetable oils. Only a few studies have reported on direct deacidification of vegetable oils using membrane system.

27.3.2.1   Direct membrane deacidification

The deacidification of crude soybean and rapeseed oils without addition of organic solvents in undiluted oils was investigated using polymeric membrane. The process produces permeate and retentate fractions containing TG and other oil constituents. The decreasing order of relative preferential permeation in the nonporous membranes is expected to be FFA, tocopherols, TG, aldehydes, peroxides, color pigments and PLs. However, during membrane processing FFA permeated preferentially compared to TG resulting in negative rejection of FFA (Subramanian *et al.*, 1998).

Similar results with negative rejection of FFA using polymeric nonporous membrane were recorded in refined sunflower oil and oleic acid model system. Oleic acid permeated preferentially over triacylglycerols in the nonporous membrane, over a wide range of concentrations (~5–70%). The differences in molecular size, solubility, diffusivity and polarity between TGs and oleic acid appear insufficient for achieving direct deacidification (Bhosle *et al.*, 2005).

Another attempt was made by using ceramic membrane and polysulfone (PSF) hollow fiber membrane for membrane ultrafiltration of crude soybean oil. Ceramic membrane resulted in higher rejection than PSF membrane with 54.45 and 34.39% FFA rejections, respectively. However, the permeate flux of the ceramic membrane ($4.16 \, kg \, m^{-2} \, h^{-1}$) was much lower than the PSF membrane ($11.58 \, kg \, m^{-2} \, h^{-1}$) (Alicieo *et al.*, 2002).

27.3.2.2 *Membrane deacidification with solvent*

Most of the previous research works on deacidification of vegetable oils were conducted in combination with solvent. This is because the difference in their molecular weights is too small and membranes alone are very difficult to be used for membrane separation process (Raman *et al.*, 1994b). If the FFA were extracted from the crude oil by a solvent that selectively dissolves the FFA, the extractant (containing the solvent + FFA) can be processed and separated through the appropriate membrane.

Hollow fiber UF membranes were used to extract FFA from oil using 1,2-butanediol as a selective extractant. The oil phase was circulated inside the fibers, and butanediol outside the fibers. Some of the membranes tested could be used successfully. However due to the high mass transfer resistance, the required membrane surface area for a given extraction was relatively high (Keurentjes *et al.*, 1992).

In other studies, the deacification of soybean oil was investigated after extracting FFA from a model crude vegetable oil with methanol. A combination of high-rejection and low-rejection membranes resulted in a retentate stream of 35% FFA and a permeate stream with less than 0.04% FFA. No alkali is required, no soap-stock is formed, and almost all streams within the membrane process are recycled with little discharged as effluent (Raman *et al.*, 1996).

Another attempt using acetone-stable nanofiltration (NF) membranes was investigated for separation of different vegetable oil/solvent mixtures. The permeate consisted almost entirely of fatty acids in acetone, and only small traces of triglycerides were found. This makes it feasible to selectively remove the fatty acids and reduce loss of triglycerides normally associated with deacidification (Zwijnenberga *et al.*, 1999).

Different types of commercial nonporous (reverse osmosis and gas separation) polymeric membranes were screened for their abilities to separate FFA, MG, DG, and TG from a lipase hydrolysate of high-oleic sunflower oil after diluting with organic solvents (ethanol and hexane). Separation of FFA, MG, and DG from TG may find an industrial application as it could replace or reduce the load on the deacidification and deodorization steps in the conventional refining process. The constituents can be separated in suitable solvents using appropriate nonporous membranes (Koike *et al.*, 2002).

In deacidification of model vegetable oils using polymeric hydrophobic nonporous and hydrophilic NF membranes, dilution with hexane had improved oil flux compared to the model undiluted system. However, membrane selectivity was completely lost as both triacylglycerols and oleic acid permeated along with the solvent. Processing of the model oil after diluting with acetone showed that oleic acid was retained less than TG by the NF membrane. However, the selectivity decreased during successive runs, owing to the gradual loss of hydrophilicity due to polarity conditioning of the membrane (Bhosle *et al.*, 2005).

Different types of commercial porous and nonporous polymeric membranes were studied to separate FFA from hydrolysate of partially hydrogenated soybean oil (PHSO). UF membrane PLAC (PLAC is a trade name of a membrane, MW cut off 1000) made from regenerated cellulose (RC) and non-porous membranes of NTR series (Nitto Denko, Japan) had recorded 95.41% FFA rejection at $27 \, kg \, m^{-2} \, h^{-1}$ and 95–96% FFA rejection at $5–10 \, kg \, m^{-2} \, h^{-1}$, respectively in the ethanol system. In spite of encouraging results, there are still some key issues to be resolved for membrane development. A good compromise between achieving high selectivity and keeping high permeate flux, and long term stabilities give promising results for scale-up operation under solvent conditions (Jala *et al.*, 2011).

27.4 CONCLUSION

The development of membrane technology in degumming and deacidification – independently or in combination with the current technology – has emerged as a new approach to overcome major drawbacks in conventional refining of vegetable oils. It is a promising tool to refine vegetable oils due to low-energy consumption, mild temperature operation, low maintenance/operation cost, easiness to scale-up, retention of nutrients and other desirable components. However, the development of suitable membranes that enable achieving higher removal of PLs and FFA at acceptable flux is necessary for industrial adoption of this technology. With this, the consumer expectation for a sustainable, mild and efficient refining process could be achieved.

ACKNOWLEDGEMENT

The authors are grateful to Sime Darby Research Sdn. Bhd. and Advanced Membrane Technology Centre (AMTEC), UTM Skudai, Johor for providing the necessary facilities and technical support for successful completion of this review chapter.

REFERENCES

Alicieo, T.V.R., Mendes, E.S., Pereira, N.C. & Motta Lima, O.C. (2002) Membrane ultrafiltration of crude soybean oil. *Desalination*, 148, 99–102.

Aly, S.M. (1992) Degumming of soybean oil. *Grasas Aceites*, 43, 284–286.

Anderson, D. (1996) A primer on oils processing technology. In: Hui, Y.H. (ed.) *Bailey's industrial oil and fat products*, Volume 4. 5th edition. John Wiley & Sons, New York, NY. pp. 1–60.

Arora, S., Manjula, S., Gopala Krishna, A.G. & Subramanian, R. (2006) Membrane processing of crude palm oil. *Desalination* 191, 454–466.

Basiron, Y. (2013) Palm oil in the global oils & fats market. *Malaysia-Myanmar Palm Oil Trade Fair & Seminar POTS Myanmar 2013, 28th June 2013, Traders Hotel, Yangon, Myanmar*. Slides 7–11.

Bhosle, B.M. & Subramanian, R. (2005) New approaches in deacidification of edible oils – a review. *Journal of Food Engineering*, 69, 481–494.

Bhosle, B.M., Subramanian, R. & Ebert, K. (2005) Deacidification of model vegetable oils using polymeric membranes. *European Journal of Lipid Science and Technology*, 107, 746–753.

Ceriani, R. & Meirelles, A.J.A. (2006) Simulation of continuous physical refiners for edible oil deacidification. *Journal of Food Engineering*, 76, 261–271.

Cheremisinoff, N.P. (2002) Membrane separation technologies. Chapter 9 in *Handbook of water and wastewater treatment technologies*. Butterworth-Heinemann, Elsevier, Oxford, UK, pp. 335–371.

Cheryan, M. (1986) *Ultrafiltration handbook*. Technomic Publishing Company, Inc., Lancaster, PA.

Cheryan, M. (1998) *Ultrafiltration and microfiltration handbook*. Technomic Publishing, Chicago, IL.

Cheryan, M. (2005) Membrane technology in the vegetable oil industry. *Membrane Technology*, 2, 5–7.

Čmolík, J. & Pokorný J. (2000) Physical refining of edible oils. *European Journal of Lipid Science and Technology*, 102, 472–486.

Cuperus, F.P. & Nijhuis, H.H. (1993) Applications of membrane technology to food processing. *Trends in Food Science and Technology*, 7, 277–282.

de Morais Coutinho, C., Chiu, M.C., Basso, R.C., Ribeiro, A.P.B., Gonçalves, L.A.G. & Viotto, L.A. (2009) State of art of the application of membrane technology to vegetable oils: a review. *Food Research International*, 42, 536–550.

de Moura, J.M.L.N., Goncalves, L.A.G., Petrus, J.C.C. &. Viotto, L.A. (2005) Degumming of vegetable oil by microporous membrane. *Journal of Food Engineering*, 70, 473–478.

De Smet (2013) Refining technology by De Smet – World leader in oils and fats technologies – extraction of oilseeds – vegetable oil. Available from: http://www.refining.be [accessed July 2015].

de Souza, M.P., Petrus, J.C.C, Goncalves, L.A.G. & Viotto, L.A. (2008) Degumming of corn oil/hexane miscella using a ceramic membrane. *Journal of Food Engineering*, 86, 557–564.

Deffense, E. (2011a) Crystallisation & Degumming S.P.R.L. *AOCS Lipid Library*. Available from: http://lipidlibrary.aocs.org/ [accessed July 2015].

Deffense, E. (2011b) Edible oil processing – chemical degumming. *AOCS Lipid Library*. Available from: http://lipidlibrary.aocs.org/processing/chem-degum/index.htm [accessed July 2015].

Dijkstra, A.J. & Opstal, V. (1989) The total degumming process. *Journal of the American Oil Chemists' Society*, 66, 1002–1009.

Faessler, P. (1998) Recent developments and improvements in palm oil stripping and fatty acid distillation. In: Leonard, E.C., Perkins, E.G. & Cahn, A. (eds.) *Proceedings of the World Conference Palm Coconut*

Oils for the 21th Century: sources, processing, applications, and competition, 1998, Denpasar Indonesia. AOCS Press, Champaign, IL. pp. 67–72.

García, A., Álvarez, S., Riera, F., Álvarez, R. & Coca, J. (2006) Sunflower oil miscella degumming with polyethersulfone membranes effect of process conditions and MWCO on fluxes and rejections. *Journal of Food Engineering,* 74, 516–522.

Gibon, V. & Tirtiaux, A. (1998) Soft degumming: the simplest route to physical refining of soft oils. *Malaysian Oil Science and Technology,* 72, 48–54.

Gibon, V., De Greyt, W. & Kellens, M. (2007) Palm oil refining. *European Journal of Lipid Science and Technology,* 109, 315–335.

Gupta, A.K.S. (1977) Process for refining crude glyceride oils by membrane filtration. US Patent 4062882.

Hodgson, A.S. (1996) *Refining and bleaching.* In Hui, Y.H. (ed.) *Bailey & industrial oil and fat products,* Volume 4. 5th editon. John Wiley & Sons, New York, NY. pp. 157–212.

Iyike, S.E., Ahmadun, F.R. & Majid, R.A. (2004) Process intensification of membrane system for crude palm oil pretreatment. *Journal of Food Process Engineering,* 27, 476–496.

Jala, R.C.R., Guo, Z. & Xu, X. (2011) Separation of FFA from partially hydrogenated soybean oil hydrolysate by means of membrane processing. *Journal of the American Oil Chemists' Society,* 88, 1053–1060.

Jamil, S., Dufour, J.-P.G. & Deffense, E.M.J. (2000) Process for degumming a fatty substance and fatty substance thus obtained. US Patent 6,015,915.

Johansson, L.N., Brimberg, U.I. & Haraldsson, G. (1988) Experience of pre-refining of vegetable oils with acids. *European Journal of Lipid Science and Technology,* 90, 447–451.

Keurentjes, J.T.F., Sluijs, J.T.M., Franssen, R.T.J. & Vant'Riet, K. (1992) Extraction and fractionation of fatty acids from oil using an ultrafiltration membrane. *Industrial Engineering and Chemistry Research,* 31, 581–587.

Kim, I.C., Kim, J.H., Lee, K.H. & Tak, T.M. (2002) Phospholipids separation (degumming) from crude vegetable oil by polyimide ultrafiltration membrane. *Journal of Membrane Science,* 205, 113–123.

Koike, S., Subramanian, R., Nabetani, H. & Nakajima, M. (2002) Separation of oil constituents in organic solvents using polymeric membranes. *Journal of the American Oil Chemists' Society,* 79, 937–942.

Koris, A. & Marki, E. (2006) Ceramic ultrafiltration membranes for non-solvent vegetable oil degumming (phospholipid removal). *Desalination,* 200, 537–539.

Koseoglu, S.S. & Engelgau, D.E. (1990) Membrane applications and research in the edible oil industry: an assessment. *Journal of the American Oil Chemists' Society,* 67, 239–249.

Ladhe, A.R. & Kumar, N.S.K. (2010) Application of membrane technology in vegetable oil processing. Chapter 5 in: Cui, Z.F. & Muralidhara, H.S. (eds.) *Membrane technology.* Elsevier, Savage, MN. pp. 63–75.

Lin, L. & Koseoglu, S.S. (2005) Membrane processing of fats and oils. In: Shahidi, F. (ed.) *Bailey's industrial oils and fats products.* Volume 6. 6th Edition. John Wiley & Sons, Hoboken, NJ. pp. 433–457.

Lin, L., Rhee, K.C. & Koseoglu, S.D. (1997) Bench-scale membrane degumming of crude vegetable oil: process optimization. *Journal of Membrane Science,* 134, 101–108.

Manjula, S. & Subramanian, R. (2009) Simultaneous degumming, dewaxing and decolorizing crude rice bran oil using nonporous membranes. *Separation and Purification Technology,* 66, 223–228.

Manjula, S., Kobayashi, I. & Subramanian, R. (2011) Characterization of phospholipid reverse micelles in nonaqueous systems in relation to their rejection during membrane processing. *Food Research International,* 44, 925–930.

Marenchino, R., Pagliero, C. & Mattea, M. (2006) Vegetable oil degumming using inorganic membranes. *Desalination,* 200, 562–564.

Muralidhara, H.S. (2010) Challenges of membrane technology in the XXI century. Chapter 2 in: Cui, Z.F. & Muralidhara, H.S. (eds.) *Membrane technology.* Elsevier, Plymouth, MN. pp. 19–32.

Ochoa, N., Pagliero, C., Marchese, J. & Mattea, M. (2001) Ultrafiltration of vegetable oils degumming by polymeric membranes. *Separation and Purification Technology,* 22–23, 417–422.

Pagliero, C., Ochoa, N., Marchese, J. & Mattea, M. (2001) Degumming of crude soybean oil by ultrafiltration using polymeric membranes. *Journal of American Oil Chemists Society,* 78, 793–796.

Pagliero, C., Ochoa, N., Marchese, J. & Mattea, M. (2004) Vegetable oil degumming with polyimide and polyvinylidenefluoride ultrafiltration membranes. *Journal of Chemical Technology and Biotechnology,* 79, 148–152.

Patterson, H.B.W. (1992) *Bleaching and purifying fats and oils, theory and practices.* AOCS Press, Champaign, IL.

Paulson, D.J., Wilson, R.L. & Spatz, D.D. (1984) Crossflow membrane technology and its applications. *Food Technology,* 38, 77–87.

PORAM (2013) Refining in Malaysia. PORAM – Palm Oil Refiners Association of Malaysia. Available from: www.poram.org.my [accessed July 2013].

R.E.A. (2013) World production of oils & fats, R.E.A Holdings PLC. Available from: http://www.rea.co.uk/rea/en/markets [accessed July 2015].

Raman, L.P., Rajagopalan, N. & Cheryan, M. (1994a) Membrane technology (in vegetable oil processing). *Oils Fats International*, 10, 28–38.

Raman, L.P., Rajagopalan, N. & Cheryan, M. (1994b) Membrane technology. *Fats & Oils International (UK)*, 10 (6), 28–34.

Raman, L.P., Cheryan, M. & Rajagopalan, N. (1996) Deacidification of soybean oil by membrane technology. *Journal of the American Oil Chemists' Society*, 73, 219–224.

Ribeiro, A.P.B., Bei, N., Goncalves, L.A.G., Petrus, J.C.C. & Viotto, L.A. (2008) The optimisation of soybean oil degumming on a pilot plant scale using a ceramic membrane. *Journal of Food Engineering*, 87, 514–521.

Rohdenburg, H., Csernitzky, K., Chikány, B., Perédi, J., Boródi, A. & Fábicsné Ruzics, A. (1993) Degumming process for plant oils. US Patent 5,239,096 (Krupp Maschinentechnik GmbH).

Sande, R.L.K.M. van de & Segers, J.C. (1989) Method of refining glyceride oils. European Patent 0 348 004 (Unilever).

Saravanan, M., Bhosle, B.M. & Subramanian, R. (2006) Processing hexane-oil miscella using a nonporous polymeric composite membrane. *Journal of Food Engineering*, 74, 529–535.

Segers, J.C. (1994) Removal of phospholipids from glyceride oil. PCT Int. Appl. WO 1994021762 A1.

Snape, J.B. & Nakajima, M. (1996) Processing of agricultural fats and technology oils using membrane technology. *Journal of Food Engineering*, 30, 1–41.

Strathmann, P.H. (1990) Synthetic membranes and their preparation. In: Porter, M.C. (ed.) *Handbook of industrial membrane technology*. Noyes Publications, Park Ridge, NJ. pp. 1–56.

Subramanian, R. & Nakajima, M. (1997) Membrane degumming of crude soybean and rapeseed oils. *Journal of the American Oil Chemists' Society*, 74, 971–975.

Subramanian, R., Nakajima, M., Kimura, T. & Maekawa, T. (1998) Membrane process for premium quality expeller-pressed vegetable oils. *Food Research International*, 31, 587–593.

Subramanian, R., Ichikawa, S., Nakajima, M., Kimura, T. & Maekawa, T. (2001) Characterization of phospholipid reverse micelles in relation to membrane processing of vegetable oils. *European Journal of Lipid Science and Technology*, 103, 93–97.

Subramaniana, R., Nakajimab, M., Raghavaraoa, K.S.M.S. & Kimura, T. (2004) Processing vegetable oils using nonporous denser polymeric composite membranes. *Journal of the American Oil Chemists' Society*, 81, 313–322.

Thiagarajan, T. & Tang, T.S. (1991) Refinery practices and oil quality. *Proceeding of PORIM International Palm Oil Conference. Progress, Prospects Challenges Towards the 21st Century. Chemistry and Technology, 9–14 September 1991, Kuala Lumpur, Malaysia*. Palm Oil Research Institute of Malaysia. (PORIM). pp. 254–267.

UN (2013) World population prospects: the 2012 revision. Population Division of the Department of Economic and Social Affairs of the United Nations Secretariat. 2013. United Nations, New York, NY.

Vintilă, I. (2009) *The physical-chemical mechanism of the edible oils depth refining*. Scientific Study & Research. Vol. X (2). pp 179–183.

Zwijnenberga, H.J., Krosse, A.M., Ebert, K., Peinemann, K.-V. & Cuperus, F.P. (1999) Acetone-stable nanofiltration membranes in deacidifying vegetable oil. *Journal of the American Oil Chemists' Society*, 76, 83–87.

CHAPTER 28

Preparation of polysulfone electrospun nanofibers: effect of electrospinning and solution parameters

Zafarullah Khan & Feras M. Kafiaha

28.1 INTRODUCTION

Electrospinning is considered as one of the unique techniques to produce fibers at micro- and nano-scales (Greiner and Wendorff, 2007; Huang *et al.*, 2003; Yao *et al.*, 2003). The process is straightforward and very simple technically, using very high electric field between nozzle and collector to derive polymeric nano-fibers onto the collector as shown in Figure 28.1.

A polymeric solution is first prepared with a specific concentration using a suitable solvent and then placed into a dispenser syringe for electrospinning at a selected applied voltage, feed rate and distance between syringe nozzle (needle) and collector. The droplet coming out of the needle tip is charged, turns into a cone (Taylor cone) and whips out the solution into jets. As the jet travels toward the collector, the solvent evaporates leaving behind nano-fibers which are deposited over the target collector.

Nano-fibers are produced due to the stretching of the solution jet by applying an electric force on free charges at the surface or inside the solution as shown in Figure 28.1. The drift velocity of charges is proportional to the electric field and mobility of carriers (Chang and Lin, 2009). The motion of the charges will transfer the electric force to the polymeric solution that will be accelerated to the field direction. At the early stages of this process, a small hemispherical droplet

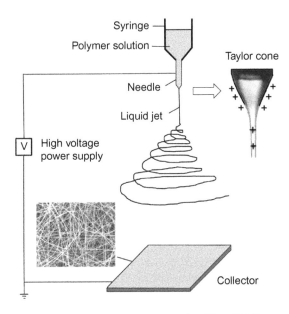

Figure 28.1. Schematic diagram for electrospinning setup (Li and Xia, 2004).

Table 28.1.   (wt/vol)% concentrations of PSF in DMF.

Concentration No.	Weight of PSF [g]	Volume of DMF [mL]
1	17	100
2	22	100
3	25	100
4	27	100
5	30	100

at spinneret tip will be produced due to electric energy and as this energy increases, it assumes a conical shape called a Taylor cone. Further increase of electric energy exceeds the surface tension of a Taylor cone and causes a jet to eject out of the Taylor cone in a whipping action. The solvent evaporates during this whipping action and dry fibers are collected on the collector plate.

Many parameters influence the resultant fibers morphology, those related to the polymeric material to be electrospun like molecular weight and molecular weight distribution and other parameters related to the polymeric solution like concentration, viscosity, surface tension and electrical conductivity, and finally the parameters related to the electrospinning process, namely, voltage, feed rate, distance between nozzle and collector, temperature and humidity (Buchko et al., 1999; Fong et al., 1999; Hajra et al., 2003; Norris et al., 2000; Shin et al., 2001).

Due to its excellent thermal properties, good resistance to inorganic acids and bases (Cha et al., 2006; Gopal et al., 2007), good solubility and low price (Yuan et al., 2004), polysulfone (PSF) has been used as one of the most important membrane materials. Electrospun nano-fibrous membranes show superior performance in separation technology because of many advantages like high porosity, sub-micron to several microns open pore structure, high permeability to gases and finally, high surface area to volume ratio (Gopal et al., 2007).

This work considers electrospinning polysulfone using N,N-dimethylformamide (DMF) as solvent to produce uniform nano-fibrous structure without beads and with minimum fiber diameters. Solution concentration and electrospinning parameters namely the applied voltage, feed rate and distance between nozzle and collector on the fiber diameter and fiber morphology are investigated.

28.2   EXPERIMENTAL

28.2.1   Materials

Polysulfone (MW $= 35,000$ g mol^{-1}) was purchased from BOC Sciences, USA and the solvent N,N-dimethylformamide (DMF) (density $= 0.944$ g cm^{-3}) was purchased from Alfa Aesar, USA. PSF was dissolved in DMF at 45°C using magnetic stirrer until uniform and homogenous solution was obtained. Table 28.1 shows the wt/vol concentration ratios used in this investigation.

28.2.2   Electrospinning and characterization

Eelctrospinning was carried out using NANON A-1 apparatus (MECC Ltd., Japan). A 5 mm hypoderemic syringe with an 8 mm inner diameter needle was used to dispense the spinning solution. Feed rate and voltage were controlled by the programmable electrospinning apparatus. The electrospun fibers were collected on aluminum foil placed on the collector plate.

PSF nano-fiber mats produced were characterized by checking their morphologies and average fibers diameter. The morphology was observed by SEM at different magnifications (1000×, 5000× and 10,000×). Average nano-fiber diameters were measured from high magnification SEM micrographs using Image J software (http://rsb.info.nih.gov/ij/) in which, averages have been taken for around 250 readings of every mat and their standard deviations, minima and

Table 28.2. Matrix of electrospun samples with variation of different parameters along with some extracted data. Note that: PSF 3, 6, 16 and 21 are with same results.

	Polysulfone in DMF samples				Extracted data			
Sample	Concentration [wt/vol]	Voltage [kV]	Distance [mm]	Feed rate [mL h^{-1}]	Average fiber diameters [nm]	Standard deviation	Min [nm]	Max [nm]
PSF 1	17.00	20.00	150.00	4.00	Fibers with droplets and beads			
PSF 2	22.00	20.00	150.00	4.00	Beaded fibers			
PSF 3	25.00	20.00	150.00	4.00	1054.3	395.8	280.6	3024.5
PSF 4	27.00	20.00	150.00	4.00	1448.6	379.7	785.6	2804.7
PSF 5	30.00	20.00	150.00	4.00	2096.9	902.3	935.6	5252.9
PSF 6	25.00	20.00	150.00	4.00	1054.3	395.8	280.6	3024.5
PSF 7	25.00	22.00	150.00	4.00	1068.9	388.1	198.2	2255.0
PSF 8	25.00	24.00	150.00	4.00	1147.2	404.8	413.2	2684.6
PSF 9	25.00	26.00	150.00	4.00	1239.3	404.3	327.2	2858.4
PSF 10	25.00	28.00	150.00	4.00	1274.8	450.1	317.0	2924.2
PSF 11	25.00	30.00	150.00	4.00	1314.1	407.1	662.1	2206.7
PSF 12	25.00	20.00	60.00	4.00	1199.4	445.6	405.2	3116.0
PSF 13	25.00	20.00	80.00	4.00	1264.0	477.0	286.0	2806.0
PSF 14	25.00	20.00	100.00	4.00	1184.0	473.0	327.0	3269.0
PSF 15	25.00	20.00	120.00	4.00	1158.7	376.1	577.8	2589.4
PSF 16	25.00	20.00	150.00	4.00	1054.3	395.8	280.6	3024.5
PSF 17	25.00	20.00	150.00	1.00	609.0	219.0	177.0	1389.0
PSF 18	25.00	20.00	150.00	1.50	699.4	219.1	158.7	1796.6
PSF 19	25.00	20.00	150.00	2.00	750.0	315.3	236.9	1748.1
PSF 20	25.00	20.00	150.00	3.00	952.0	360.0	335.0	2857.0
PSF 21	25.00	20.00	150.00	4.00	1054.3	395.8	280.6	3024.5

maxima were calculated. Table 28.2 shows the electrospinning parameters that have been studied on the left side and extracted data like average fibers diameter, standard deviation, minima and maxima on the right side. PSF concentrations of 17, 22, 25, 27 and 30% (wt/vol) (Samples PSF 1 to PSF 5) were studied keeping voltage at 20 kV, feed rate at 4 mL h^{-1} and distance between nozzle to collector at 150 mm. PSF concentration was then fixed at 25% (wt/vol) for reasons to be discussed in Section 28.3.1. Voltages from 20 kV to 30 kV with a 2 kV increments (Samples PSF 6 to PSF 11) were tested keeping feed rate at 4 mL h^{-1} and distance at 150 mm. Feed rates of 1, 1.5, 2, 3 and 4 mL h^{-1} (Samples PSF 17 to PSF 21) were selected after fixing voltage at 20 kV and distance at 150 mm. Finally, effect of distance between nozzle to collector was studied by adopting 60, 80, 100, 120 and 150 mm distances (Samples PSF 12 to PSF 16) keeping voltage at 20 kV and feed rate at 4 mL h^{-1}.

28.3 RESULTS AND DISCUSSION

28.3.1 *Effect of PSF concentration on fiber morphology*

Concentration or the amount of polymer in the solution is an important factor determining fiber morphology (Jalili et al., 2005). Too high concentration will hinder the solution from being pumped (Dhakate et al., 2010; Jun et al., 2003), and on the other hand, too low concentration will lower the amount of polymer chain entanglements that are required to keep the jet from breaking down (Chang and Lin, 2009).

At the concentration of 17% (wt/vol), there was insufficient polymer entanglement to stabilize the jet and the electrospinning at this concentration resulted in the formation of droplets and beaded fiber morphology as shown in SEM micrograph (Fig. 28.2a). As the concentration

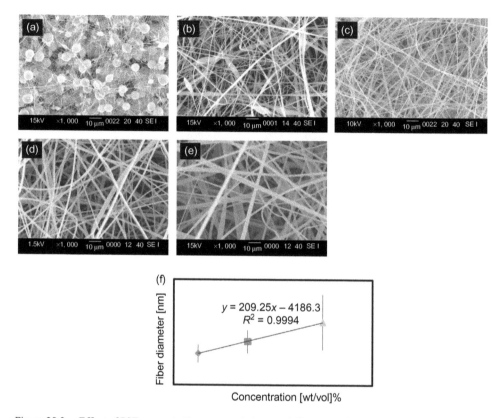

Figure 28.2.   Effect of PSF concentration on morphology and diameters of nano-fibers: SEM micrographs for (a) 17% (wt/vol), (b) 22% (wt/vol), (c) 25% (wt/vol), (d) 27% (wt/vol) and (e) 30% (wt/vol). (f) Variation of average fiber diameters with concentration (feed rate 4 mL h^{-1}, distance distance between nozzle and the collector plate: 150 mm, voltage: 20 kV).

increased to 22% (wt/vol), jet was stabilized but many beads still appear in the fibers as shown in (Fig. 28.2b). The beaded morphology results from high solution surface tension which at this concentration is more controlling due to the fewer chain entanglements and higher amounts of solvent (Ramakrishna *et al.*, 2005) than electric force, which is responsible for stretching fibers. Yuan *et al.* (2004) suggested that lower viscosity and conductivity of the lower concentration PSF solutions could be the possible reasons leading to beaded fiber morphology. These beads started to disappear at the concentration of 25% (wt/vol) PSF/DMF solution. At this concentration uniform fibers without beads were obtained (Fig. 28.2c). Further increase in the solution concentration was responsible for increasing average fiber diameter (Fig. 28.2d,e). Figure 28.2f shows the relation between average fiber diameter with solution concentration. As the concentration increases the average fiber diameter increases and becomes broader. The figure has taken only three concentrations (25, 27 and 30% (wt/vol)) and excluded 17% and 22% in which the presence of droplets and beads did not permit accurate fiber diameter measurements. Demir *et al.* (2002) studied this relation during polyurethane electrospinning and they found the average fiber diameter to increase with concentration according to the third power law. Many researchers who studied the effect of concentration on the average fiber diameter (Baumgarten, 1971; Beachley and Wen, 2009; Chowdhury and Stylios, 2010; Demir *et al.*, 2002; Fong *et al.*, 1999; Heikkilä and Harlin, 2008; Larrondo and St John Manley, 1981; Li *et al.*, 2007; Megelski *et al.*, 2002; Mit-Uppatham *et al.*, 2004; Morota *et al.*, 2004; Ramakrishna *et al.*, 2005; Supaphol *et al.*, 2005; Sutasinpromprae *et al.*, 2006) found the same trend as the present study. These authors report

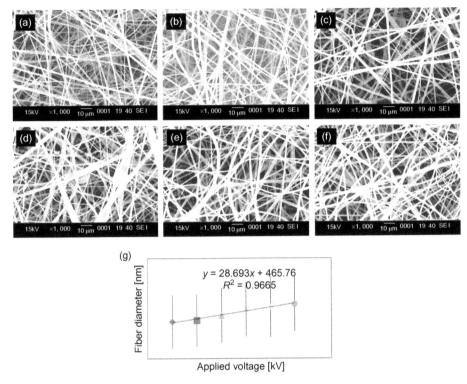

Figure 28.3.   Effect of electrospinning voltage on the morphology and diameters of nano-fibers: SEM micrographs for (a) 20 kV, (b) 22 kV, (c) 24 kV, (d) 26 kV, (e) 28 kV and (f) 30 kV. (g) Variation of average fiber diameters with applied voltage (feed rate 4 mL h^{-1}, distance between nozzle and the collector plate: 150 mm).

that increasing concentration will in some cases increase fiber length (Beachley and Wen, 2009), decreases tendency of bead formation, and finally fibers become more uniform and smooth. Fong *et al.* (1999) found that as beads become bigger, the distance between them becomes larger and they change from spherical to spindle-like with increasing solution concentration (solution viscosity) before acquiring a uniform fiber morphology.

28.3.2   *Effect of applied voltage on fiber morphology*

Applied voltage is considered to be the most contradictory parameter in affecting nano-fibers morphology and average fiber diameter. The researchers find that the fiber diameter is either directly (Li and Hsieh, 2005; Li *et al.*, 2007; Morota *et al.*, 2004; Pornsopone *et al.*, 2005; Supaphol *et al.*, 2005; Zhang *et al.*, 2005; Zhao *et al.*, 2004; Zong *et al.*, 2002) or inversely (Buchko *et al.*, 1999; Chang and Lin, 2009; Deitzel *et al.*, 2001; Ding *et al.*, 2002; Gu and Ren, 2005; Hsu and Shivkumar, 2004; Lee *et al.*, 2004; Li *et al.*, 2006; Megelski *et al.*, 2002; Wang *et al.*, 2006a; Yuan *et al.*, 2004) proportional to the applied voltage and in some cases report a mixed trend (Chowdhury and Stylios, 2010; Jalili *et al.*, 2005; Jun *et al.*, 2003; Lee *et al.*, 2002; Li *et al.*, 2006; Sencadas *et al.*, 2012; Wang *et al.*, 2006b) with variation in the applied voltage.

In this study, the average fiber diameter was found to increase with increasing voltage for solution concentration of 25% (which provides uniform fibers with lowest fiber diameter), 4 mL h^{-1} feed rate and 150 mm distance between nozzle and the collector plate. Figure 28.3 shows SEM micrographs for PSF nano-fibers with variation of voltages from 20 kV (Fig. 28.3a) to 30 kV (Fig. 28.3f) with the 2 kV increments.

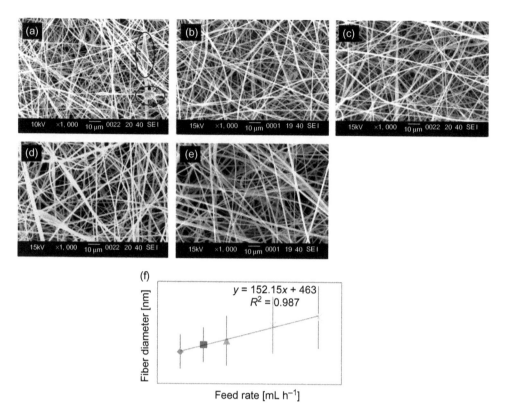

Figure 28.4.   Effect of solution feed rate on the morphology and diameters of nano-fibers. SEM micrographs for (a) 1 mL h^{-1}, (b) 1.5 mL h^{-1}, (c) 2 mL h^{-1}, (d) 3 mL h^{-1} and (e) 4 mL h^{-1}. (f) Variation of average fiber diameters with feed rate (distance between nozzle and the collector plate: 150 mm, voltage: 20 kV).

Although increasing voltage is expected to increase elongational force, which is responsible for fiber thinning, the time for the jet to travel toward the collector is shortened and as a result there is insufficient time for full solvent evaporation. In addition, the higher voltage causes more mass flow and all of these favor an increase in the fiber diameter as corroborated by other researchers (Demir *et al.*, 2002; Li and Hsieh, 2005).

Some researchers found presence of beads with increasing voltages (Cui *et al.*, 2007; Deitzel *et al.*, 2001; Krishnappa *et al.*, 2003; Megelski *et al.*, 2002), and they attribute this to the instability of the jet at the needle in which the Taylor cone is receding (Zong *et al.*, 2002). The other reason is the steep increase in spinning current (Subbiah *et al.*, 2005). However, in this study, beads at high voltages were not observed.

28.3.3   *Effect of solution feed rate on fiber morphology*

The feed rate controls the amount of solution to be electrospun (Khayet and Matsuura, 2011; Ramakrishna *et al.*, 2005) and it is an important electrospinning parameter which is related to Taylor cone stabilization. Feed rate in this study was varied from 1 to 4 mL h^{-1}, keeping the spinning voltage at 20 kV, needle tip to collector distance at 150 mm and solution concentration at 25% (wt/vol).

Average fiber diameters were found to increase with increasing feed rate (Fig. 28.4f) and this was proven by many research groups (Chang and Lin, 2009; Chowdhury and Stylios, 2010;

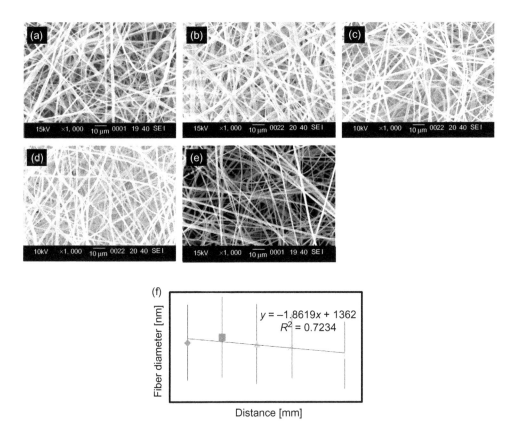

Figure 28.5.   Effect of distance between nozzle and collector on the morphology and diameters of nano-
fibers. SEM micrographs for (a) 60 mm, (b) 80 mm, (c) 100 mm, (d) 120 mm and (e) 150 mm.
(f) Variation of average fiber diameters with nozzle to collector distance (f) Variation of average
fiber diameters with concentration (feed rate 4 mL h^{-1}, voltage: 20 kV).

Feng *et al.*, 2011; Homayoni *et al.*, 2009; Ramakrishna *et al.*, 2005; Touny *et al.*, 2010). These
researchers attributed this to the lesser time for the jet to reach the collector and hence lower
solvent evaporation rate which leaves thicker fibers.

At lower feed rates of 1 mL h^{-1} some beads (red marks) were noticed (as shown in Fig. 28.4a),
which start to disappear with increasing feed rate (Fig. 28.4b–e). This could be due to the smaller
volume of polymer coming out of needle which at high applied voltage makes the jet unstable
and discontinuous resulting in an increase in the surface tension and favors formation of beads.

It was also noted (as shown in Fig. 28.4f and Table 28.2) that an increase in the fiber diameter
occurs with increasing feed rate in which standard deviation of readings at 1 mL h^{-1} is almost
half of 4 mL h^{-1} feed rate. This could again be attributed to the increase in time for complete
solvent evaporation at lower feed.

28.3.4   *Effect of distance between nozzle and collector on fiber morphology*

The distance between nozzle and collector influences two important parameters related to fiber
morphology. First is the jet travelling time that is coined with solvent evaporation rate and second
is the field strength which describes the voltage over distance (Chang and Lin, 2009; Subbiah
et al., 2005). Decreasing this distance will shorten the time for jet travel and increase the field
strength, which will result in higher fiber diameters (Chowdhury and Stylios, 2010; Hekmati
et al., 2013; Kameoka and Craighead, 2003; Zhao *et al.*, 2004). Figure 28.5a–e shows SEM

micrographs for 25% (wt/vol) PSF nano-fibers electrospun at 20 kV voltage, 4 mL h⁻¹ feed rate
and different needle to collector distances varied from 60 to 150 mm.

A slight decrease in average fiber diameter was noticed by changing the needle to collector distance. In the case of 60 mm distance fiber diameters were measured to be equal to 1199.4 ± 445.6 nm whereas at 150 mm distance, the fiber diameters decreased to 1054.3 ± 395.8 nm. This slight decrease in the fiber diameter could be attributed to the extra time for solvent evaporation that causes fiber thinning on one hand and to the lower field strength (lower stretching force) on the other hand.

Unlike some results found in the literature (Hsu and Shivkumar, 2004; Megelski *et al.*, 2002), distance between nozzle and collector was unrelated with the appearance of beads. Megelski *et al.* (2002) found that average fiber diameter doesn't change significantly with needle to collector distance and their observations are consistent with the findings of this study. But on the other hand, these authors found some elongated beads along Polystyrene (PS) fibers.

28.4   CONCLUSION

In this study, the effect of solution concentration and process parameters on the electrospun fiber morphology for Polysulfone/DMF was investigated. It was found that droplets and beads disappeared and fibers became uniform after reaching 25% (wt/vol) PSF concentration. Increasing concentration above this value causes an increase in average fiber diameter. Increasing electrospinning applied voltage and feed rate was found to increase average fiber diameter as well. On the other hand, increasing the nozzle to collector distance resulted in a slight decrease in average fiber diameters.

The smallest fiber diameters (around 700 nm) with uniform structure and without beads were found in the case of PSF/DMF solution concentration of 25% (wt/vol) at the electrospinning voltage of 20 kV, feed rate of 1.5 mL h⁻¹ and needle to collector distance of 150 mm.

ACKNOWLEDGEMENT

The authors wish to gratefully acknowledge the support of King Fahd University of Petroleum and Minerals, Dhahran, Saudi Arabia for supporting this research under the Center of Excellence in Research Collaboration with MIT grant CW-R5-08.

REFERENCES

Baumgarten, P.K. (1971) Electrostatic spinning of acrylic microfibers. *Journal of Colloid and Interface Science*, 36, 71–79.

Beachley, V. & Wen, X. (2009) Effect of electrospinning parameters on the nanofiber diameter and length. *Materials Science and Engineering C*, 29, 663–668.

Buchko, C.J., Chen, L.C., Shen, Y. & Martin, D.C. (1999) Processing and microstructural characterization of porous biocompatible protein polymer thin films. *Polymer*, 40, 7397–7407.

Buruaga, L., Gonzalez, A. & Iruin, J.J. (2009) Electrospinning of poly (2-ethyl-2-oxazoline). *Journal of Materials Science*, 44, 3186–3191.

Cha, D.I., Kim, K.W., Chu, G.H., Kim, H.Y., Lee, K.H. & Bhattarai, N. (2006) Mechanical behaviors and characterization of electrospun polysulfone/polyurethane blend nonwovens. *Macromolecular Research*, 14, 331–337.

Chang, K.-H. & Lin, H.-L. (2009) Electrospin of polysulfone in N,N'-dimethyl acetamide solutions. *Journal of Polymer Research*, 16, 611–622.

Chowdhury, M. & Stylios, G. (2010) Effect of experimental parameters on the morphology of electrospun Nylon 6 fibres. *International Journal of Basic & Applied Sciences*, 10, 116–131.

Cui, W., Li, X., Zhou, S. & Weng, J. (2007) Investigation on process parameters of electrospinning system through orthogonal experimental design. *Journal of Applied Polymer Science*, 103, 3105–3112.

Deitzel, J., Kleinmeyer, J., Harris, D.E.A. & Beck Tan, N. (2001) The effect of processing variables on the morphology of electrospun nanofibers and textiles. *Polymer*, 42, 261–272.

Demir, M.M., Yilgor, I., Yilgor, E. & Erman, B. (2002) Electrospinning of polyurethane fibers. *Polymer*, 43, 3303–3309.

Dhakate, S., Singla, B., Uppal, M. & Mathur, R. (2010) Effect of processing parameters on morphology and thermal properties of electrospun polycarbonate nanofibers. *Advanced Materials Letters*, 1, 200–204.

Ding, B., Kim, H.-Y., Lee, S.-C., Lee, D.-R. & Choi, K.-J. (2002) Preparation and characterization of nanoscaled poly (vinyl alcohol) fibers via electrospinning. *Fibers and Polymers*, 3, 73–79.

Feng, Y., Meng, F., Xiao, R., Zhao, H. & Guo, J. (2011) Electrospinning of polycarbonate urethane biomaterials. *Frontiers of Chemical Science and Engineering*, 5, 11–18.

Fong, H., Chun, I. & Reneker, D. (1999) Beaded nanofibers formed during electrospinning. *Polymer*, 40, 4585–4592.

Gopal, R., Kaur, S., Feng, C.Y., Chan, C., Ramakrishna, S., Tabe, S. & Matsuura, T. (2007) Electrospun nanofibrous polysulfone membranes as pre-filters: particulate removal. *Journal of Membrane Science*, 289, 210–219.

Greiner, A. & Wendorff, J.H. (2007) Electrospinning: a fascinating method for the preparation of ultrathin fibers. *Angewandte Chemie International Edition*, 46, 5670–5703.

Gu, S.Y. & Ren, J. (2005) Process optimization and empirical modeling for electrospun poly (D,L-lactide) fibers using response surface methodology. *Macromolecular Materials and Engineering*, 290, 1097–1105.

Hajra, M.G., Mehta, K. & Chase, G.G. (2003) Effects of humidity, temperature, and nanofibers on drop coalescence in glass fiber media. *Separation and Purification Technology*, 30, 79–88.

Heikkilä, P. & Harlin, H. (2008) Parameter study of electrospinning of polyamide-6. *European Polymer Journal*, 44, 3067–3079.

Hekmati, A.H., Rashidi, A., Ghazisaeidi, R. & Drean, Y.-D. (2013) Effect of needle length, electrospinning distance, and solution concentration on morphological properties of polyamide-6 electrospun nanowebs. *Textile Research Journal*, 83, 1452–1466.

Homayoni, H., Ravandi, S.A.H. & Valizadeh, N. (2009) Electrospinning of chitosan nanofibers: processing optimization. *Carbohydrate Polymers*, 77, 656–661.

Hsu, C.-M. & Shivkumar, S. (2004) Nano-sized beads and porous fiber constructs of poly (ε-caprolactone) produced by electrospinning. *Journal of Materials Science*, 39, 3003–3013.

Huang, Z.-M., Zhang, Y.Z., Kotaki, M. & Ramakrishna, S. (2003) A review on polymer nanofibers by electrospinning and their applications in nanocomposites. *Composites Science and Technology*, 63, 2223–2253.

Jalili, R., Hosseini, S.A. & Morshed, M. (2005) The effects of operating parameters on the morphology of electrospun polyacrilonitrile nanofibers. *Iranian Polymer Journal*, 14, 1074.

Jun, K., Reid, O., Yanou, Y., David, C., Robert, M., Geoffrey, W.C. & Craighead, H.G. (2003) A scanning tip electrospinning source for deposition of oriented nanofibers. *Nanotechnology*, 14, 1124.

Kameoka, J. & Craighead, H. (2003) Fabrication of oriented polymeric nanofibers on planar surfaces by electrospinning. *Applied Physics Letters*, 83, 371–373.

Khayet, M. & Matsuura, T. (2011) *Membrane distillation: principles and applications.* Elsevier, Amsterdam, The Netherlands.

Krishnappa, R., Desai, K. & Sung, C. (2003) Morphological study of electrospun polycarbonates as a function of the solvent and processing voltage. *Journal of Materials Science*, 38, 2357–2365.

Larrondo, L. & St John Manley, R. (1981) Electrostatic fiber spinning from polymer melts. I. Experimental observations on fiber formation and properties. *Journal of Polymer Science* B: *Polymer Physics*, 19, 909–920.

Lee, C.K., Kim, S.I. & Kim, S.J. (2005) The influence of added ionic salt on nanofiber uniformity for electrospinning of electrolyte polymer. *Synthetic Metals*, 154, 209–212.

Lee, J.S., Choi, K.H., Ghim, H.D., Kim, S.S., Chun, D.H., Kim, H.Y. & Lyoo, W.S. (2004) Role of molecular weight of atactic poly (vinyl alcohol) (PVA) in the structure and properties of PVA nanofabric prepared by electrospinning. *Journal of Applied Polymer Science*, 93, 1638–1646.

Lee, K.H., Kim, H.Y., La, Y.M., Lee, D.R. & Sung, N.H. (2002) Influence of a mixing solvent with tetrahydrofuran and *N,N*-dimethylformamide on electrospun poly (vinyl chloride) nonwoven mats. *Journal of Polymer Science* B: *Polymer Physics*, 40, 2259–2268.

Li, D. & Xia, Y. (2004) Electrospinning of nanofibers: reinventing the wheel? *Advanced Material*, 16, 1151–1170.

Li, L. & Hsieh, Y.-L. (2005) Ultra-fine polyelectrolyte fibers from electrospinning of poly (acrylic acid). *Polymer*, 46, 5133–5139.

Li, Q., Jia, Z., Yang, Y., Wang, L. & Zhicheng, G. (2007) Preparation and properties of poly(vinyl alcohol) nanofibers by electrospinning. *9th International Conference on Solid Dielectrics (ICSD 2007), July 8–13, 2007, Winchester, UK.* pp. 215–218.

Li, Y., Huang, Z. & Lü, Y. (2006) Electrospinning of nylon-6, 66, 1010 terpolymer. *European Polymer Journal*, 42, 1696–1704.

Liu, F., Guo, R., Shen, M., Wang, S. & Shi, X. (2009) Effect of processing variables on the morphology of electrospun poly [(lactic acid)-co-(glycolic acid)] nanofibers. *Macromolecular Materials and Engineering*, 294, 666–672.

Megelski, S., Stephens, J.S., Chase, D.B. & Rabolt, J.F. (2002) Micro-and nanostructured surface morphology on electrospun polymer fibers. *Macromolecules*, 35, 8456–8466.

Mit-Uppatham, C., Nithitanakul, M. & Supaphol, P. (2004) Effects of solution concentration, emitting electrode polarity, solvent type, and salt addition on electrospun polyamide-6 fibers: a preliminary report. *Macromolecular Symposia.* Vol. 216, *Wiley Online Library.* pp. 293–300.

Morota, K., Matsumoto, H., Mizukoshi, T., Konosu, Y., Minagawa, M., Tanioka, A., Yamagata, Y. & Inoue, K. (2004) Poly (ethylene oxide) thin films produced by electrospray deposition: morphology control and additive effects of alcohols on nanostructure. *Journal of Colloid and Interface Science*, 279, 484–492.

Norris, I.D., Shaker, M.M., Ko, F.K. & MacDiarmid, A.G. (2000) Electrostatic fabrication of ultrafine conducting fibers: polyaniline/polyethylene oxide blends. *Synthetic Metals*, 114, 109–114.

Pornsopone, V., Supaphol, P., Rangkupan, R. & Tantayanon, S. (2005) Electrospinning of methacrylate-based copolymers: effects of solution concentration and applied electrical potential on morphological appearance of as-spun fibers. *Polymer Engineering & Science*, 45, 1073–1080.

Ramakrishna, S., Fujihara, K., Teo, W.-E., Lim, T.-C. & Ma, Z. (2005) *An introduction to electrospinning and nanofibers.* World Scientific.

Sencadas, V., Correia, D., Areias, A., Botelho, G., Fonseca, A., Neves, I., Gomez Ribelles, J. & Lanceros Mendez, S. (2012) Determination of the parameters affecting electrospun chitosan fiber size distribution and morphology. *Carbohydrate Polymers*, 87, 1295–1301.

Shin, Y.M., Hohman, M.M., Brenner, M.P. & Rutledge, G.C. (2001) Experimental characterization of electrospinning: the electrically forced jet and instabilities. *Polymer*, 42, 09955–09967.

Subbiah, T., Bhat, G., Tock, R., Parameswaran, S. & Ramkumar, S. (2005) Electrospinning of nanofibers. *Journal of Applied Polymer Science*, 96, 557–569.

Supaphol, P., Mit-Uppatham, C. & Nithitanakul, M. (2005) Ultrafine electrospun polyamide-6 fibers: effect of emitting electrode polarity on morphology and average fiber diameter. *Journal of Polymer Science* B: *Polymer Physics*, 43, 3699–3712.

Sutasinpromprae, J., Jitjaicham, S., Nithitanakul, M., Meechaisue, C. & Supaphol, P. (2006) Preparation and characterization of ultrafine electrospun polyacrylonitrile fibers and their subsequent pyrolysis to carbon fibers. *Polymer International*, 55, 825–833.

Touny, A.H., Lawrence, J.G., Jones, A.D. & Bhaduri, S.B. (2010) Effect of electrospinning parameters on the characterization of PLA/HNT nanocomposite fibers. *Journal of Materials Research*, 25, 857–865.

Wang, C., Hsu, C.-H. & Lin, J.-H. (2006a) Scaling laws in electrospinning of polystyrene solutions. *Macromolecules*, 39, 7662–7672.

Wang, C., Zhang, W., Huang, Z., Yan, E. & Su, Y. (2006b) Effect of concentration, voltage, take-over distance and diameter of pinhead on precursory poly (phenylene vinylene) electrospinning. *Pigment & Resin Technology*, 35, 278–283.

Yao, L., Haas, T.W., Guiseppi-Elie, A., Bowlin, G.L., Simpson, D.G. & Wnek, G.E. (2003) Electrospinning and stabilization of fully hydrolyzed poly (vinyl alcohol) fibers. *Chemistry of Materials*, 15, 1860–1864.

Yuan, X., Zhang, Y., Dong, C. & Sheng, J. (2004) Morphology of ultrafine polysulfone fibers prepared by electrospinning. *Polymer International*, 53, 1704–1710.

Zhang, C., Yuan, X., Wu, L., Han, Y. & Sheng, J. (2005) Study on morphology of electrospun poly (vinyl alcohol) mats. *European Polymer Journal*, 41, 423–432.

Zhao, S., Wu, X., Wang, L. & Huang, Y. (2004) Electrospinning of ethyl-cyanoethyl cellulose/tetrahydrofuran solutions. *Journal of Applied Polymer Science*, 91, 242–246.

Zong, X., Kim, K., Fang, D., Ran, S., Hsiao, B.S. & Chu, B. (2002) Structure and process relationship of electrospun bioabsorbable nanofiber membranes. *Polymer*, 43, 4403–4412.

Index

Sustainable Water Developments

Book Series Editor: Jochen Bundschuh

ISSN: 2373-7506

Publisher: CRC Press/Balkema, Taylor & Francis Group

1. Membrane Technologies for Water Treatment: Removal of Toxic Trace
   Elements with Emphasis on Arsenic, Fluoride and Uranium
   Editors: Alberto Figoli, Jan Hoinkis & Jochen Bundschuh
   2016
   ISBN: 978-1-138-02720-6 (Hbk)

2. Innovative Materials and Methods for Water Treatment:
   Solutions for Arsenic and Chromium Removal
   Editors: Marek Bryjak, Nalan Kabay, Bernabé L. Rivas & Jochen Bundschuh
   2016
   ISBN: 978-1-138-02749-7 (Hbk)

3. Membrane Technology for Water and Wastewater Treatment,
   Energy and Environment
   Editors: Ahmad Fauzi Ismail & Takeshi Matsuura
   2016
   ISBN: 978-1-138-02901-9 (Hbk)